T0335877

Geometrical Properties of Differential Equations

Applications of Lie Group Analysis in Financial Mathematics

Geometrical Properties of Differential Equations

Applications of Lie Group Analysis in Financial Mathematics

Ljudmila A. Bordag

University of Applied Sciences Zittau/Görlitz, Germany

NEW JERSEY • LONDON • SINGAPORE • BEIJING • SHANGHAI • HONG KONG • TAIPEI • CHENNAI

Published by

World Scientific Publishing Co. Pte. Ltd.

5 Toh Tuck Link, Singapore 596224

USA office: 27 Warren Street, Suite 401-402, Hackensack, NJ 07601

UK office: 57 Shelton Street, Covent Garden, London WC2H 9HE

Library of Congress Cataloging-in-Publication Data

Bordag, Ljudmila A.

Geometrical properties of differential equations : applications of Lie group analysis in financial mathematics / by Ljudmila A. Bordag (University of Applied Sciences, Zittau/Görlitz, Germany).

pages cm

Includes bibliographical references and index.

ISBN 978-9814667241 (hardcover : alk. paper)

1. Lie groups. 2. Differential equations, Partial. 3. Geometry, Differential. 4. Business mathematics. I. Title.

QA387.B625 2015

512'.482--dc23

2015002797

British Library Cataloguing-in-Publication Data

A catalogue record for this book is available from the British Library.

Printed in Singapore

Preface

This textbook is devoted to Lie group analysis of differential equations with applications to financial mathematics. The new models arising in financial mathematics are often presented in the form of nonlinear partial differential equations (PDEs). Lie group analysis was earlier successfully applied to the studies of algebraic and group structures of linear and nonlinear PDEs which describe physical or chemical processes. It is to be expected that this method will also be widely applied to new types of PDEs appearing in financial mathematics.

The content of this book was developed for the lecture courses *Geometrical properties of differential equations* given in the framework of the master program "Master in financial mathematics" at Halmstad University, Sweden for the academic years 2006/2007 till 2010/2011. The one-year program was oriented towards mathematicians, physicists and computer sciences engineers who wanted to learn actual methods and modern models in financial mathematics.

The goal of this and other lecture courses in the master program was to provide a set of practical skills which allowed the students to use the studied methods in industry. So the main focus of the program was on a clear understanding of the main ideas and self-confident use of learned tools instead of deep theoretical proofs and theorems. We had about 20-30 students per year. The program was very international and the students came from Armenia, China, Germany, Lithuania, Pakistan, Poland, Russia, Serbia, Singapore, Slovakia, Sweden, Tunisia and Turkey. The atmosphere in the master program was very friendly and students supported each other. As far as I know many students stayed in touch years after finishing their studies, working in finance institutions in different countries all over the world. Despite different skills and educational backgrounds the students in my lecture course were really enthusiastic to learn new ideas and apply them in practice.

At Halmstad University the complete academic year was divided into four periods with eight teaching weeks each and one or two examination weeks in between. During the first period I provided four academic hours of lectures and four academic hours of classroom tutorials every week. Students were also given assignments with problems to solve every week. These assignments and classroom tutorials, as well as exams are included in this book accompanied by detailed solutions. In the last, fourth, teaching period of the study year all students wrote a master project (15 ECTS points) devoted to one of the actual problems in financial mathematics. These projects are published on the homepage of the library of Halmstad University. Part of the results of the master projects were later published as journal papers. Some students used Lie group analysis and studied new financial models with this method.

Later in Germany I used these lecture notes during the *Compact course on Lie group analysis* that took place in Zittau, 2013, in the framework of FP 7 Marie Curie Initial Training Network (ITN) STRIKE (see www.itn-strike.eu). The aim of the project ITN STRIKE is to understand complex (mostly nonlinear) financial models and to develop effective and robust numerical schemes for solving linear and nonlinear problems arising from the mathematical theory of pricing financial derivatives and related financial products. This two-week intensive course was conducted for PhD students of different European countries working in the ITN STRIKE. The lectures were presented by me, while the practical exercises and assignments were led by Ivan P. Yamshchikov. Because the participants were interested in applications of the new knowledge to numerical schemes we emphasized this part. We discussed how the admitted Lie algebraic structure of studied equations can be used to provide improved numerical schemes, how to get and use the invariant solutions and to get in touch with new models. In two weeks the participants learned a tool which they can use practically.

In my opinion this book with its large number of problems with detailed solutions can be used as a textbook for a regular lecture course, or for a compact lecture course of a few weeks or as a book for convenient self-study.

The lecture notes were prepared with the strong help of my previous students: Tony Huschto, Anna Mikaelyan, Ivan P. Yamshchikov and Dmitry Zhelezov. I am very grateful for their help and efforts to make this text better.

Zittau, Germany

Ljudmila A. Bordag
May 11, 2015

Abstract

This textbook is a short comprehensive and intuitive introduction to Lie group analysis of ordinary and partial differential equations. This practical oriented material contains a large number of examples and problems accompanied by detailed solutions and figures. In comparison with known beginner guides to Lie group analysis, this book is oriented towards students who are interested in financial mathematics and mathematical economics. Avoiding reference to physical intuition, ideas, such as the following, are developed step-by-step: point transformation, meaning of a continuous one-parameter group, infinitesimal action, invariants, Lie algebra, symmetry reductions of differential equations. In this book we look at differential equations from a geometrical point of view, and explain the idea of invariant solutions and symmetries. The admitted symmetries will be used to find the reductions of given differential equations and invariant solutions accompanied by a large number of examples.

The book contains nine chapters. The first few chapters are devoted to the development of the ideas and tools of Lie group analysis. All notations, ideas and methods are explained first with very simple examples of ordinary differential equations and only then with partial differential equations. This structure allows the students to get familiar with the main tools of Lie group analysis as quickly and as easily as possible.

In Chapter 8 we study the famous Black-Scholes model for option pricing. Its algebraic structure is compared with the structure of the heat equation. The explicit analytic solution for a call option is derived simply as an invariant solution of Black-Scholes equation.

In Chapter 9 we provide the results of the Lie group analysis of actual models in financial mathematics using recent publications. These models are usually formulated as nonlinear partial differential equations and are rather difficult for investigations. With the help of Lie group analysis it is possible to describe some important properties of these models and get some interesting reductions in a clear and understandable algorithmic way.

The material in this book has been used for the academic years 2006/2007 till 2010/2011 as part of the Master in Financial Mathematics program at Halmstad University, Sweden and in a two-week intensive compact course given 2013 in Zittau in the framework of FP 7 Marie Curie Initial Training Network (ITN) STRIKE (www.itn-strike.eu). The participants of the compact course were PhD students primarily interested in new effective numeric schemes for advanced models in financial mathematics.

As prerequisites for this textbook one needs to know some basics of the theory of differential equations which corresponds usually to the bachelor level in mathematics or applied mathematics. This book can be a short introductory for a further study of modern geometrical analysis applied to models in financial mathematics. It can also be used as a textbook in a master program, in an intensive compact course or for self-study.

Contents

Chapter 1

Introduction

Lie group analysis of differential equations was developed by Sophus Lie during his work at Leipzig University from 1886 to 1898. Sophus Lie was deeply unsatisfied with the theory of differential equations at that time. The textbooks concerning differential equations were to some extent similar to cook books: take this substitution, apply it to this equation and you get the solution or obtain a reduction of the equation to a simpler one. Sophus Lie decided to find an internal structure which could allow one to uncover the procedure of discovering all these substitutions, if they exist. His ideas on point transformations and their algebraic structure are represented in the books [51], [52], [53]. His books are written in a clear and understandable way. Sophus Lie was a Norwegian but he used German language brilliantly, though it was not his mother tongue.

Sophus Lie provided the complete classification of all ordinary differential equations (ODEs) up to the third order. Additionally he studied the group structure of two-dimensional linear partial differential equations (PDEs) in detail. The quantity and quality of Lie's work was so impressive and powerful, that for many years most mathematicians assumed that there was no room for any further development of the Lie group analysis of ODEs or PDEs. The main focus was placed upon the development of the theory of Lie algebras, Lie groups and later on upon the quantum-group theory.

The idea to use Lie group analysis to study PDEs came back into practice during the 1960's, with the development of new models in mathematical physics. Using the modern achievements in the theory of Lie algebras and Lie groups it was possible to get new results and renovate the theory of Lie group analysis of differential equations.

The scientific group of Ovsiannikov [60] and later Ibragimov [40] played a leading role in this process. In the last decades of the 20th century this method was widely used in mathematical physics. Many books devoted to the Lie group theory with applications in physics and mathematical physics were written in this time. There are textbooks for beginners like [38], [21], [27] and more sophisticated ones like [59], [10], [8], [9]. This topic was included in education curricula for mathematics and physics students in many universities over the wold.

Financial mathematics rapidly developed in the 20th century. Since the famous works of Black and Scholes [7] and Merton [55], financial mathematics is moving from pure econometric and stochastic methods to more sophisticated tools of mathematics. The main subject of the Black-Scholes (BS) model was a linear parabolic PDE. In the framework of this model it was possible to get directly the simple and useful solutions for European call and put options in an exact analytic form [7]. In a language of the Lie group analysis the famous Black-Scholes (BS) formula represents an invariant solution of this equation. With the work of Black and Scholes the rich arsenal of methods of theory of PDEs was suddenly available for the mathematicians working in the area of financial mathematics. The properties of parabolic PDEs were very well studied before in the framework of mathematical physics; one just needed to reformulate and interpret the results for the new applications.

These circumstances not only made the BS model famous and the authors Nobel Prize winners, but also allowed the model to be used in financial companies. On the other hand it attracted a lot of other specialists in the area of PDEs with their own instruments and experience in the PDE theory to financial mathematics. The BS model was developed for an ideal financial market, it was transparent and easy to use. However different improvements are necessary if we take into account deviations of a real market from the ideal one. A lot of models given in the form of nonlinear PDEs which take into account different frictions on the financial market have appeared in the last 40 years.

After the financial crisis in 2008 practitioners all over the world realized that illiquidity is one of the most demanding problems at present. Financial mathematics applied in the industry was mainly focused on applications of statistical methods to the models implying the linearity of the corresponding processes. Yet it is known that the modeling of hedging and pricing options in non-ideal markets with illiquidity problems or transaction costs leads to strong nonlinearities in partial stochastic differential equations and correspondingly in PDEs. In many cases it is impossible to regard these PDEs simply as a perturbation of a linear PDE allowing a proper approximation. Moreover, known numerical methods may break down due to the strong nonlinearity. Because of that, financial mathematics demand new mathematical methods in application to PDEs and numerics to handle these phenomena.

Hence, one needs to apply a broader spectrum of the advanced mathematical methods to nonlinear equations. One of these methods is the method of geometrical analysis, especially Lie group analysis. This method is new in this developing area and so far, Lie group analysis applied to financial models is not reflected at all in the textbooks.

Only 25 years after the famous paper of Black and Scholes was the first paper [33] written where the Lie group analysis method was applied to the BS model and to similar PDEs appearing in financial mathematics. A bit later the group of Leach started to use this method in applications to linear and quasi-linear PDE models in financial mathematics [57]. We also used this method to study strongly nonlinear models in [14], [11] [20], [15], [12], [16], [13], [18], [19].

Some of these models and results of their analyses are presented in Chapter 9. Nowadays one can see a growing interest in Lie group analysis and more and more papers which use this method in applications to new models in financial mathematics appear every year. On one hand the Lie algebraic structure of the studied equations is used to get invariant solutions and to apply them as benchmarks for instance. On the other hand this structure is used to develop new numeric schemes [25], [26], [6] which are more efficient and stable [66], [64].

This textbook is suitable for students of mathematical departments at the master level or the higher bachelor level who are interested in applications of modern mathematical methods in financial mathematics, mathematical economics or applied mathematics with the focus on economic aspects. As a prerequisite, the reader needs the usual bachelor level courses in ODEs (a lecture course on the theory of PDEs would be rather useful as well). Unlike well-known textbooks devoted to Lie group analysis applied to differential equations we do not use any physical intuition. Most of these textbooks were written when Lie group analysis was mainly addressing the needs of mathematical physics and such analogies were very welcome. Nowadays students who are interested in economics or financial mathematics do not have any physics background and are usually not able to use any hints linked to physical phenomena. This textbook with its numerous examples will be useful not only for students who are interested in financial mathematics but also for people who are working in other areas of research that are not directly connected with physics (for instance in areas of applied mathematics like mathematical economics, bio systems, coding theory etc.).

The book has the following structure.

In Chapter 2 and Chapter 3, the ideas of point transformations on $\mathbb{R}^2$, one-parameter group and invariants are introduced on the intuitive level with a large number of examples, problems which are accompanied by very detailed solutions and exercises. In a similar way, in Chapter 4 the idea of first order ODEs, Lie point symmetries and solutions of these ODEs using their internal algebraic structure are introduced. Chapter 4 also contains a lot of examples and problems with extensively described solutions that allow students to gain experience with this material.

The next two chapters, Chapter 5 and Chapter 6, are of a more technical nature. In Chapter 5 the idea of a prolongation is introduced. In order to apply Lie group analysis to higher order differential equations one needs to be familiar with the concept of a prolongation. In Chapter 6 some properties of Lie algebras are introduced and explained. We also present some optimal systems of subalgebras for Lie algebras which will be covered later in Chapter 9. These properties are used in the subsequent chapters to describe the different families of invariant solutions. A lot of examples, problems accompanied by solutions and exercises are included in all the chapters.

Chapter 7 is devoted to Lie group analysis of high order ODEs. This chapter is an important step towards Lie group analysis in application to financial mathematics. Since new models in financial mathematics are often represented by nonlinear PDEs, after using a reduction procedure they are usually transformed to high order ODEs. So, it is important that the reader is familiar with the equations of such type and can work with them. We provide in Chapter 7 a lot of examples, problems with detailed solutions and some exercises so that the reader can always check his level of understanding. We also provide an example of the usage of the software package **IntroToSymmetry** which is an additional package to the computer algebra system **Mathematica**. We give a hint on how to use other typical software packages like **ReLie** based on **REDUCE** and **SymboLie** based on **Mathematica** [58].

Technical tools are introduced and adjusted to the needs of Lie group analysis of PDEs in Chapter 8. The applications of Lie group analysis are demonstrated with two examples. For two linear PDEs, the heat equation and the Black-Scholes equation, we provide admitted Lie algebras, describe the symmetry group structure and discuss the idea of invariant solutions using many examples. It is shown that both equations have isomorphic and similar Lie algebras and may be reduced to each other with simple transformations. These transformations are often used in different papers or even textbooks devoted to financial mathematics but one rarely mentions the nature and background concepts that stand behind this similarity. In this chapter we obtain an explicit analytical solution for a call option that was first presented in the work of Black and Scholes as an invariant solution of this PDE.

After the technical tools are developed in the earlier chapters, in Chapter

9 we continue to study different models describing processes in real financial markets with Lie group analysis. In the beginning we give some short introduction to the theory of financial markets, define the ideal market, give a formulation of Ito's Lemma and list the assumptions under which the BS model is valid. Further we discuss how this model can be improved and list some modern models. These models were developed under different economical and financial assumptions and the corresponding PDEs look also rather different. But after Lie group analysis one can prove that the models are often connected to each other and have isomorphic algebraic structures.

This book contains a list of references to the most important monographs and textbooks as well as some papers which are directly connected to the book material. The short subject index makes the book more convenient to work with.

Chapter 2

Point transformations on $\mathbb{R}^2$

This is the first chapter where we introduce some elementary ideas and try to explain them intuitively. We briefly discuss point transformations on the plane, a group, group representations, an infinitesimal action of the group, and a group orbit. We also introduce the ideas of a direction field provided by an ordinary differential Equation (ODE) first order and by a one-parameter group, an integral curve and a relation between a one-parameter family of curves and a first order ODE. These ideas are probably known for the majority of the readers of this book but they are very important for our further material. That is why we pay a lot of attention to the details and clarify every single one with examples. We also discuss the relations between these ideas in order to make them practically accessible for our reader.

Section 2.1 *Groups of transformations* introduces a general idea of a one-parameter continuous group. This concept is crucial for applications of Lie group analysis to differential equations. We compare continuous and discontinuous groups of transformations on $\mathbb{R}^2$ using two simple examples. They explain two crucial aspects of the theory that are needed later: the idea of an infinitesimal transformation and the idea of a fundamental region. Though discontinuous groups have many applications to the theory of differential equations as well in this textbook we focus just on a special type of continuous groups namely on Lie groups. We introduce the idea of an orbit of the group and discuss how the properties of the orbits depend on the properties of the studied group. The ideas of infinitesimal transformation and direction field arising from the action of transformation group on $\mathbb{R}^2$ will help us later understand a connection between a Lie group and a corresponding Lie algebra.

Section 2.2 *The general form of a one-parameter group of point transformations in* $\mathbb{R}^2$ starts with a general form of a set of point transformations on a plane. Then the idea of a group in application to point transformations is introduced step-by-step. Using the example of a rotation group in a closed form both in Cartesian and polar coordinates we prove all group properties. We define a group representation, a group action, a one-parameter group of transformations and a flow. Using the example of the rotation group we introduce an idea of an infinitesimal action of a point transformation group. This infinitesimal action of the point transformation group on the plane can be described by two functions playing a major role in Sophus Lie theory. We provide definitions of a direction field, an integral curve to a direction field and an orbit of a point transformation group. We also introduce the Lie equations which are the key relation between infinitesimal and closed form of a point transformation group. They also describe the correspondence between a direction field and orbits of a point transformation group.

Section 2.3 *Problems with solutions* contains eleven problems with detailed solutions. These practical tasks should help the reader to understand the ideas from previous sections. We recommend that the reader tries to solve the problems himself before looking at the solutions provided. The examples address some basic questions of the theory of the first order ODEs such as: how can one describe the family of integral curves? How are the direction field and integral curves connected? How can one find a most general form of a first order ODE with a given one-parameter family of integral curves? From our experience, these ideas are not new for many students, but many do not realize that these are just different points of view on the same problem. They also lack a certain amount of scientific intuition concerning elementary ODEs of the first order. Working with such basic tasks can seem rather easy yet is crucial for development of this intuition. A number of similarly solved problems with slightly different emphasis gives the student the chance to understand the material better and thereafter operate all the needed concepts with ease.

The chapter ends with some exercises which are very similar to the problems and examples given in the text of the chapter.

2.1 Groups of transformations

Let us begin this course on geometrical properties of differential equations by studying two very simple, yet important, examples. They will help us to see the principle difference between discontinuous and continuous groups. These two examples are extremely important since in this book we will only study continuous groups.

Example 2.1.1 Let us consider a plane $\mathbb{R}^2$ with Cartesian coordinates denoted by x and y. We take a point $(x, y) \in \mathbb{R}^2$ and a transformation of the following type

$$x_\varepsilon = x + \varepsilon, \qquad (2.1.1)$$
$$y_\varepsilon = y,$$

where ε is an arbitrary parameter, $\varepsilon \in \mathbb{R}$.

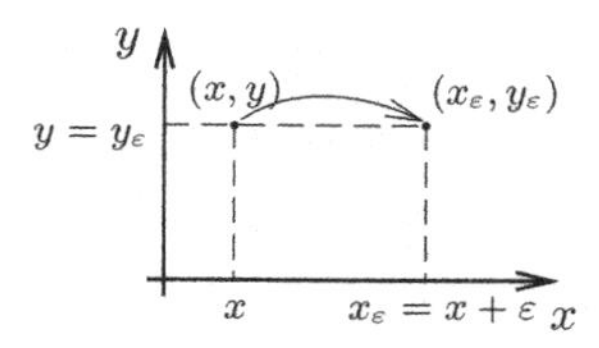

Figure 2.1: A translation of the point (x, y) on the plane parallel to the x-axes

We see that each source point $(x, y) \in \mathbb{R}$ has exactly one image point $(x_\varepsilon, y_\varepsilon)$ for every value of the parameter ε. We can interpret this situation in three different ways:

1. We have one $\mathbb{R}^2$-plane and map each point (x, y) into a point $(x_\varepsilon, y_\varepsilon)$, see Figure 2.1.

2. We have two copies of the plane $\mathbb{R}^2$, where $(x, y) \in \mathbb{R}^2$ and $(x_\varepsilon, y_\varepsilon) \in \mathbb{R}^2$, and establish a one-to-one correspondence between different copies of $\mathbb{R}^2$, see Figure 2.2.

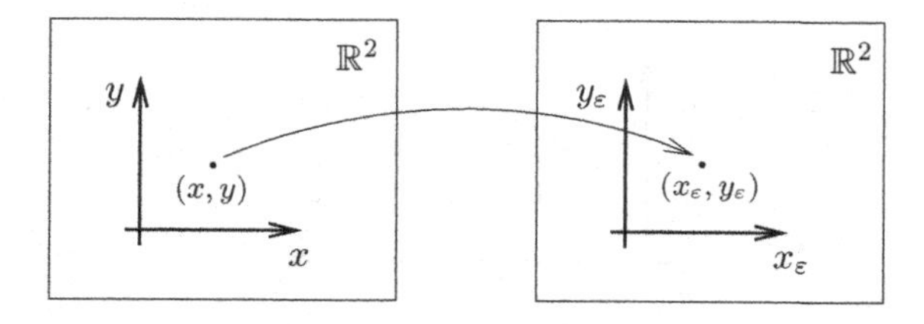

Figure 2.2: A map of the sourse point (x, y) from one plane to an image point $(x_\varepsilon, y_\varepsilon)$ on other plane

3. We have $\mathbb{R}^2$ and the point (x, y) that stays the same in a physical sense, but has different coordinates as we change the coordinate system from (x, y)-coordinates to $(x_\varepsilon, y_\varepsilon)$-coordinates and obtain the correspondence $(x, y) \to (x_\varepsilon, y_\varepsilon)$, see Figure 2.3.

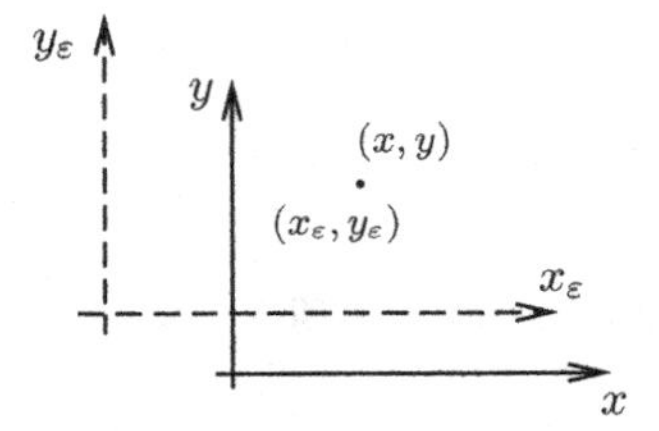

Figure 2.3: The point on the plane is fixed, the coordinate system is changed

The most important aspect for us now is that we have a smooth one-to-one point transformation. Let us look back at our example (2.1.1). The transformation is one-to-one and if we use this transformation with $\varepsilon_1 = -\varepsilon$, we obtain the source point again

$$\begin{aligned}
\tilde{x} &= x_\varepsilon - \varepsilon = (x + \varepsilon) - \varepsilon = x, \\
\tilde{y} &= y_\varepsilon = y.
\end{aligned}$$

As we can chose ε arbitrarily, we can get an image point $(x_\varepsilon, y_\varepsilon)$ of the source point (x, y) in every arbitrary small neighborhood of (x, y). In this case we say that the transformation family (2.1.1) has an infinitesimal transformation. When we look for all possible images of the point (x, y), we get a continuous line – a straight line which is parallel to the x-axis, see Figure 2.4. This line consists of all possible images of one point on this line and is called an orbit.

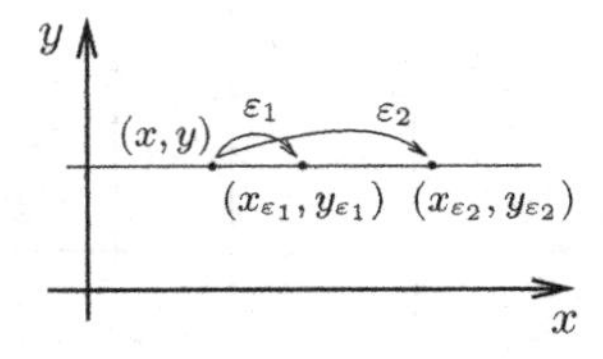

Figure 2.4: One of the orbits of the point transformation (2.1.1)

Intuitively we can understand the orbit of the transformation group as the locus of all the images of (x, y) under the action of the group. A full and detailed definition of the orbit is given in the next chapter but this intuitive understanding is sufficient for our current example. Another important fact is that if we take two different source points they either generate the same orbit or their orbits do not intersect, because otherwise we would not have a one-to-one map.

Let us now consider a composition of two point transformations: at first a translation with a parameter ε_1, then with ε_2. As in (2.1.1) we obtain

$$\tilde{x} = x_{\varepsilon_1} + \varepsilon_2 = (x + \varepsilon_1) + \varepsilon_2 = x + \underbrace{(\varepsilon_1 + \varepsilon_2)}_{\varepsilon_3},$$
$$\tilde{y} = y_{\varepsilon_1} = y.$$

We can see that two successive actions are equivalent to one map with a parameter equal to the sum of the two initial parameters $\varepsilon_3 = \varepsilon_1 + \varepsilon_2$, i.e. $(\tilde{x}, \tilde{y}) = (x_{\varepsilon_3}, y_{\varepsilon_3})$. We proved that the family of point transformations (2.1.1) has a group structure:

1. For each transformation with a parameter ε the family of point transformations (2.1.1) has an inverse transformation with the parameter equal to $-\varepsilon$.

2. The identity transformation $\tilde{x} = x$, $\tilde{y} = y$ belongs to the family of transformations (2.1.1)(with the parameter $\varepsilon = 0$).

3. The successive actions of two point transformations are equivalent to one point transformation with a parameter equal to the sum of the original parameters.

Thus, the point transformations (2.1.1) form a group, called translation group. The group structure is equivalent to the group structure of real numbers $\mathbb{R}$ together with the addition operation. We can represent the action of this group geometrically as point transformations on a plane.

On the other hand, we have seen that this transformation group also has topological properties. It means that for each source point we can introduce a fixed arbitrary neighborhood and say if an image point belongs to this neighborhood or not. In any neighborhood of every source point we can find infinitely many image points for the family of point transformations (2.1.1). In other words, we cannot separate the source point from the image points. It means that there exists an infinitesimal transformation in the family of transformations (2.1.1). We have also seen that the geometrical properties of the orbits are defined by the group, not by the source point of an orbit. All orbits of the translation group (2.1.1) are continuous, straight lines independent of the chosen source point. We can conclude that the family of transformations (2.1.1) fulfills the group properties: there exists an infinitesimal transformation, the orbits are continuous lines and the group has only one-parameter ε. $\diamond$

Example 2.1.2 In the second example we consider a point transformation of the following kind

$$x_r = x + \omega r \qquad (2.1.2)$$
$$y_r = y,$$

where $\omega \in \mathbb{R}$ is fixed and $r \in \mathbb{Z}$.

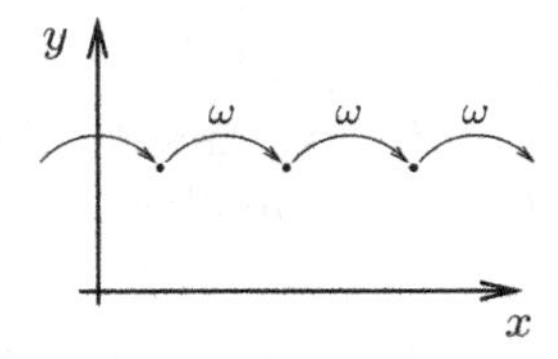

Figure 2.5: A discrete point transformation with the step $\omega > 0$, translation of a point parallel to the x-axes

If we apply the transformation (2.1.2) with the parameter $r = 1$ a second time to the image point (x_1, y_1) then we obtain

$$\tilde{x} = x_1 + \omega = x + 2\omega,$$
$$\tilde{y} = y_1 = y,$$

i.e. the image point $(\tilde{x}, \tilde{y})$ coincide with the point (x_2, y_2) which we obtain if we apply the transformation (2.1.2) one time to the point (x, y) with $r = 2$. The inverse transformation, it means the transformation (2.1.2) applied to $(\tilde{x}, \tilde{y})$ with $r = -1$ leads to

$$x = \tilde{x} - \omega,$$
$$y = \tilde{y}.$$

So, the transformation maps the image point $(\tilde{x}, \tilde{y})$ back to the source point (x, y). In this case we also have a group structure: this group structure is equivalent to the group of integers $\mathbb{Z}$. This group as well as the previous one describes translations of a source point parallel to the x-axis. But there are very important differences between those translation groups (2.1.1) and (2.1.2). We can see easily that in the second example all images are separated from each other. We can find small neighborhoods of each source point (x, y) which do not contain any image point $(\tilde{x}, \tilde{y})$. The orbit of each point is a discrete set of points.

Hence, we can chose a non-empty region on the plane with the property that each point of this region does not have any of its image points in exactly the same region. In our case this region is a vertical strip of width ω with a right or left boundary. We could take a thinner one as well, but this is the maximal region of points that does not contain any of its images. Such a region is called a fundamental region of the group.

The family of point transformations (2.1.2) in our second example

1. has the group properties,

2. has a non-trivial fundamental region,

3. all orbits are discrete sets of points.

We see that this group of point transformations does not admit an infinitesimal transformation. Between every source and the corresponding image point we have a minimal distance equal to ω. Such a group of point transformations (that does not possess any infinitesimal transformation) is called a discontinuous group. A group of point transformations that has only discrete sets of points as orbits is called a discrete group. A discontinuous group is always discrete, but the opposite statement is false. In our example (2.1.2) we have both properties, i.e. the group is a discontinuous one. $\diamond$

Let us compare the two examples of translations group studied before. We can unify the notations for the translations groups (2.1.1) and (2.1.2) as follows

$$\begin{aligned} x_r &= x + r\omega, \\ y_r &= y, \end{aligned} \tag{2.1.3}$$

where $\omega \in \mathbb{R}$ is non-zero and fixed, whereas in the first example we have $r \in \mathbb{R}$ and in the second one $r \in \mathbb{Z}$. This means that in the first case (2.1.1) we denote $r\omega = \varepsilon$. We see that we have a one group parameter in both cases. The properties of the group parameter define the properties of the corresponding group.

In the first case of point transformations (2.1.1) we have to do with a continuous, in the second case (2.1.2) with a discontinuous group. For the first example we also have an infinitesimal transformation which we do not have in the second one. In comparison with the second example, the continuous group (2.1.1) does not have any fundamental region.

Now we have two very simple examples for groups of one-parameter point transformations: one for a continuous group and one for a discontinuous group depending on the properties of the group parameter. Later we will define the idea of a Lie group and will see that the first example is in fact a Lie group.

In this book we study only continuous groups and their applications to differential equations and models in financial mathematics. We would like to emphasize here that the existence of an infinitesimal transformation is extremely important for these applications.

Let us look at the transformation group (2.1.1) once again.

We have seen that all orbits are straight lines parallel to the x-axis. If we take one point (x, y) on the orbit, the action of the group can be considered as a motion of the point along the orbit. Let us examine a piece of a continuous curve c. Under the action of the group each point of this curve will move along its orbit and for a fixed value of the parameter ε, so in the end we obtain another curve $\tilde{c}$.

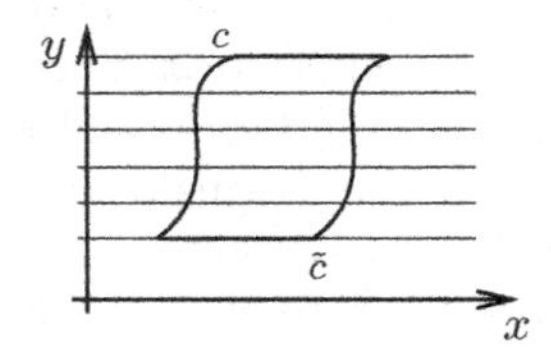

Figure 2.6: A piece c of a continuous curve under the action of a continuous one-parameter transformation group (2.1.1)

In the case (2.1.1) we have a parallel translation of the curve c. If we look at the region between c, its image $\tilde{c}$, upper and lower orbits, we can see that all points inside this region are image points of points on the curve c. Therefore, we cannot get any gaps in this region which are free from image points of c, because the curve c and the translation group (2.1.1) are both continuous and the transformations are defined in all points of the plane. This is a very important property of continuous groups and we will discuss it in the general case by application to differential equations.

We notice that every point is moved along its own orbit. This motion is defined by the parameter of the group. We can differentiate the formulas for the point transformations for each point on the orbit and obtain the rate of change of the point position with respect to the change of the parameter value

$$\frac{d\tilde{y}}{d\varepsilon} = 0, \quad \frac{d\tilde{x}}{d\varepsilon} = 1.$$

We have a constant rate of change in every point on the plane. The rate of change for the coordinate y is equal to zero and for the coordinate x to one. These values define also the direction of transformation in each point. Furthermore, we can associate with every point an infinitesimal piece of a straight line parallel to the x-axis and obtain a direction field corresponding to the group (2.1.1).

Consequently, now we have three possibilities to analyze the action of the group:

1. We use the global rule

$$\tilde{x} = x + \varepsilon,$$
$$\tilde{y} = y,$$

2. we have a family of orbits and use them,

3. we have a direction field associated with this action of a one-parameter group.

All these representations of the action of the same group are equivalent up to a scaling of the parameter ε. We will show it for a general situation a bit later.

2.2 The general form of a one-parameter group of point transformations in $\mathbb{R}^2$

We consider a set of point transformations on the plane

$$\begin{aligned} x_\epsilon &= X_\varepsilon(x, y) \\ y_\epsilon &= Y_\varepsilon(x, y) \end{aligned} \tag{2.2.4}$$

that depends on a continuous parameter $\varepsilon \in \mathbb{R}$. Any particular value of ε determines one transformation of the set and each transformation maps a point (x, y) to the image point (x_ϵ, y_ϵ). Every transformation has its own region where it is well-defined.

We suppose now that for $\varepsilon = \varepsilon_0$ we get the identical transformation

$$X_{\varepsilon_0}(x, y) = x,$$
$$Y_{\varepsilon_0}(x, y) = y.$$

Further on, we suppose that all transformations are defined in a small neighborhood of the identical transformation and that the transformations depend on a continuous parameter ε and both functions X and Y are continuous, at least in the neighborhood of the point (x, y). For two arbitrary transformations from the set we can find a common region where both of them are defined and apply these two transformations one after another.

We demand that the set of the transformations is large enough. It means that for any transformation of the set we have an inverse transformation which also belongs to the set. We demand as well that this set fulfills the other group properties, for example that two transformations, executed in succession, are equivalent to another transformation from the set.

Let us write down all the properties of the transformation set (2.2.4) that we need:

1. We have a set of one-to-one point transformations with one parameter $\varepsilon \in \mathbb{R}$.

2. The identity transformation belongs to the set with the fixed value of the parameter $\varepsilon = \varepsilon_0$.

3. The transformations are well defined and continuous, at least in a neighborhood of ε_0.

4. Every transformation possesses an inverse transformation which belongs to the set.

5. Several transformations, carried out in succession, are equivalent to another transformation of the set.

All of these properties allow us to speak about a local continuous, one-parameter group. The locality condition can be removed if all point transformations are defined on the whole space $\mathbb{R}^2$ for all parameter values ε from a fixed interval.

Let us consider an example of a continuous group, where all transformations of the group (2.2.4) are well defined on $\mathbb{R}^2$.

Example 2.2.1 Let the global rule for a transformation be

$$\begin{aligned} x_\varepsilon &= X_\varepsilon(x, y) = x\cos\varepsilon - y\sin\varepsilon, \\ y_\varepsilon &= Y_\varepsilon(x, y) = x\sin\varepsilon + y\cos\varepsilon, \quad \forall\varepsilon \in \mathbb{R}. \end{aligned} \tag{2.2.5}$$

Now let us see if this set of one-parameter point transformations has the properties of the continuous one-parameter group:

1. All transformations are well defined on the whole plane $\mathbb{R}^2$ and for any value of a parameter ε.

2. We have a one-to-one correspondence between a source and an image point.

3. The identity transformation belongs to the set (for $\varepsilon = \varepsilon_0 = 0$). Indeed, if we take $\varepsilon = 0$ equations (2.2.5) reduce to $x_\varepsilon = x, \quad y_\varepsilon = y$.

4. Every transformation from the set (2.2.5) possesses an inverse transformation which belongs to the same set of transformations. We denote the transformation with the value of parameter ε as g_ε: $(x, y) \to (x_\epsilon, y_\epsilon)$ then

$$g_\varepsilon : \begin{array}{l} x_\varepsilon = x\cos\varepsilon - y\sin\varepsilon, \\ y_\varepsilon = x\sin\varepsilon + y\cos\varepsilon, \end{array} \qquad g_{-\varepsilon} : \begin{array}{l} x_{-\varepsilon} = x\cos\varepsilon + y\sin\varepsilon, \\ y_{-\varepsilon} = -x\sin\varepsilon + y\cos\varepsilon. \end{array}$$

We prove that after the successive action (denoted by "$\circ$") of these two transformations $(g_\varepsilon \circ g_{-\varepsilon})(x, y) = g_\varepsilon(g_{-\varepsilon}(x, y))$ we obtain the source point (x, y), i.e. $g_\varepsilon(g_{-\varepsilon}(x, y)) = (x, y)$. Indeed

$$\begin{aligned} (x_\varepsilon)_{-\varepsilon} &= x_\varepsilon \cos\varepsilon + y_\varepsilon \sin\varepsilon \\ &= (x\cos\varepsilon - y\sin\varepsilon)\cos\varepsilon + (x\sin\varepsilon + y\cos\varepsilon)\sin\varepsilon \\ &= x(\cos^2\varepsilon + \sin^2\varepsilon) = x, \\ (y_\varepsilon)_{-\varepsilon} &= -x_\varepsilon \sin\varepsilon + y_\varepsilon \cos\varepsilon \\ &= -(x\cos\varepsilon - y\sin\varepsilon)\sin\varepsilon + (x\sin\varepsilon + y\cos\varepsilon)\cos\varepsilon \\ &= y(\sin^2\varepsilon + \cos^2\varepsilon) = y. \end{aligned}$$

It means that the inverse transformation we obtain by $\varepsilon \to -\varepsilon$.

5. We prove that the composition of any two transformations $g_{\varepsilon_2} \circ g_{\varepsilon_1}$ yields a transformation $g_{\varepsilon_1+\varepsilon_2}$ which belongs to the same family of transformations

$$\begin{aligned} (x_{\varepsilon_1})_{\varepsilon_2} &= x_{\varepsilon_1}\cos\varepsilon_2 - y_{\varepsilon_1}\sin\varepsilon_2 \\ &= (x\cos\varepsilon_1 - y\sin\varepsilon_1)\cos\varepsilon_2 - (x\sin\varepsilon_1 + y\cos\varepsilon_1)\sin\varepsilon_2 \\ &= x(\cos\varepsilon_1\cos\varepsilon_2 - \sin\varepsilon_1\sin\varepsilon_2) - y(\sin\varepsilon_1\cos\varepsilon_2 + \cos\varepsilon_1\sin\varepsilon_2) \\ &= x\cos(\varepsilon_1 + \varepsilon_2) - y\sin(\varepsilon_1 + \varepsilon_2), \\ (y_{\varepsilon_1})_{\varepsilon_2} &= x_{\varepsilon_1}\sin\varepsilon_2 + y_{\varepsilon_1}\cos\varepsilon_2 \\ &= x(\cos\varepsilon_1\sin\varepsilon_2 + \sin\varepsilon_1\cos\varepsilon_2) + y(\cos\varepsilon_1\cos\varepsilon_2 - \sin\varepsilon_1\sin\varepsilon_2) \\ &= x\sin(\varepsilon_1 + \varepsilon_2) + y\cos(\varepsilon_1 + \varepsilon_2). \end{aligned}$$

This means that the one-parameter family of transformations (2.2.5) forms a continuous group of transformations defined on $\mathbb{R}^2$. This group is called a rotation group on $\mathbb{R}^2$. The action of an abstract group $\mathcal{G}$ which is equivalent to the group of real numbers $\mathbb{R}$ together with the addition operation is realized as a rotation of a point (x, y) on the plane $\mathbb{R}^2$. ◇

We have seen in the previous examples that the action of an abstract group $\mathcal{G}$ can be represented as an action of a transformation group on the plane. We introduce now a general definition of a group representation.

Definition 2.2.1 (Representation of a group, action, exact representation) *A representation of a group $\mathcal{G}$ ($\mathcal{G}$ is here equivalent to the group of real numbers $\mathbb{R}$) as a group of point transformations $G(\mathbb{R}^2)$ on the plane is a group homomorphism from $\mathcal{G}$ to G, i.e. a map $g : \mathcal{G} \to G(\mathbb{R}^2)$. It means that for any value of a parameter $\varepsilon \in \mathbb{R}$ the image $g \circ \varepsilon = g(\varepsilon)$ is a transformation $g_\varepsilon \in G(\mathbb{R}^2)$ and $g(\varepsilon_1) \circ g(\varepsilon_2) = g(\varepsilon_1 + \varepsilon_2)$.*
The map $((x, y), \varepsilon) \to g_\varepsilon(x, y) = (x_\varepsilon, y_\varepsilon)$, where $((x, y), \varepsilon) \in \mathbb{R}^2 \times \mathcal{G} \to \mathbb{R}^2$, is called action of the group $\mathcal{G}$ on $\mathbb{R}^2$.
If the map g is an isomorphism, we have to do with an exact representation. The action of the group $\mathcal{G}$ defines the group structure on $\mathbb{R}^2$.

Now we can define also a one-parameter group of point transformations on the plane. We provided some practical examples of such groups before and hope that they helped the reader to understand this idea.

Definition 2.2.2 (One-parameter group of point transformations) *The (local) exact representation $g : \mathbb{R} \to G(\mathbb{R}^2)$ is called a one-parameter group of point transformations on $\mathbb{R}^2$.*

In many books especially oriented on physical applications authors used the term flow, let us here provide this definition as well.

Definition 2.2.3 (Flow) *A continuous, one-parameter group of smooth point transformations is called a flow on $\mathbb{R}^2$.*

It is important for us that the group parameter $\varepsilon \in \mathbb{R}$ and the functions $X_\varepsilon(x, y), Y_\varepsilon(x, y)$ in (2.2.4) are well defined continuous functions of parameter ε, moreover, later we use mostly smooth $X_\varepsilon(x, y), Y_\varepsilon(x, y)$ functions. We can look at an infinitesimal change of the group parameter now. In this way we can get a description of the infinitesimal action of the group in the neighborhood of the identical transformation. Let us look at an example of the rotation group on the plane.

Example 2.2.2 We consider the rotation group on the plane given by the closed form transformation rules (2.2.5). Differentiation with respect to ε

yields in the point $\varepsilon_0 = \varepsilon = 0$

$$\left.\frac{dx_\varepsilon}{d\varepsilon}\right|_{\varepsilon=0} = x(-\sin\varepsilon)|_{\varepsilon=0} - y\cos\varepsilon|_{\varepsilon=0} = -y,$$
$$\left.\frac{dy_\varepsilon}{d\varepsilon}\right|_{\varepsilon=0} = x\cos\varepsilon|_{\varepsilon=0} - y\sin\varepsilon|_{\varepsilon=0} = x.$$

Using the Taylor series we obtain in a general case

$$x_\varepsilon = x + \left.\frac{\partial X_\varepsilon}{\partial\varepsilon}\right|_{\varepsilon=\varepsilon_0} (\varepsilon - \varepsilon_0) + \mathcal{O}\left((\varepsilon-\varepsilon_0)^2\right),$$
$$y_\varepsilon = y + \left.\frac{\partial Y_\varepsilon}{\partial\varepsilon}\right|_{\varepsilon=\varepsilon_0} (\varepsilon - \varepsilon_0) + \mathcal{O}\left((\varepsilon-\varepsilon_0)^2\right).$$

If we denote $\Delta x = x_\varepsilon - x$ and $\Delta\varepsilon = \varepsilon - \varepsilon_0$, for the rotation group we get

$$\Delta x = -y\Delta\varepsilon,$$
$$\Delta y = x\Delta\varepsilon,$$

or, by considering infinitesimal changes,

$$dx = -y\,d\varepsilon, \qquad (2.2.6)$$
$$dy = x\,d\varepsilon.$$

The system (2.2.6) is the pair of coupled ordinary differential equations which can also be written in the form

$$\frac{dx}{-y} = \frac{dy}{x} = \frac{d\varepsilon}{1}.$$

We can solve these equations in order to get the closed form (2.2.5) of the group transformations again.

Instead we study the first of these two equations more carefully

$$\frac{dy}{x} = -\frac{dx}{y} \quad \Rightarrow \quad \frac{dy}{dx} = -\frac{x}{y}.$$

We look at the variable y as a dependent variable and x as an independent one. The solution of this ordinary differential Equation (ODE) is a one-parameter family of curves. From a geometrical point of view, the value of the fraction $\frac{dy}{dx}$ describes a slope (defined by this ODE) in every point of the plane $\mathbb{R}^2$. We can take an arbitrary point $((x, y) \neq (0, 0))$, calculate this slope and easily get the direction field (i.e. in each point of the plane except $(0, 0)$, we have a prescribed slope or direction) for our transformation group.

Thus, we can "solve" this equation step by step: We start with an arbitrary point and move along the piece of the straight line defined by the direction field in this point. Then we go to another point and move along the next straight line defined by the same direction field and so on. In the end we obtain an approximation of an integral curve of this equation. We provide a definition of the integral curve of the given direction field.

Definition 2.2.4 (Integral curve) *An integral curve of a direction field is a curve whose tangent line at every point coincides with the image of the direction field at that point.*

In our case (2.2.6) for the rotation group we obtain circles with the center in the point $(0, 0)$, see Figure 2.7. The point $(0, 0)$ is also a circle but with a zero radius. The position of this point is unaltered under the action of the rotation group. Those circles with different radii are orbits of the rotation group on $\mathbb{R}^2$.

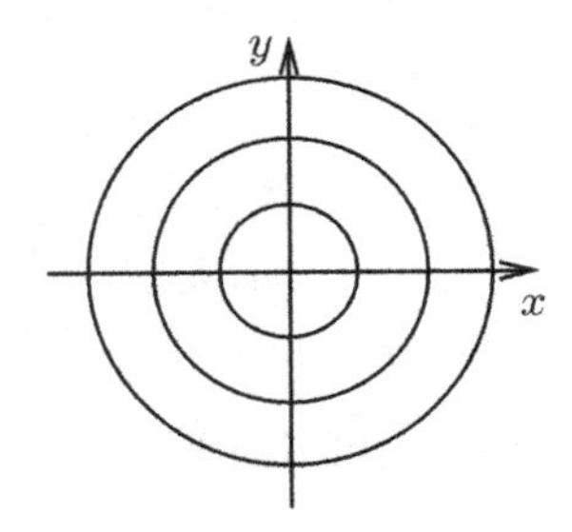

Figure 2.7: Some orbits of the rotation group on the plane

Let us, at last, examine the transformation rule of (2.2.5) once again. Instead of the original Cartesian coordinates we introduce polar coordinates

$$\begin{aligned} x &= r \cdot \cos\theta, \qquad r \in [0, \infty) \\ y &= r \cdot \sin\theta, \qquad \theta \in [0, 2\pi). \end{aligned}$$

The transformation rules (2.2.5) change to

$$\begin{aligned} r_\varepsilon \cos\theta_\varepsilon &= r\cos\theta\cos\varepsilon - r\sin\theta\sin\varepsilon, \\ r_\varepsilon \sin\theta_\varepsilon &= r\cos\theta\sin\varepsilon + r\sin\theta\cos\varepsilon, \end{aligned}$$

or, in a simplified form

$$\begin{aligned} r_\varepsilon \cos\theta_\varepsilon &= r\cos(\theta + \varepsilon), \\ r_\varepsilon \sin\theta_\varepsilon &= r\sin(\theta + \varepsilon). \end{aligned}$$

If we introduce the complex variables then this yields

$$r_\varepsilon e^{i\theta_\varepsilon} = re^{i(\theta+\varepsilon)} \to r_\varepsilon = r, \quad \theta_\varepsilon = \theta + \varepsilon.$$

Immediately we see that in polar coordinates the plane $\mathbb{R}^2$ transforms in a positive half-plane $r \geq 0$ and the action of the rotation group in polar coordinates looks like a simple translation along the θ-axis. It is evident that it is much more convenient to work with a translation group than with a rotation group. We obtained such a complicated closed form of the original transformation rules (2.2.5) because we used Cartesian coordinates. In polar coordinates the transformations look much simpler. ◇

Can we find convenient coordinates for every one-parameter group?

Yes, locally every continuous, one-parameter group of point transformations is equivalent to a translation group. We will prove this theorem a bit later.

Now we want to get an infinitesimal representation of a one-parameter group of point transformations $G(\mathbb{R}^2)$ in a general case. We have the closed form for the family of transformations given by

$$\begin{aligned} x_\varepsilon &= X_\varepsilon(x, y), \quad \varepsilon \in \mathbb{R} \\ y_\varepsilon &= Y_\varepsilon(x, y). \end{aligned}$$

We suppose that this family of transformation builds a group of point transformations on $\mathbb{R}^2$ with the identical transformation $\varepsilon = \varepsilon_0$ and the functions $(X_\varepsilon, Y_\varepsilon)$ are smooth enough.

Let us analyze a small neighborhood of the source point (x, y). The representation of the action of the group in a small neighborhood of an identical transformation is given by the Taylor series

$$\begin{aligned} x_\varepsilon &= x + \left.\frac{\partial X_\varepsilon}{\partial \varepsilon}\right|_{\varepsilon=\varepsilon_0} (\varepsilon - \varepsilon_0) + \mathcal{O}\left((\varepsilon - \varepsilon_0)^2\right), \\ y_\varepsilon &= y + \left.\frac{\partial Y_\varepsilon}{\partial \varepsilon}\right|_{\varepsilon=\varepsilon_0} (\varepsilon - \varepsilon_0) + \mathcal{O}\left((\varepsilon - \varepsilon_0)^2\right). \end{aligned}$$

The functions $\left.\frac{\partial X_\varepsilon}{\partial \varepsilon}\right|_{\varepsilon=\varepsilon_0}$ and $\left.\frac{\partial Y_\varepsilon}{\partial \varepsilon}\right|_{\varepsilon=\varepsilon_0}$ are the functions of the coordinates of the source point (x, y) only.

We introduce a more convenient notation for them

$$\left.\frac{\partial X_\varepsilon}{\partial \varepsilon}\right|_{\varepsilon=\varepsilon_0} = \xi(x, y), \quad \left.\frac{\partial Y_\varepsilon}{\partial \varepsilon}\right|_{\varepsilon=\varepsilon_0} = \eta(x, y). \tag{2.2.7}$$

This notation was introduced by Sophus Lie himself.

Exactly in the same way as in the previous example, we get a system of coupled ODEs corresponding to the one-parameter group of point transformations on $\mathbb{R}^2$

$$\begin{aligned} dx &= \xi(x,y)\,d\varepsilon, \\ dy &= \eta(x,y)\,d\varepsilon, \end{aligned} \tag{2.2.8}$$

respectively we can re-write this system of equations in the form

$$\frac{dy}{\eta(x,y)} = \frac{dx}{\xi(x,y)} = \frac{d\varepsilon}{1}. \tag{2.2.9}$$

This system of differential equations is called **Lie equations** and can be used to define the direction field corresponding to the group

$$\frac{dy}{dx} = \frac{\eta(x,y)}{\xi(x,y)}.$$

The integral curves of this direction field are the orbits of the group. If we can integrate exactly the system of ODEs, we obtain the closed form representation of the group action, i.e. the functions $X_\varepsilon(x,y)$ and $Y_\varepsilon(x,y)$ respectively. In general it is a difficult procedure. Usually we are able to find the functions $\xi(x,y)$ and $\eta(x,y)$, but deriving the functions $X_\varepsilon(x,y)$ and $Y_\varepsilon(x,y)$ sometimes can be rather complicated. We do not really need to know the closed form of point transformations in order to describe the main properties of the one-parameter group. We will see that we can get all this information from functions $\xi(x,y)$ and $\eta(x,y)$. In the next section we provide some problems with solutions which the reader can use to get familiar with introduced ideas. We pay special attention to the connection between a closed form and an infinitesimal form of a one-parameter group of point transformations on the plane.

2.3 Problems with solutions

This section is devoted to practical exercises. In Problems 2.3.1, 2.3.2, and 2.3.4 we discuss conditions under which a one-parameter set of point transformation build a group.

In the Problems 2.3.3 and 2.3.6 one should find the closed form using the infinitesimal form of a one-parameter group of point transformations. In the Problem 2.3.5 one has a closed form of a one-parameter group and should find an infinitesimal form.

In Problems 2.3.7 to 2.3.11 a one-parameter family of curves on a plane is given and one has to find a first order ODE with integral curves described by the given family.

Problem 2.3.1 *Let a one-parameter family of point transformations be given by the rules*

$$x_\epsilon = x,$$
$$y_\epsilon = \epsilon y + \epsilon^2 y^2, \quad \epsilon \in \mathbb{R}.$$

Do these transformations build a group?

Solution To check if these one-parameter family of transformations build a group or not, we should verify that all three group properties (existence of the identity element, an inverse element and group property) hold. At first we show for which value ϵ_0 we obtain identical transformation, i.e. we solve the equations for ϵ_0

$$\begin{aligned} x_{\epsilon_0} &= x, \\ \epsilon_0 y + {\epsilon_0}^2 y^2 &= y. \end{aligned}$$

The first equation puts no constraints on the value ϵ_0. From the second equation we get ϵ_0 as a solution of the second order algebraic equation, which depends on y. The value ϵ_0 should be a constant. Consequently the one-parameter family of transformations do not build any group.

Problem 2.3.2 *Let the group of* **affine transformations** *be given by*

$$\begin{aligned} x_\epsilon &= e^\epsilon x, \\ y_\epsilon &= y, \qquad \epsilon \in \mathbb{R}. \end{aligned}$$

Prove the group properties and sketch the orbits of the group.

Solution Let us check the group properties:

1. We prove at first that the identity transformation is included in the family of the transformations. We solve the equations

$$\begin{aligned} x &= e^{\epsilon_0} x, \\ y &= y_{\epsilon_0} \end{aligned}$$

for ϵ_0. The system has a unique solution $\epsilon_0 = 0$ for any value of x, y.

2. We prove now that for any transformation $g_\epsilon : (x, y) \to (x_\epsilon, y_\epsilon)$ there exist an inverse transformation $g_{\epsilon_1}(x_\epsilon, y_\epsilon) \to (x, y)$ with some value ϵ_1. We take the point (x_ϵ, y_ϵ) and apply the transformation g_{ϵ_1} with an unknown value ϵ_1 and solve the system of equations $g_{\epsilon_1}(x_\epsilon, y_\epsilon) = (x, y)$

$$\begin{aligned} g_{\epsilon_1} x_\epsilon &= e^{\epsilon_1} x_\epsilon = e^{\epsilon_1} e^\epsilon x = e^{\epsilon_1 + \epsilon} = x, \quad \epsilon_1 = -\epsilon, \\ g_{\epsilon_1} y_\epsilon &= y_\epsilon = y. \end{aligned}$$

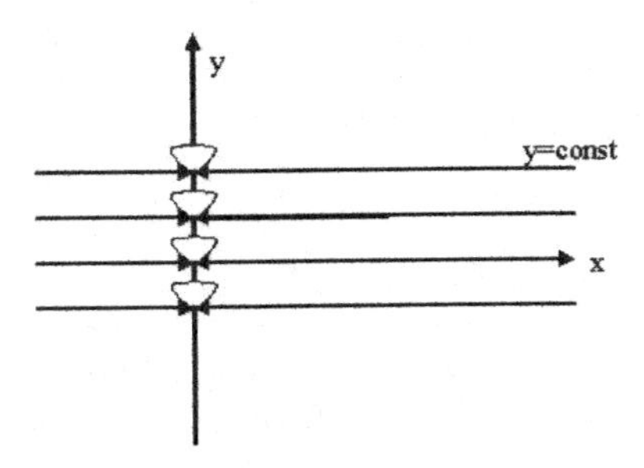

Figure 2.8: The sketch of the affine transformation group orbits.

If we choose the value $\epsilon_1 = -\epsilon$ then for any value ϵ we obtain the transformation which maps the image point (x_ϵ, y_ϵ) to the source point (x, y) for any $(x, y) \in \mathbb{R}^2$. Because the value $-\epsilon \in \mathbb{R}$ for any value $\epsilon \in \mathbb{R}$ we proved in this way that the inverse transformation is included in the family of transformations.

3. We prove the last group property $g_{\epsilon_1+\epsilon_2} = g_{\epsilon_2} g_{\epsilon_1}$.

$$\begin{aligned} g_{\epsilon_2} g_{\epsilon_1} x &= e^{\epsilon_2} x_{\epsilon_1} = e^{\epsilon_2} e^{\epsilon_1} x = e^{\epsilon_1+\epsilon_2} x = g_{\epsilon_1+\epsilon_2} x, \\ g_{\epsilon_2} g_{\epsilon_1} y &= y_{\epsilon_2} = y_{\epsilon_1} = y = g_{\epsilon_1+\epsilon_2} y. \end{aligned}$$

Indeed the family of transformations build a group of point transformations on $\mathbb{R}^2$.

The one-dimensional orbits of this group are half lines that lie parallel to the x-axis. The y-axis is the locus of all invariant points of this group (in other words the null-dimensional orbits). The sketch of the orbits is presented in Figure 2.8.

Problem 2.3.3 *Find the closed form of transformation equations if the infinitesimal representation of the group transformations is given by*

$$\xi(x, y) = x - y, \quad \eta(x, y) = x + y.$$

Solution To find the closed representation of the group we use the Lie equations

$$\frac{dx}{x-y} = \frac{dy}{x+y} = \frac{d\epsilon}{1}. \tag{2.3.10}$$

In order to solve the first equation of the system, we change the Cartesian coordinates (x, y) to polar coordinates (r, θ), i.e. $x = r\cos\theta$, $y = r\sin\theta$, where $r \in \mathbb{R},\ \theta \in [0, 2\pi)$

$$\begin{aligned} dx &= dr\cos\theta - r\sin\theta d\theta, \\ dy &= dr\sin\theta + r\cos\theta d\theta. \end{aligned} \tag{2.3.11}$$

If we insert the expressions (2.3.11) in the first equation of the system (2.3.10), we get

$$\frac{dr\cos\theta - r\sin\theta d\theta}{r(\cos\theta - \sin\theta)} = \frac{dr\sin\theta + r\cos\theta d\theta}{r(\cos\theta + \sin\theta)}.$$

Multiplying the equations above by denominators and collecting the terms with dr and $d\theta$, we obtain

$$\begin{gathered} ((\cos\theta + \sin\theta)\cos\theta - \sin\theta(\cos\theta - \sin\theta))\,dr = \\ r\,(\sin\theta(\cos\theta + \sin\theta) + \cos\theta(\cos\theta - \sin\theta))\,d\theta, \end{gathered}$$

or using the trigonometry we obtain $dr = rd\theta$. Now the Lie equations take the form

$$\frac{dr}{r} = d\theta = d\epsilon.$$

The solution of the system of equations above in polar coordinates is given by

$$r_\epsilon = re^\epsilon, \quad \theta_\epsilon = \theta + \epsilon. \tag{2.3.12}$$

In Cartesian coordinates, i.e. in variables (x, y) this solution takes the form

$$\begin{aligned} x_\epsilon = r_\epsilon\cos(\theta + \epsilon) = e^\epsilon\sqrt{x^2 + y^2}\cos\left(\arctan\frac{y}{x} + \epsilon\right), \\ y_\epsilon = r_\epsilon\sin(\theta + \epsilon) = e^\epsilon\sqrt{x^2 + y^2}\sin\left(\arctan\frac{y}{x} + \epsilon\right). \end{aligned}$$

After a simplification of these expressions we obtain the closed form for transformations of this group

$$\begin{aligned} x_\epsilon &= e^\epsilon\sqrt{x^2 + y^2}\left(\cos\left(\arctan\frac{y}{x}\right)\cos\epsilon - \sin\left(\arctan\frac{y}{x}\right)\sin\epsilon\right) \\ &= e^\epsilon\left(x\cos\epsilon - y\sin\epsilon\right), \\ y_\epsilon &= \sqrt{x^2 + y^2}e^\epsilon\left(\sin\left(\arctan\frac{y}{x}\right)\cos\epsilon + \cos\left(\arctan\frac{y}{x}\right)\sin\epsilon\right) \\ &= e^\epsilon\left(y\cos\epsilon + x\sin\epsilon\right), \end{aligned}$$

where we used the transformations

$$\frac{y}{x} = \tan\theta,\ r^2 = x^2 + y^2, \quad \text{or} \quad \theta = \arctan\frac{y}{x},\ r = \sqrt{x^2 + y^2}.$$

Finally we get

$$x_\epsilon = e^\epsilon \left(x \cos \epsilon - y \sin \epsilon\right),$$
$$y_\epsilon = e^\epsilon \left(y \cos \epsilon + x \sin \epsilon\right).$$

From these equations we see that the orbits of this group are logarithmic spirals.

Problem 2.3.4 *Let a one-parameter family of point transformations be given by the rules*

$$x_\varepsilon = xe^\varepsilon, \quad y_\varepsilon = ye^\varepsilon, \quad \varepsilon \in \mathbb{R}. \tag{2.3.13}$$

Do these transformations build a group? If it is the case, which orbits has this group?

Solution Let us check the group properties:

1. We prove that the identity transformation is included in the family of the transformations. We solve the equations

$$x = e^{\varepsilon_0} x,$$
$$y = e^{\varepsilon_0} y$$

   for ε_0. The system has a unique solution $\varepsilon_0 = 0$ for any value of x, y.

2. We prove now that for any transformation $g_\varepsilon : (x, y) \to (x_\varepsilon, y_\varepsilon)$ there exists an inverse transformation $g_{\varepsilon_1} : (x_\varepsilon, y_\varepsilon) \to (x, y)$, i.e. we can find the corresponding value ε_1.
   We take the point $(x_\varepsilon, y_\varepsilon)$ and apply the transformation g_{ε_1} with an unknown value ε_1 and solve the system of equations

$$g_{\varepsilon_1} x_\varepsilon = e^{\varepsilon_1} x_\varepsilon = e^{\varepsilon_1} e^\varepsilon x = e^{\varepsilon_1 + \varepsilon} x = x, \quad \varepsilon_1 = -\varepsilon,$$
$$g_{\varepsilon_1} y_\varepsilon = e^{\varepsilon_1} e^\varepsilon y = e^{\varepsilon_1 + \varepsilon} y = y, \quad \varepsilon_1 = -\varepsilon.$$

   If we choose the value $\varepsilon_1 = -\varepsilon$ then we obtain the transformation which maps the image point $(x_\varepsilon, y_\varepsilon)$ to the source point (x, y) for any $(x, y) \in \mathbb{R}^2$. Since the value $-\varepsilon \in \mathbb{R}$ for any value $\varepsilon \in \mathbb{R}$ we proved that the inverse transformation is included in the family of transformations for any value $\varepsilon \in \mathbb{R}$.

3. We prove the group property $g_{\varepsilon_1 + \varepsilon_2} = g_{\varepsilon_2} g_{\varepsilon_1}$, i.e. that the result of an action of two successive transformations with parameters $\varepsilon_1, \varepsilon_2$ is

equivalent to action of one transformation of the set with the parameter equal to $\varepsilon_1 + \varepsilon_2$. After (2.3.13) we have

$$\begin{aligned} g_{\varepsilon_2} g_{\varepsilon_1} x &= e^{\varepsilon_2} x_{\varepsilon_1} = e^{\varepsilon_2} e^{\varepsilon_1} x = e^{\varepsilon_1+\varepsilon_2} x = g_{\varepsilon_1+\varepsilon_2} x, \\ g_{\varepsilon_2} g_{\varepsilon_1} y &= e^{\varepsilon_2} y_{\varepsilon_1} = e^{\varepsilon_2} e^{\varepsilon_1} y = e^{\varepsilon_1+\varepsilon_2} y = g_{\varepsilon_1+\varepsilon_2} y. \end{aligned}$$

We have proved that we have to do with a group of point transformations. The group (2.3.13) is called the group of **uniform dilation**. The one-dimensional orbits of this group are half lines starting in the origin of coordinate system. The point $(0,0)$ is the invariant point of this group (in other words - a null-dimensional orbit).

Problem 2.3.5 *Let the group of* **linear fractional transformations** *be given by the closed form transformation equations*

$$\begin{aligned} x_\varepsilon &= \frac{x}{1-\varepsilon x}, \\ y_\varepsilon &= \frac{y}{1-\varepsilon y}, \quad \varepsilon \in \mathbb{R}. \end{aligned} \tag{2.3.14}$$

Find the infinitesimal representation of this group (i.e. the functions $\xi(x,y)$ and $\eta(x,y)$). Where are the point transformations well defined?

Solution The group (2.3.14) is a local group, because each transformation has its own region where it is defined $\varepsilon \in (-a,a)$ where $a = \min\left(\frac{1}{|x|}, \frac{1}{|y|}\right)$ and all transformations are well defined just in a infinitesimal small neighborhood of the identical transformation.
The linear part of a transformation is given by

$$\begin{aligned} \xi(x,y) &= \left.\frac{dx_\varepsilon}{d\varepsilon}\right|_{\varepsilon=0} = -\left.\frac{-x^2}{(1-\varepsilon x)^2}\right|_{\varepsilon=0} = x^2, \\ \eta(x,y) &= \left.\frac{dy_\varepsilon}{d\varepsilon}\right|_{\varepsilon=0} = -\left.\frac{-y^2}{(1-\varepsilon y)^2}\right|_{\varepsilon=0} = y^2. \end{aligned}$$

Problem 2.3.6 *Find the closed form of transformation equations for the* **Lorentz group** *if the infinitesimal representation is given by*

$$\xi(x,y) = y, \quad \eta(x,y) = x.$$

Solution Let us write Lie equations for this group

$$\frac{dx}{y} = \frac{dy}{x} = d\varepsilon. \tag{2.3.15}$$

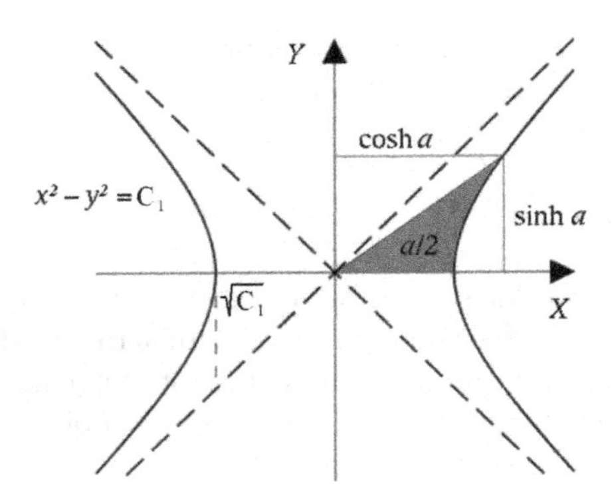

Figure 2.9: The sketch of the Lorentz group's orbit.

The first equation of this system defines the orbits of the group. For the orbits we obtain the expression

$$x^2 - y^2 = C_1, \quad C_1 \in \mathbb{R}. \tag{2.3.16}$$

From this expression we obtain the value x as a function of y and C_1 and substitute into the second equation of (2.3.15) so we obtain an equation on y

$$\frac{dy}{\sqrt{y^2 + C_1}} = d\varepsilon.$$

After integration we obtain the following expression

$$y = \sqrt{C_1} \sinh(\varepsilon - C_2), \quad C_2 \in \mathbb{R}. \tag{2.3.17}$$

Using the same method we can easily get the expression for x as a function of ε.

Now we obtain the coordinates of the point (x, y) after the point transformation with parameter ε ,i.e. the values $(x_\varepsilon, y_\varepsilon)$. We substitute in the last equations the expression for the constant C_1 and get

$$y_\varepsilon = \sqrt{x^2 - y^2}(\sinh\varepsilon \cosh C_2 - \cosh\varepsilon \sinh C_2),$$
$$x_\varepsilon = \sqrt{x^2 - y^2}(\cosh\varepsilon \cosh C_2 + \sinh\varepsilon \sinh C_2).$$

Looking at Figure 2.9 we can understand that if we denote $C_2 = a$ then the closed form of Lorenz group of point transformations will be

$$x_\varepsilon = x\cosh\varepsilon + y\sinh\varepsilon,$$
$$y_\varepsilon = y\cosh\varepsilon - x\sinh\varepsilon. \tag{2.3.18}$$

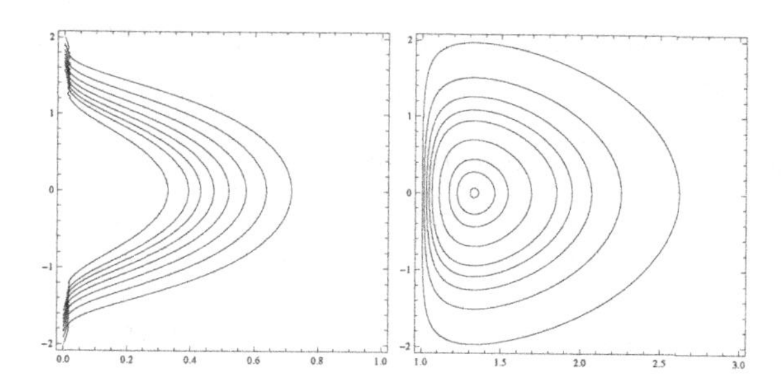

Figure 2.10: The integral curves for (2.3.20) $0 < x < 1$ and for $x > 1$.

Problem 2.3.7 *Let a one-parameter family of curves be defined by the formulas*

$$x^2 - 2y \ln x = Cy, \quad y = 0, \quad x = 0. \tag{2.3.19}$$

Find a first order ordinary differential equation for which this family of curves represents the integral curves of the equation.

Solution We rewrite the given expression for the family of curves in the form

$$C = \frac{x^2}{y} - 2 \ln x.$$

Calculating a total differential for both sides of the equation we obtain

$$0 = \frac{2xydx - x^2dy}{y^2} - 2\frac{dx}{x},$$

separating the terms with dx from the terms with dy we obtain

$$\frac{2x^2ydx - 2y^2}{xy^2}dx = \frac{x^3}{xy^2}dy.$$

We look at this ODE more attentively. We should remember that the lines $y = 0$ and $x = 0$ also belongs to the family of curves (2.3.19) and they should be integral curves of the equation which we look for. But the equation we have obtained does not have these solutions. Therefore we should multiply both sides of the equation by xy^2 and only this new equation will be the correct one

$$(2x^2y - 2y^2)dx = x^3dy. \tag{2.3.20}$$

Problem 2.3.8 *Let a one-parameter family of curves be given by*

$$x = Cy + ye^x, \quad y = 0, \quad x = 0. \tag{2.3.21}$$

Find a first order ordinary differential equation for which this family of curves represents the integral curves of the equation.

Solution Let us find the expression for C

$$\frac{x}{y} - e^x = C. \tag{2.3.22}$$

Calculating a total differential of both sides of the equation we obtain

$$\frac{ydx - xdy}{y^2} - e^x dx = 0,$$

separating the terms with dx from the terms with dy we obtain

$$\left(\frac{1}{y} - e^2\right) dx = \frac{x}{y^2} dy.$$

We know that $y = 0$ and $x = 0$ should also solve the differential equation, but the equation we have obtained does not have the integral curve $y = 0$. Therefore we should multiply both sides of the equation by y^2 and only this new equation will be the ODE which we are looking for

$$y(1 - ye^x)dx = xdy.$$

Problem 2.3.9 *Let a one-parameter family of curves be given by*

$$x^2 = (C + y)e^y.$$

Find a first order ordinary differential equation for which this family of curves represents the integral curves of the equation.

Solution Let us find the expression for C

$$C = x^2 e^{-y} - y.$$

Calculating a total differential of both sides of the equation we obtain

$$2xe^{-y} - x^2 e^{-y} dy - dy = 0,$$

separating the terms with dx from the terms with dy we obtain the solution

$$2xe^{-y} dx = (x^2 e^{-y} + 1)dy.$$

Problem 2.3.10 *Let a one-parameter family of curves be described by the formula*

$$y^3 - x^3 - xy + x = C.$$

Find a first order ordinary differential equation for which this family of curves represents the integral curves of the equation.

Solution Let us calculate a total differential of both sides of the equation

$$3y^2dy - 3x^2dx - ydx - xdy + dx = 0.$$

Separating the terms with dx from the terms with dy leads to the ODE

$$(3x^2 + y - 1)dx = (3y^2 - x)dy. \tag{2.3.23}$$

Problem 2.3.11 *Let a one-parameter family of curves be given by a formulas*

$$x = Cy - y^2, \;\; y = 0. \tag{2.3.24}$$

Find a first order ordinary differential equation for which this family of curves represents the integral curves of the equation.

Solution Let us find the expression for the parameter C of the family

$$C = \frac{x + y^2}{y}.$$

Differentiating both sides gives us

$$\frac{ydx + 2y^2dy - xdy - y^2dy}{y^2} = 0.$$

Separating the terms with dx from the terms with dy we obtain an ODE

$$\frac{1}{y}dx = \left(\frac{x}{y^2} - 1\right) dy.$$

We remember that $y = 0$ also solves the equation to find, while the equation we have obtained does not have this solution. Therefore we should multiply both sides of the equation by y^2 and only this new equation will be the correct answer to the problem

$$ydx = \left(x - y^2\right) dy. \tag{2.3.25}$$

2.4 Exercises

1. Let the group of projective transformations be given by the closed transformation equations

$$x_\epsilon = \frac{x}{1-\epsilon x},$$
$$y_\epsilon = \frac{y}{1-\epsilon x}, \quad \epsilon \in \mathbb{R}.$$

   Find the infinitesimal representation of this group (i.e. the functions $\xi(x,y)$ and $\eta(x,y)$).

2. Find the closed form of the one-parameter point transformations for a group having the following infinitesimal representation

$$\xi = x + y, \quad \eta = 3y.$$

Chapter 3

Invariants of a one-parameter group of point transformations

In this chapter we introduce the ideas of invariant points, invariant curves, and invariant family of curves on the plane. These ideas build the basic background needed to apply Lie group theory to differential equations. We also discuss a connection between a closed form and infinitesimal form of point transformations. We show that a first order differential operator called *infinitesimal generator* is deeply connected with the studied group of transformations. This general concept of an infinitesimal generator, invented by Sophus Lie, contributed a lot to the future success of his theory. In fact it is much more convenient to work with linear differential operators than with nonlinear and complicated formulas describing the action of the group in closed form. Using infinitesimal generators one can describe the action of the group in a small neighborhood of the source point. It is also an effective tool to describe invariance properties and all main features of the transformation group. So, the infinitesimal generators are a natural way to apply the Lie group theory to differential equations.

3.1 Group invariants

In the previous chapter we have seen that an orbit of a one-parameter group represented either as a curve on a plane or as a function in $\mathbb{R}^2$ remains unaltered under the action of the group. In this chapter we give a general definition of an invariant and study different types of invariants.

We give a definition of an orbit which was intuitively introduced in the previous chapter.

Definition 3.1.1 (Orbit) *An orbit of a one-parameter continuous group of point transformations on* $\mathbb{R}^2$ *is a subset of* $\mathbb{R}^2$*, defined for each* $(x, y) \in \mathbb{R}^2$ *by*

$$O_{(x,y)} = \{g_\varepsilon(x, y) | g_\varepsilon \in G(\mathbb{R}^2)\}. \tag{3.1.1}$$

Now let us describe the idea of a group invariant in an intuitively clear way and later we will develop this idea.

Definition 3.1.2 (Group invariants) *A group invariant (or a group invariant function) is a function* $u(x, y)$ *whose value remains unaltered under the action of the group* $G(\mathbb{R}^2)$

$$u(x_\varepsilon, y_\varepsilon) = u(x, y). \tag{3.1.2}$$

This definition is easy to use in the following way. If we have some function $u(x, y)$ and a one-parameter group of point transformations is given in the closed form, then we can just insert transformed variables x and y in the function $u(x, y)$ and see if the form of function stays unaltered. However it is not so easy to find an invariant function using this definition. Now we develop a more convenient step-by-step procedure that allows us to find invariant functions using infinitesimal representation of this group.

In the previous chapter we introduced the infinitesimal representation of a one-parameter group on a plane

$$x_\varepsilon = x + \left.\frac{\partial X_\varepsilon}{\partial \varepsilon}\right|_{\varepsilon=0} \cdot \varepsilon + \mathcal{O}(\varepsilon^2),$$
$$y_\varepsilon = y + \left.\frac{\partial Y_\varepsilon}{\partial \varepsilon}\right|_{\varepsilon=0} \cdot \varepsilon + \mathcal{O}(\varepsilon^2),$$

and denoted the linear parts of the Taylor series by $\xi(x, y)$ and $\eta(x, y)$. If we examine the limit of $x_\varepsilon - x$ by $\varepsilon \to 0$, we get

$$\lim_{\varepsilon \to 0} \frac{x_\varepsilon - x}{\varepsilon} = \left.\frac{\partial X_\varepsilon}{\partial \varepsilon}\right|_{\varepsilon=0} + \mathcal{O}(\varepsilon) \;\Rightarrow\; dx = \xi(x, y) d\varepsilon$$

and by analog $dy = \eta(x, y) d\varepsilon$. The Lie equations

$$\frac{dx}{\xi(x, y)} = \frac{dy}{\eta(x, y)} = \frac{d\varepsilon}{1} \tag{3.1.3}$$

can be used to get the closed form representation of the one-parameter group of transformations if we have the functions $\xi(x, y), \eta(x, y)$, i.e. the infinitesimal representation.

Let $u(x,y)$ be an invariant function under the action of the one-parameter group of point transformations which is given in the closed form by

$$x_\varepsilon = X_\varepsilon(x,y)$$
$$y_\varepsilon = Y_\varepsilon(x,y),$$

or, in the infinitesimal form by

$$\begin{aligned} dx &= \xi(x,y)d\varepsilon, \\ dy &= \eta(x,y)d\varepsilon. \end{aligned} \tag{3.1.4}$$

Taking the equality (3.1.2) and differentiating it with respect to ε we obtain

$$\frac{du(x_\varepsilon, y_\varepsilon)}{d\varepsilon} = \frac{\partial u}{\partial x_\varepsilon}\frac{dx_\varepsilon}{d\varepsilon} + \frac{\partial u}{\partial y_\varepsilon}\frac{dy_\varepsilon}{d\varepsilon} = 0, \tag{3.1.5}$$

because the right part of (3.1.2) does not depend on ε. The Equation (3.1.5) holds for any value of ε in particular for the value $\varepsilon = 0$. Using the equations (3.1.4) we rewrite the middle part of (3.1.5) and obtain that any invariant function $u(x,y)$ solves the first order PDE

$$\xi(x,y)u_x + \eta(x,y)u_y = 0. \tag{3.1.6}$$

The characteristic equation associated with this first order PDE is a first order ODE and has the form

$$\frac{dx}{\xi(x,y)} = \frac{dy}{\eta(x,y)}. \tag{3.1.7}$$

The solution of this ODE is called a first integral and we denote it as $\phi(x,y) =$ const. From the theory of the first order PDEs it follows that the general solution $u(x,y)$ of the linear PDE (3.1.6) is an arbitrary function $F(\cdot)$ of the first integral $\phi(x,y)$ of the characteristic Equation (3.1.7). It means that any solution of (3.1.6) has a form $u(x,y) = F(\phi(x,y))$.

If we compare the Equation (3.1.7) with the Lie equations (3.1.3) we see that it is the first of these equations. In other words it is the equation which describes the orbits of the Lie group. The first integral $\phi(x,y) =$ const. of this ODE represent the orbits of the group. It means that any function which is invariant under action of the studied group is a function of orbits $u(x,y) = F(\phi(x,y))$.

With the PDE (3.1.6) we have found a convenient tool to find all functions which are invariant under the action of a one-parameter group of point transformations. It is enough to know the functions $\xi(x,y), \eta(x,y)$, i.e. the infinitesimal representation of the group to be able to prove if a function is an invariant of this group or to find all invariant functions as solutions of (3.1.6). We formulate and prove the corresponding theorem in the next section.

Let us look at two examples of invariant functions.

Example 3.1.1 (Translation group) We take a translation group which provides a homogeneous translation of a point (x, y) along a line parallel to the x-axis, i.e., in the infinitesimal form the transformation is described by $\xi(x, y) = 1$, $\eta(x, y) = 0$. We can say that we move the point (x, y) with constant horizontal velocity.

We insert the values $\xi(x, y) = 1$, $\eta(x, y) = 0$ in the characteristic Equation (3.1.7) and obtain $1 \cdot u_x + 0 \cdot u_y = 0$. To obtain the most general function $u(x, y)$ that is invariant under the action of this group we have to solve the ODE

$$\frac{dx}{1} = \frac{dy}{0}.$$

Remark 3.1.1 *The fraction on the right part of the equation does not mean that we divide by zero. It is just a formal expression which means that y is an invariant. We will explain it in the next lines.*

This equation is equivalent to $0 \cdot dx = 1 \cdot dy$. We obtain the first integral in the form $\phi(x, y) = \int dy = \text{const}$. It means that $\phi(x, y) = y$ is the first integral of the characteristic equation. The function $u(x, y) = F(\phi(x, y)) = F(y)$, where $F(\cdot)$ is an arbitrary function, is the most general function $u(x, y)$ which is invariant under the action of this group. It means that $u(x, y) = F(y)$ depends only on the variable y and does not depend on the variable x.

We can verify that this function $u(x, y)$ solves the equation (3.1.6) for the given translation group. We calculate that $u_x = 0$, $u_y = F'(y)$ and therefore, $1 \cdot u_x + 0 \cdot u_y = 1 \cdot 0 + 0 \cdot F'(y) = 0$.

We now give a geometrical interpretation of the invariant function which will be used often later. Let us take a space $\mathbb{R}^3$ with coordinates (x, y, u). We obtain a surface $u = u(x, y)$ in this space, which is invariant under the action of the translation group, if we choose any curve $u = F(y)$ in the (y, u)-plane and move it parallel to the x-axis it will stay on the same surface. All such surfaces represent invariants of the translation group. ◇

Example 3.1.2 (Rotation group) In Chapter 2 we have found the infinitesimal form of the rotation group on the plain. We got $\xi(x, y) = -y$ and

$\eta(x,y) = x$. If a function $u(x,y)$ is invariant under the action of the rotation group, then it is a solution of the equation

$$-yu_x + xu_y = 0.$$

The corresponding characteristic equation for this linear PDE of the first order is

$$\frac{dx}{-y} = \frac{dy}{x}.$$

Looking for the first integral of this equation, we conclude that

$$\int x\,dx = \int -y\,dy \;\Rightarrow\; \frac{x^2}{2} + \frac{y^2}{2} = \tilde{\phi}(x,y) = \text{const.}$$

Since the first integral multiplied by a constant is still a constant we simplify a bit the expression for the first integral and take the expression $\phi(x,y) = x^2 + y^2$ as a first integral. An arbitrary function $F(\cdot)$ of the first integral, i.e. $F(\phi(x,y))$ will be invariant under the action of the rotation group. So we obtain that an invariant function of the rotation group has the form

$$u(x,y) = F(x^2 + y^2).$$

Let us verify that the function $u(x,y)$ is invariant. We insert $u(x,y)$ in (3.1.6) and obtain $u_x = 2xF'$, $u_y = 2yF'$, and $-y \cdot 2xF' + x \cdot 2yF' = 0$.

The above expression for the invariant function $u(x,y)$ has the following geometrical interpretation. If we take a space $\mathbb{R}^3$ with coordinates (x,y,u) and choose an arbitrary curve in the plane (x,u) or in the plane (y,u) and rotate it around the u-axis we obtain a surface which is invariant under the action of a rotation group on the plane. ◇

3.2 An infinitesimal generator U

Let us look once more at Equation (3.1.6) where each solution $u(x,y)$ is an invariant function under the action of the one-parameter group of point transformations. We can rewrite this equation in the following way

$$\left(\xi(x,y)\frac{\partial}{\partial x} + \eta(x,y)\frac{\partial}{\partial y}\right) u(x,y) = 0.$$

We notice that the action of the first order differential operator $\mathbf{U}$ on the function $u(x,y)$ must be equal to zero if it is an invariant. The first order linear differential operator in question is

$$\mathbf{U} = \xi(x,y)\frac{\partial}{\partial x} + \eta(x,y)\frac{\partial}{\partial y}. \tag{3.2.8}$$

It represents locally the action of the group. By using this operator we can obtain the direction field which defines the infinitesimal motion of this point for each point on the plane, the orbits which are integral curves of this direction field and the closed form of the group transformations if we solve the corresponding Lie equations. In other words, we can generate the whole one-parameter point transformation group with all its features. This is the reason why this operator has a special name - the **infinitesimal generator U** of the one-parameter continuous group.

We can study the action of this operator on different functions, not only on an invariant function $u(x, y)$.

Let us study the result of the action of $\mathbf{U}$ on the variable x

$$\mathbf{U}x = \left(\xi(x, y)\frac{\partial}{\partial x} + \eta(x, y)\frac{\partial}{\partial y} \right) x = \xi(x, y),$$

and on the variable y

$$\mathbf{U}y = \left(\xi(x, y)\frac{\partial}{\partial x} + \eta(x, y)\frac{\partial}{\partial y} \right) y = \eta(x, y).$$

It was to be expected that we get the functions $\xi(x, y)$ and $\eta(x, y)$.

Now we can recall how the Taylor series for the coordinates $(x_\varepsilon, y_\varepsilon)$ of an image point looks like

$$x_\varepsilon = x + \left.\frac{\partial X_\varepsilon}{\partial \varepsilon}\right|_{\varepsilon=0} \varepsilon + \frac{1}{2} \left.\frac{\partial^2 X_\varepsilon}{\partial \varepsilon^2}\right|_{\varepsilon=0} \varepsilon^2 + \mathcal{O}(\varepsilon^3),$$
$$y_\varepsilon = y + \left.\frac{\partial Y_\varepsilon}{\partial \varepsilon}\right|_{\varepsilon=0} \varepsilon + \frac{1}{2} \left.\frac{\partial^2 Y_\varepsilon}{\partial \varepsilon^2}\right|_{\varepsilon=0} \varepsilon^2 + \mathcal{O}(\varepsilon^3).$$

We rewrite these expressions using the operator $\mathbf{U}$

$$x_\varepsilon = x + \xi(x, y) \cdot \varepsilon + \mathcal{O}(\varepsilon^2) = x + \mathbf{U}x \cdot \varepsilon + \mathcal{O}(\varepsilon^2),$$
$$y_\varepsilon = y + \eta(x, y) \cdot \varepsilon + \mathcal{O}(\varepsilon^2) = y + \mathbf{U}y \cdot \varepsilon + \mathcal{O}(\varepsilon^2).$$

Here we represented the functions $\xi(x, y)$ and $\eta(x, y)$ as the result of the action of the infinitesimal generator $\mathbf{U}$ on variables x and y correspondingly.

Can we describe the next term in the Taylor series, the expression $\frac{\partial^2 X_\varepsilon}{\partial \varepsilon^2}$, as a result of an action of $\mathbf{U}$ on some function as well?

The second derivative $\frac{\partial^2 X_\varepsilon}{\partial \varepsilon^2}$ is a derivative of the first derivative, i.e. we can use the previous results and obtain

$$\begin{aligned}
\frac{\partial}{\partial \varepsilon}\left(\frac{\partial X_\varepsilon(x,y)}{\partial \varepsilon}\right)\bigg|_{\varepsilon=0} &= \frac{\partial}{\partial \varepsilon}\left(\xi(x_\varepsilon, y_\varepsilon)\right)\bigg|_{\varepsilon=0} \\
&= \left(\frac{\partial \xi(x_\varepsilon, y_\varepsilon)}{\partial x_\varepsilon}\cdot\frac{\partial X_\varepsilon}{\partial \varepsilon} + \frac{\partial \xi(x_\varepsilon, y_\varepsilon)}{\partial y_\varepsilon}\cdot\frac{\partial Y_\varepsilon}{\partial \varepsilon}\right)\bigg|_{\varepsilon=0} \\
&= \left(\xi(x_\varepsilon, y_\varepsilon)\frac{\partial \xi(x_\varepsilon, y_\varepsilon)}{\partial x_\varepsilon} + \eta(x_\varepsilon, y_\varepsilon)\frac{\partial \xi(x_\varepsilon, y_\varepsilon)}{\partial y_\varepsilon}\right)\bigg|_{\varepsilon=0} \\
&= \mathbf{U}\xi(x,y) = \mathbf{U}(\mathbf{U}x) = \mathbf{U}^2 x.
\end{aligned}$$

Consequently we represent the Taylor series for $(x_\varepsilon, y_\varepsilon)$ in the following form

$$x_\varepsilon = x + \mathbf{U}x\cdot\varepsilon + \frac{1}{2}\mathbf{U}^2 x\cdot\varepsilon^2 + \mathcal{O}(\varepsilon^3) = \sum_{n=0}^{\infty}\frac{\mathbf{U}^n x}{n!}\varepsilon^n = \mathrm{e}^{\varepsilon\mathbf{U}}x, \tag{3.2.9}$$

$$y_\varepsilon = y + \mathbf{U}y\cdot\varepsilon + \frac{1}{2}\mathbf{U}^2 y\cdot\varepsilon^2 + \mathcal{O}(\varepsilon^3) = \sum_{n=0}^{\infty}\frac{\mathbf{U}^n y}{n!}\varepsilon^n = \mathrm{e}^{\varepsilon\mathbf{U}}y.$$

The expression $\mathbf{U}^n$ is defined as a successive application of the operator $\mathbf{U}$. The expressions $x_\varepsilon = \mathrm{e}^{\varepsilon\mathbf{U}}x$ and $y_\varepsilon = \mathrm{e}^{\varepsilon\mathbf{U}}y$ are defined by the Taylor series above. The expression $\mathrm{e}^{\varepsilon\mathbf{U}}$ is called an **exponential map**. We can use the exponential map to reproduce the closed form of the group of one-parameter point transformations instead of solving the Lie equations for this group using $\xi(x,y)$ and $\eta(x,y)$.

We illustrate these ideas with two examples. We take the rotation and translation group and use the exponential map to get the transformations of these groups in a closed form.

Example 3.2.1 (Rotation group) For the rotation group we got before that $\xi(x,y) = -y$ and $\eta(x,y) = x$. The infinitesimal generator of the rotation group has correspondingly the form

$$\mathbf{U} = -y\frac{\partial}{\partial x} + x\frac{\partial}{\partial y}.$$

The terms in the Taylor series for the transformed point $(x_\varepsilon, y_\varepsilon)$ in the neighborhood of the point (x,y), i.e. for small ε are

$$\begin{array}{l|l}
\mathbf{U}\,x = \left(-y\frac{\partial}{\partial x} + x\frac{\partial}{\partial y}\right)x = -y, & \mathbf{U}\,y = x, \\
\mathbf{U}^2 x = \left(-y\frac{\partial}{\partial x} + x\frac{\partial}{\partial y}\right)(-y) = -x, & \mathbf{U}^2 y = -y, \\
\mathbf{U}^3 x = y, & \mathbf{U}^3 y = -x, \\
\mathbf{U}^4 x = x, & \mathbf{U}^4 y = y, \\
\vdots & \vdots
\end{array}$$

Thus, we obtain

$$\begin{aligned}
x_\varepsilon &= x + (-y) \cdot \varepsilon - x \cdot \frac{\varepsilon^2}{2} + y \cdot \frac{\varepsilon^3}{3!} + x \cdot \frac{\varepsilon^4}{4!} + \mathcal{O}(\varepsilon^5) \\
&= x \left(1 - \frac{\varepsilon^2}{2} + \frac{\varepsilon^4}{4!} + \mathcal{O}(\varepsilon^6)\right) - y \left(\varepsilon - \frac{\varepsilon^3}{3!} + \mathcal{O}(\varepsilon^5)\right) \\
&= x \cos \varepsilon - y \sin \varepsilon,
\end{aligned}$$

$$\begin{aligned}
y_\varepsilon &= x \left(\varepsilon - \frac{\varepsilon^3}{3!} + \mathcal{O}(\varepsilon^5)\right) + y \left(1 - \frac{\varepsilon^2}{2} + \mathcal{O}(\varepsilon^4)\right) \\
&= x \sin \varepsilon + y \cos \varepsilon.
\end{aligned}$$

Actually, we got the closed form for the transformations of the rotation group on a plane. We have seen that the action of the group indeed can be described as the action of a first order differential operator $\mathbf{U}$ (the infinitesimal generator of the group) on the source point (x, y) in an infinitesimal small neighborhood of this point. ◇

Example 3.2.2 (Translation group) The translation group (along the line parallel to the x-axis) has the infinitesimal generator

$$\mathbf{U} = \frac{\partial}{\partial x}$$

because of $\xi(x, y) = 1$ and $\eta(x, y) = 0$. The corresponding terms in the Taylor series for the transformed point $(x_\varepsilon, y_\varepsilon)$ in a small neighborhood of the source point (x, y) are

$$\begin{array}{l|l}
\mathbf{U}\, x = 1, & \mathbf{U}\, y = 0, \\
\mathbf{U}^2 x = 0, & \mathbf{U}^2 y = 0, \\
\vdots & \vdots \\
\mathbf{U}^n x = 0, & \mathbf{U}^n y = 0.
\end{array}$$

Using the exponential map we obtain the closed form for the transformations of the translation group: $x_\varepsilon = x + \varepsilon, \quad y_\varepsilon = y$. ◇

Using the idea of an infinitesimal generator we can formulate the following theorem about an invariant function.

Theorem 3.2.1 *A function $u(x, y)$ is an invariant of the group $G(\mathbb{R}^2)$ with the infinitesimal generator $\mathbf{U} = \xi(x, y)\frac{\partial}{\partial x} + \eta(x, y)\frac{\partial}{\partial y}$ if and only if it solves the homogeneous linear PDE*

$$\mathbf{U}u = \xi(x, y)\frac{\partial u}{\partial x} + \eta(x, y)\frac{\partial u}{\partial y} = 0. \qquad (3.2.10)$$

Proof. Let us take an arbitrary invariant function $u(x, y)$. According to the definition of an invariant function the value of the function is kept unaltered after the transformation of the variables. From the other side, we obtain

$$u(x_\varepsilon, y_\varepsilon) = u(x + \xi\varepsilon + \mathcal{O}(\varepsilon^2), y + \eta\varepsilon + \mathcal{O}(\varepsilon^2)) = u(x, y) + \varepsilon\xi u_x + \varepsilon\eta u_y + \mathcal{O}(\varepsilon^2).$$

From (3.1.2) we get that all terms except the first one on the right side of this equation should be equal zero, and the second term is also equal to zero. It means that the function $u(x, y)$ should be a solution of (3.2.10).

So let $u(x, y)$ be a solution of (3.2.10) now, i.e. $\mathbf{U}u = 0$. Then (3.2.10) is valid at any point of the plane $\mathbb{R}^2$. Considering an image point $(x_\varepsilon, y_\varepsilon)$, we obtain

$$\xi(x_\varepsilon, y_\varepsilon)\frac{\partial u(x_\varepsilon, y_\varepsilon)}{\partial x_\varepsilon} + \eta(x_\varepsilon, y_\varepsilon)\frac{\partial u(x_\varepsilon, y_\varepsilon)}{\partial y_\varepsilon} = 0.$$

If we calculate the first derivative of the function $u(x_\varepsilon, y_\varepsilon)$ with respect to ε we obtain

$$\frac{du(x_\varepsilon, y_\varepsilon)}{d\varepsilon} = \frac{\partial u}{\partial x_\varepsilon}\frac{dx_\varepsilon}{d\varepsilon} + \frac{\partial u}{\partial y_\varepsilon}\frac{dy_\varepsilon}{d\varepsilon}.$$

Using the formulas (3.1.4) we get

$$\frac{\partial u}{\partial x_\varepsilon}\frac{dx_\varepsilon}{d\varepsilon} + \frac{\partial u}{\partial y_\varepsilon}\frac{dy_\varepsilon}{d\varepsilon} = \xi(x_\varepsilon, y_\varepsilon)u_{x_\varepsilon} + \eta(x_\varepsilon, y_\varepsilon)u_{y_\varepsilon} = 0$$

because the function $u(x, y)$ is a solution of (3.2.10) in any point of the plane also in the point $(x_\varepsilon, y_\varepsilon)$. Therefore, we obtain

$$\frac{du(x_\varepsilon, y_\varepsilon)}{d\varepsilon} = 0, \quad \rightarrow \quad u(x_\varepsilon, y_\varepsilon) = u(x, y).$$

Alternatively, for the proof we can use the exponential map and calculate

$$u(x_\varepsilon, y_\varepsilon) = \mathrm{e}^{\varepsilon\mathbf{U}}u(x, y) = \left(1 + \varepsilon\mathbf{U} + \frac{\varepsilon^2}{2!}\mathbf{U}^2 + \ldots\right)u(x, y) = u(x, y).$$

Because of $\mathbf{U}u(x, y) = 0$ we also have that $\mathbf{U}^n u(x, y) = 0$ for any $n > 1$, thus $u(x_\varepsilon, y_\varepsilon) = u(x, y)$. □

3.3 Invariant curves and points

Let us look at the invariance problem from a geometrical point of view. Let us define zero- or one–dimensional subsets on a plane which stay unaltered under the action of a one-parameter group of point transformations.

Definition 3.3.1 (Invariant curves, invariant points) *A curve c whose points, considered as source points, map into other points of the same curve c for all transformations of the group, is called an invariant curve. If the point* (x_0, y_0) *is unaltered after any transformation of the group then it is called an invariant point.*

As we can see in Figure 3.1, there are two possibilities.

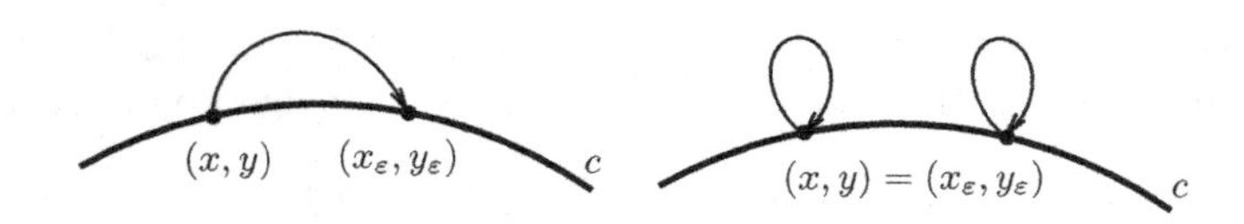

Figure 3.1: Invariant curve c and invariant points of a point transformation on $\mathbb{R}^2$

In the first case the curve c is an orbit of the group G and in the second case both functions $\xi(x, y)$ and $\eta(x, y)$ are equal to zero in all points of c. This means that the curve c is a locus of invariant points.

Example 3.3.1 (Rotation group)
As we have calculated before, the orbits for the rotation group are given by the equation $c = x^2 + y^2$. Orbits are circles with centrum in the point $(0, 0)$, all of them are invariant curves of the rotation group. We obtain the invariant points, when $\xi(x, y)$ and $\eta(x, y)$ are simultaneously equal to zero, i.e. $\xi(x, y) = -y = 0$ and $\eta(x, y) = x = 0$. Both conditions are fulfilled only at one point of the plane, i.e. at the point $(x, y) = (0, 0)$. It is the invariant point of the rotation group. ◇

Example 3.3.2 (Translation group) The orbits for this group are given by the equation $c = y$, so they are straight lines parallel to the x-axis. Because of $\eta(x, y) = 0$ and $\xi(x, y) = 1$, we can never get $\xi(x, y) = \eta(x, y) = 0$ simultaneously. Thus the translation group has no invariant points. ◇

3.4 Invariant families of curves

Now we introduce the idea of an invariant family of curves and develop here the methods that were first introduced by Sophus Lie. In general, a one-parameter family of curves is given by the equation

$$\tilde{\Phi}(x, y, c) = 0, \text{ or in an explicit form by } \Phi(x, y) = c, \tag{3.4.11}$$

where the constant c labels the curves of the family. For simplicity we suppose that the function $\Phi(x, y)$ is a smooth function of its arguments. A one-parameter family of curves is called invariant if the image of each curve of the family is again the curve of the same family.

Definition 3.4.1 (Invariant family of curves) *The one-parameter family of the curves $\Phi(x, y) = c$ is invariant under the action of the one-parameter group of point transformations $G(\mathbb{R}^2)$ if for any fixed value c and for any fixed ε the image points $(x_\varepsilon, y_\varepsilon)$ satisfy*

$$\Phi(x_\varepsilon, y_\varepsilon) = \Phi\left(X_\varepsilon(x, y), Y_\varepsilon(x, y)\right) = C(c, \varepsilon), \tag{3.4.12}$$

with the same $C(c, \varepsilon)$ (here $\Phi(x, y) = c$).

If the closed form of the point transformations is known, then it is easy to prove if the given family of curves is invariant under the action of the group or not. It is much more difficult to find the family of invariant curves using this definition. Similar to the case of invariant curves we develop some criterium for a one-parameter family of curves to be invariant under the action of the one-parameter continuous group.

We reformulate the condition (3.4.12) using the infinitesimal generator $\mathbf{U}$ of the one-parameter group $G(\mathbb{R})$. We differentiate (3.4.12) with respect to ε. Since this relation is fulfilled for any ε we can also take $\varepsilon = 0$ and obtain

$$\left.\frac{\partial X_\varepsilon}{\partial \varepsilon}\frac{\partial \Phi}{\partial x}\right|_{\varepsilon=0} + \left.\frac{\partial Y_\varepsilon}{\partial \varepsilon}\frac{\partial \Phi}{\partial y}\right|_{\varepsilon=0} = \left.\frac{\partial C}{\partial \varepsilon}\right|_{\varepsilon=0} = \mathcal{F}(c).$$

Thus, we obtain a non-homogeneous PDE of the first order on the function $\Phi(x, y)$

$$\xi(x, y)\frac{\partial \Phi}{\partial x} + \eta(x, y)\frac{\partial \Phi}{\partial y} = \mathcal{F}(c) = \mathcal{F}(\Phi).$$

It contains an arbitrary function $\mathcal{F}(\cdot)$. We will simplify this equation and exclude the arbitrary function $\mathcal{F}(\cdot)$, moreover we will replace it by a constant equal to one. In fact we described before the one-parameter family of curves by $\Phi(x, y) = c$ but this parametrization of the curves in the family is not unique. If we take other parametrization we obtain an equation for the same family of curves in the form $\Psi(x, y) = c_1$. Let us take such parametrization that $\Psi(x, y) = c_1 = \mathcal{H}(\Phi(x, y))$. Now we can choose the function $\mathcal{H}(\cdot)$ in special form to simplify the PDE. We insert the new parametrization of the family of curves in the equation and obtain

$$\begin{aligned}
\xi(x, y)\frac{\partial \Psi}{\partial x} + \eta(x, y)\frac{\partial \Psi}{\partial y} &= \xi(x, y)\frac{d\mathcal{H}}{d\Phi}\frac{\partial \Phi}{\partial x} + \eta(x, y)\frac{d\mathcal{H}}{d\Phi}\frac{\partial \Phi}{\partial y} \\
&= \left(\xi(x, y)\frac{\partial \Phi}{\partial x} + \eta(x, y)\frac{\partial \Phi}{\partial y}\right)\frac{d\mathcal{H}}{d\Phi} = \mathcal{F}(\Phi)\frac{d\mathcal{H}}{d\Phi} = 1.
\end{aligned}$$

The last equation defines the new function $\mathcal{H}(\Phi)$. We get

$$\frac{d\mathcal{H}}{d\Phi} = \frac{1}{\mathcal{F}(\Phi)} \Rightarrow \mathcal{H}(\Phi) = \int \frac{d\Phi}{\mathcal{F}(\Phi)}.$$

It means that under this convenient parametrization of the curves in the family we obtain the following equation

$$\xi(x,y)\Psi_x + \eta(x,y)\Psi_y = \mathbf{U}\Psi = 1. \tag{3.4.13}$$

This equation defines the invariant family of curves under the action of the one-parameter group of point transformations $G(\mathbb{R}^2)$ with the infinitesimal generator $\mathbf{U}$. We provide now an example how to use this equation to find the invariant family of curves for the rotation group.

Example 3.4.1 We take the rotation group with $\xi(x,y) = -y$ and $\eta(x,y) = x$ and solve the equations

$$\frac{dx}{-y} = \frac{dy}{x} = \frac{d\Psi}{1} \tag{3.4.14}$$

to get the invariant family of curves. In the first step we solve the homogeneous equation

$$x\,dx + y\,dy = 0 \Rightarrow u^2 = x^2 + y^2,$$

and obtain the expression for orbits in the form $x^2 = u^2 - y^2$ (the invariant function $u(x,y) = \text{const.}$). Then the second equation in (3.4.14) yields to

$$\frac{dy}{x} = \frac{d\Psi}{1} \Rightarrow \frac{dy}{\sqrt{u^2 - y^2}} = d\Psi.$$

We solve this ODE and obtain

$$\int \frac{dy}{\sqrt{u^2 - y^2}} = \Psi + \text{const.} \Rightarrow \arcsin\left(\frac{y}{u}\right) = \Psi + \text{const.}$$

It leads to the following expression for the function $\Psi(x,y)$

$$\frac{y}{u} = \sin(\Psi + \text{const.}), \quad \cos(\Psi + \text{const.}) = \sqrt{1 - \frac{y^2}{u^2}} = \frac{\sqrt{u^2 - y^2}}{u}.$$

In the end we get

$$\tan(\Psi + \text{const.}) = \frac{\sin(\Psi + \text{const.})}{\cos(\Psi + \text{const.})} = \frac{yu}{u\sqrt{u^2 - y^2}} = \frac{y}{x},$$

or

$$\Psi + \text{const.} = \arctan\left(\frac{y}{x}\right).$$

We denote const. $= c$ and because of the relation const. $= c = F(u) = F(x^2 + y^2)$, we obtain the most general solution of a linear non-homogeneous Equation (3.4.14) in the form

$$\Psi(x, y) = \arctan\left(\frac{y}{x}\right) + F(x^2 + y^2), \quad \Psi = \text{const.},$$

where $F(\cdot)$ is an arbitrary smooth function. This expression looks complicated because we used the Cartesian coordinates. Let us rewrite it in polar coordinates. As we proved it before in polar coordinates we have the following closed form for transformations of the rotation group $r_\varepsilon = r$ and $\theta_\varepsilon = \theta + \varepsilon$. The family of the curves which is invariant under the action of the rotation group in polar coordinates is given by

$$\Psi(r, \theta) = F(x^2 + y^2) + \arctan\left(\frac{x}{y}\right) = F(r) + \theta = c_1,$$

where $F(\cdot)$ is an arbitrary smooth function of its argument and c_1 is a constant. ◇

3.5 Canonical variables, the normal form of an infinitesimal generator

In Example 3.4.1 we demonstrated that the choice of coordinate system can reasonably simplify formulas. In this section we describe the method that helps to find *convenient* variables describing the action of the one-parameter point transformation group in the most transparent way. Most of the examples in this book are devoted to one-parameter point transformations group on the plane. This is done to make our arguments clear and understandable even for an unexperienced reader. We formulate some of the definitions or theorems for the n–dimensional case, for instance, for one-parameter point transformation groups on $\mathbb{R}^n$ or on a part of $\mathbb{R}^n$, when we can give a generalization of the method without complicating the text.

Definition 3.5.1 (Canonical variables, normal form) *For a given one-parameter group $G(\mathbb{R}^n)$ of point transformations on $\mathbb{R}^n$ canonical variables denoted as $s, t_i, i = 1, \ldots, n-1$, are defined by the condition that $G(\mathbb{R}^n)$ is* *reduced to the translation group along the s-axis in $\mathbb{R}^n$, i.e. the transformation rules for the coordinates are*

$$s_\varepsilon = s + \varepsilon,\ t_{i,\varepsilon} = t_i, \quad i = 1, \ldots, n-1. \tag{3.5.15}$$

The infinitesimal generator **U** *takes the so-called* **normal form** *in these coordinates*

$$\mathbf{U} = \frac{\partial}{\partial s}. \tag{3.5.16}$$

Theorem 3.5.1 *Canonical variables exist for any continuous one-parameter group $G(\mathbb{R}^n)$ of point transformations on $\mathbb{R}^n$. (Any infinitesimal generator* **U** *can be reduced to its normal form.)*

Proof: To demonstrate the procedure reducing an infinitesimal operator to the normal form transparently we provide the proof of the theorem for the case $G(\mathbb{R}^2)$, i.e. $n = 2$, the proof for higher dimensions is very similar. We also assume for simplicity that the functions $\xi(x,y)$ and $\eta(x,y)$ which define the generator **U** are smooth functions.

Let us change the variables $(x,y) \to (s,t)$, i.e. introduce new coordinates

$$s = s(x,y),$$
$$t = t(x,y).$$

How does the form of the infinitesimal generator **U** change under the transformation of variables? We insert after usual differentiation rules

$$\frac{\partial}{\partial x} = \frac{\partial s}{\partial x}\frac{\partial}{\partial s} + \frac{\partial t}{\partial x}\frac{\partial}{\partial t}$$

because we have to replace the differentiation with respect to x and y by differentiation with respect to the new variables s, t. In the same way we obtain

$$\frac{\partial}{\partial y} = \frac{\partial s}{\partial y}\frac{\partial}{\partial s} + \frac{\partial t}{\partial y}\frac{\partial}{\partial t}.$$

Consequently the formula for an infinitesimal generator **U** related to $G(\mathbb{R}^2)$ takes the form

$$\begin{aligned}
\mathbf{U} &= \xi(x,y)\frac{\partial}{\partial x} + \eta(x,y)\frac{\partial}{\partial y} \\
&= \xi(x,y)\left(\frac{\partial s}{\partial x}\frac{\partial}{\partial s} + \frac{\partial t}{\partial x}\frac{\partial}{\partial t}\right) + \eta(x,y)\left(\frac{\partial s}{\partial y}\frac{\partial}{\partial s} + \frac{\partial t}{\partial y}\frac{\partial}{\partial t}\right) \\
&= \left(\xi(x,y)\frac{\partial s}{\partial x} + \eta(x,y)\frac{\partial s}{\partial y}\right)\frac{\partial}{\partial s} + \left(\xi(x,y)\frac{\partial t}{\partial x} + \eta(x,y)\frac{\partial t}{\partial y}\right)\frac{\partial}{\partial t} \\
&= (\mathbf{U}s)\frac{\partial}{\partial s} + (\mathbf{U}t)\frac{\partial}{\partial t}.
\end{aligned} \tag{3.5.17}$$

Now we can look for more convenient coordinates for our one-parameter group $G(\mathbb{R}^2)$. If we can find functions $s(x, y)$ and $t(x, y)$ such that $\mathbf{U}s = 1$ and $\mathbf{U}t = 0$, the infinitesimal generator will take the normal form, or in other words, it will take the form of the infinitesimal generator of the translation group.

The theorem says that it is possible to find canonical variables for any arbitrary one-parameter group of point transformations, which means that it is always possible to solve the following coupled system of linear PDEs

$$\begin{aligned} \xi(x,y)\frac{\partial s}{\partial x} + \eta(x,y)\frac{\partial s}{\partial y} &= 1, \\ \xi(x,y)\frac{\partial t}{\partial x} + \eta(x,y)\frac{\partial t}{\partial y} &= 0. \end{aligned} \tag{3.5.18}$$

We find the solution of the system if we solve its characteristic systems of ODEs

$$\begin{aligned} \frac{dx}{\xi(x,y)} &= \frac{dy}{\eta(x,y)} = \frac{d\Psi}{1}, \\ \frac{dx}{\xi(x,y)} &= \frac{dy}{\eta(x,y)} = \frac{du}{0}, \end{aligned} \tag{3.5.19}$$

where Ψ and respectively u denote the first integrals of the corresponding equations. The system (3.5.19) has a solution for any smooth functions $\xi(x, y)$ and $\eta(x, y)$. Based on the theory of linear PDEs the solution of (3.5.18) has the form

$$\begin{cases} t(x,y) = H(u), \\ s(x,y) = F(u) + \Psi, \end{cases} \tag{3.5.20}$$

where $F(\cdot)$ and $H(\cdot)$ are arbitrary smooth functions. So we have found canonical variables s, t for the group $G(\mathbb{R}^2)$ and proved the theorem.

We notice that the second equation in (3.5.19) is equivalent to the equation for orbits of the group of point transformations, and the first equation in (3.5.19) is equivalent to the equation of the curves which are locally orthogonal to the orbits. We also should note that the first equation in (3.5.19) is the equation for an invariant family of curves to $G(\mathbb{R}^2)$. $\square$

3.6 A geometrical meaning of canonical coordinates

Let us now analyze the geometrical meaning of canonical coordinates. If we look at the infinitesimal generator

$$\mathbf{U} = \xi(x,y)\frac{\partial}{\partial x} + \eta(x,y)\frac{\partial}{\partial y},$$

we see that this differential operator locally describes the action of the group in given coordinates (x, y). We can realize this action in the following way: to each point (x, y) in $\mathbb{R}^2$ we attach a tangent plane spanned by the coordinate vectors $(e_1, e_2) = \left(\frac{\partial}{\partial x}, \frac{\partial}{\partial y}\right)$. Then we describe the infinitesimal action of the group in the point (x, y) by a vector with the coordinates (ξ, η)

$$\mathbf{U} = \xi e_1 + \eta e_2 = (\xi, \eta) \bullet \nabla = (\xi, \eta) \bullet \begin{pmatrix} \frac{\partial}{\partial x} \\ \frac{\partial}{\partial y} \end{pmatrix} = \xi\frac{\partial}{\partial x} + \eta\frac{\partial}{\partial y},$$

where $\bullet$ denotes usual scalar product of two vectors. The action of the one-parameter group produces the motion of the point (x, y) along the orbit of the group. This motion does not coincide with arbitrary chosen directions of the x- and y-axes of the coordinate system. But in every case we can change the coordinate system $(e_1, e_2) \to (\tilde{e}_1, \tilde{e}_2)$ in a way that the direction of the action of the group coincides with the direction of one of the new axes $\tilde{e}_1$. After the change of coordinates the group will act as a translation group in the $\tilde{e}_1$-direction. The coordinates are denoted as s and t. In the next two examples we look into this method in detail.

Example 3.6.1 Again we take the rotation group on the plane with $\xi = -y$ and $\eta = x$ and look for $s = s(x, y)$ and $t = t(x, y)$ which solve the coupled system of linear PDEs (3.5.18). The characteristic system (3.5.19) in this case has the form

$$\frac{dx}{-y} = \frac{dy}{x} = \frac{d\Psi}{1}, \tag{3.6.21}$$

$$\frac{dx}{-y} = \frac{dy}{x} = \frac{du}{0}. \tag{3.6.22}$$

We should find the first integrals u and Ψ of those equations, then the most general solution is

$$s = \Psi + F(u),$$
$$t = H(u).$$

Here F and H are arbitrary functions of the first integral u of the homogeneous Equation (3.6.22) and Ψ is an arbitrary solution of the non-homogeneous Equation (3.6.21).

The solution of the characteristic system is presented in detail in Example 3.4.1. Here we provide just a summary of that. From (3.6.22) we derive

$$x\,dx + y\,dy = 0 \;\Rightarrow\; u^2 = x^2 + y^2,$$

or $x^2 = u^2 - y^2$. Equation (3.6.21) yields

$$\frac{dy}{x} = \frac{d\Psi}{1} \;\Rightarrow\; \frac{dy}{\sqrt{u^2 - y^2}} = d\Psi$$

and eventually we get

$$\Psi + c = \arctan\left(\frac{y}{x}\right).$$

Because $c = F(u) = F(x^2 + y^2)$, we obtain a general form of canonical variables

$$\begin{cases} s(x,y) = \arctan\left(\frac{y}{x}\right) + F(x^2 + y^2) \\ t(x,y) = H(x^2 + y^2). \end{cases}$$

In applications we are interested in the simplest form of canonical variables. For the rotation group we choose

$$s(x,y) = \arctan\left(\frac{y}{x}\right), \quad t(x,y) = x^2 + y^2.$$

We prove that the new variables s and t reduce the infinitesimal generator to its normal form. We can easily show that

$$\mathbf{U}s = \left(-y\frac{\partial}{\partial x} + x\frac{\partial}{\partial y}\right)\left(\arctan\left(\frac{y}{x}\right)\right) = \frac{y^2}{x^2 + y^2} + \frac{x^2}{x^2 + y^2} = 1,$$

$$\mathbf{U}t = \left(-y\frac{\partial}{\partial x} + x\frac{\partial}{\partial y}\right)(x^2 + y^2) = 0.$$

◇

Example 3.6.2 Let us take the scaling group. The coordinates of the generator $\mathbf{U}$ have the form $\xi((x,y) = x$ and $\eta(x,y) = y$.

To find the canonical variables we have to solve the system of linear PDEs

$$x\frac{\partial s}{\partial x} + y\frac{\partial s}{\partial y} = 1, \tag{3.6.23}$$

$$x\frac{\partial t}{\partial x} + y\frac{\partial t}{\partial y} = 0, \tag{3.6.24}$$

respectively the characteristic system has the form

$$\frac{dx}{x} = \frac{dy}{y} = \frac{d\Psi}{1},$$
$$\frac{dx}{x} = \frac{dy}{y} = \frac{du}{0}.$$

From (3.6.24) we get the expression for the orbits

$$\frac{dx}{x} = \frac{dy}{y} \Rightarrow \ln x = \ln y + c \Rightarrow \ln\left(\frac{x}{y}\right) = c$$
$$\Rightarrow u = \frac{x}{y},$$

and from (3.6.23)

$$\frac{dx}{x} = d\Psi \Rightarrow \ln x = \Psi + \tilde{c} \Rightarrow \Psi = \ln x + \tilde{c} = \ln x + F\left(\frac{x}{y}\right).$$

Thus one of the most simplest solutions which describes the canonical variables for the scaling group is given by

$$s(x, y) = \ln x,$$
$$t(x, y) = \frac{x}{y}.$$

In these variables the infinitesimal generator of the scaling group takes its normal form. ◇

3.7 Problems with solutions

In Problems 3.7.1 and 3.7.2, one should find the form of the infinitesimal generator if the transformations of the group are given in the closed form. In Problems 3.7.2 and 3.7.5, the reader should find the orbits of the one-parameter group of point transformations. In Problems 3.7.3 and 3.7.4, we discuss how to find canonical variables and reduce the infinitesimal generator to the normal form.

Problem 3.7.1 *Let the group of linear fractional transformations be given by*

$$x_\epsilon = \frac{x}{1 - \epsilon x},$$
$$y_\epsilon = \frac{y}{1 - \epsilon y}, \quad \epsilon \in \mathbb{R}.$$

Find the infinitesimal representation of this group (i.e. the functions $\xi(x, y)$ and $\eta(x, y)$) and infinitesimal generator $\mathbf{U}$ *of this group.*

Solution The group of the linear fractional transformations is a local group, because each transformation has its own region of definition $\epsilon \in (-a, a)$ where $a = \min[|x|, |y|]$ and all transformations are defined just in a small neighborhood of the identical transformation.
The linear part of the transformation is given by

$$\xi(x, y) = \left.\frac{dx_\epsilon}{d\epsilon}\right|_{\epsilon=0} = -\left.\frac{-x^2}{(1-\epsilon x)^2}\right|_{\epsilon=0} = x^2,$$
$$\eta(x, y) = \left.\frac{dy_\epsilon}{d\epsilon}\right|_{\epsilon=0} = -\left.\frac{-y^2}{(1-\epsilon y)^2}\right|_{\epsilon=0} = y^2.$$

Correspondingly the infinitesimal generator of the group has the form

$$\mathbf{U} = x^2\frac{\partial}{\partial x} + y^2\frac{\partial}{\partial y}.$$

Problem 3.7.2 *Let a one-parameter family of point transformations be given by the rules*

$$x_\varepsilon = x + \varepsilon, \quad y_\varepsilon = \frac{xy}{x+\varepsilon}, \quad \varepsilon \in \mathbb{R}.$$

Do these point transformations build a group? If it is the case, which orbits does this group have? Is it a global or local group on $\mathbb{R}^2$*? If it is a local group where are the transformations well defined?*

Solution It is easy to prove that the transformations build a group. This group is called the group of **hyperbolic transformations**. The hyperbolic transformation group is a local group of transformations because each transformation is defined for $\varepsilon \in (-x, x)$.
First we find the infinitesimal generator $\mathbf{U}$

$$\xi(x, y) = \left.\frac{dx_\varepsilon}{d\varepsilon}\right|_{\varepsilon=0} = 1,$$
$$\eta(x, y) = \left.\frac{dy_\varepsilon}{d\varepsilon}\right|_{\varepsilon=0} = \left.\frac{-xy}{(x+\varepsilon)^2}\right|_{\varepsilon=0} = -\frac{y}{x}.$$

The infinitesimal generator of the hyperbolic group has the form

$$\mathbf{U} = \frac{\partial}{\partial x} - \frac{y}{x}\frac{\partial}{\partial x}.$$

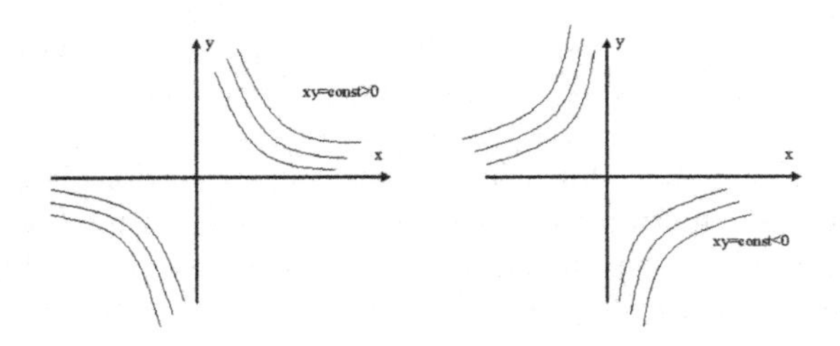

Figure 3.2: The sketch of some orbits of the hyperbolic group of transformations in the cases $c > 0$ and $c < 0$.

The equation for the orbits is now

$$\frac{dx}{x} = -\frac{dy}{y}.$$

The solutions of this equation are the hyperbolas

$$y = \frac{c}{x}, \quad c = \text{constant}.$$

The set of all orbits contains the hyperbolas $y = \frac{c}{x}$, $c = $ constant (the sketch is presented in Figure 3.2). On the x-axis (in the case $y = 0$) and on the y-axis (i.e. for $x = 0$) the transformations of the group are not well defined.

Problem 3.7.3 *Reduce the infinitesimal operator with the following coefficients*

$$\xi(x, y) = x + y, \quad \eta(x, y) = 3y$$

to the normal form.

Solution Using the coefficients $\xi(x, y)$ and $\eta(x, y)$ we write an infinitesimal generator of the one-parameter group of point transformations $G(R^2)$ in the form

$$\mathbf{U} = (x + y)\frac{\partial}{\partial x} + 3y\frac{\partial}{\partial y}.$$

We should reduce the operator $\mathbf{U}$ to the normal form, i.e. to the translation operator. We are looking for the new canonical variables $s = s(x, y)$ and $t = t(x, y)$. They solve the coupled system of equations

$$\begin{aligned} \mathbf{U}t &= 0, \\ \mathbf{U}s &= 1. \end{aligned} \tag{3.7.25}$$

Correspondingly, the system of characteristic equations has the form

$$\frac{dx}{x+y} = \frac{dy}{3y} = \frac{dt}{0}, \tag{3.7.26}$$

$$\frac{dx}{x+y} = \frac{dy}{3y} = \frac{ds}{1}. \tag{3.7.27}$$

Let us solve now the first Equation (3.7.26)

$$\frac{dx}{x+y} = \frac{dy}{3y},$$

after transformation we get

$$\frac{dy}{dx} = \frac{3y}{x+y} = \frac{3}{\frac{x}{y}+1}.$$

After the substitution $\frac{x}{y} = z$, i.e $x = zy$ and $dx = zdy + ydz$, we obtain

$$\frac{dy}{zdy + ydz} = \frac{3}{z+1}.$$

We rewrite this equation in the following way.

$$\frac{dy}{3y} = \frac{dz}{1-2z}.$$

Integrating, we get

$$y^2(1-2z)^3 = \text{const}.$$

Thus we obtain that the most general form of the new variable t is given by the expression

$$t = F\left(\frac{(y-2x)^3}{y}\right).$$

Now we solve the Equation (3.7.27) and obtain the most general form for the second canonical variable

$$s = \frac{1}{3}\ln y + \text{const}.$$

Because we can add the most general solution of the corresponding homogeneous equation to a special solution of the non-homogeneous linear partial differential equation we replace $c = H(\frac{(y-2x)^3}{y})$ in the equation above and obtain expressions for the canonical variables

$$t = F\left(\frac{(y-2x)^3}{y}\right),$$

$$s = \frac{1}{3}\ln y + H\left(\frac{(y-2x)^3}{y}\right).$$

To reduce our infinitesimal generator to the normal form it is enough to take new variables in the most simple form, for instance, in the form

$$\begin{aligned} t &= \frac{(y-2x)^3}{y}, \\ s &= \frac{1}{3}\ln y. \end{aligned}$$

We verify that the new variables s and t indeed are solutions of the system (3.7.25) (and the infinitesimal generator $\mathbf{U}$ is reduced to its normal form $\mathbf{U} = \frac{\partial}{\partial s}$). We can easily show that

$$\begin{aligned} \mathbf{U}t &= \left((x+y)\frac{\partial}{\partial x} + 3y\frac{\partial}{\partial y}\right)\left(\frac{(y-2x)^3}{y}\right) \\ &= \frac{(y-2x)^2}{y}(-6x-6y+9y-3y+6x) = 0, \\ \mathbf{U}s &= \left((x+y)\frac{\partial}{\partial x} + 3y\frac{\partial}{\partial y}\right)\left(\frac{1}{3}\ln y\right) = 3y\frac{1}{3y} = 1. \end{aligned}$$

Problem 3.7.4 *Find the canonical variables for the one-parameter point transformation group defined by the infinitesimal operator with the following coefficients*

$$\xi(x,y) = \frac{1}{x}, \quad \eta(x,y) = \frac{1}{y}.$$

Solution Since we know the coefficients $\xi(x,y)$ and $\eta(x,y)$, we can write the infinitesimal generator

$$\mathbf{U} = \frac{1}{x}\frac{\partial}{\partial x} + \frac{1}{y}\frac{\partial}{\partial y}.$$

Similar to the solution of the previous problem we have to solve a system of linear PDEs (3.5.19) to find the canonical variables

$$\begin{aligned} \frac{1}{x}\frac{\partial s}{\partial x} + \frac{1}{y}\frac{\partial s}{\partial y} &= 1, \\ \frac{1}{x}\frac{\partial t}{\partial x} + \frac{1}{y}\frac{\partial s}{\partial y} &= 0. \end{aligned}$$

Respectively the characteristic system of equations takes the form

$$xdx = ydy = ds, \tag{3.7.28}$$

$$xdx = ydy = \frac{dt}{0}. \tag{3.7.29}$$

From Equation (3.7.29) we get the expression for the orbits and correspondingly for the variable t

$$\frac{1}{2}x^2 = \frac{1}{2}y^2 + \text{const.} \quad \Rightarrow \quad x^2 = y^2 + \text{const.} \quad \Rightarrow \quad t = F(x^2 - y^2).$$

Analogously, from Equation (3.7.28) we obtain the expression for the invariant family of curves and for the variable s

$$s = \frac{1}{2}x^2 + \text{const.} = \frac{1}{2}x^2 + H(x^2 - y^2).$$

We got the general form of canonical variables t, s. Let us look for one of the most simplest expressions for the new variables t and s. We can take the canonical variables in the form

$$\begin{aligned} t &= x^2 - y^2, \\ s &= \frac{1}{2}x^2. \end{aligned}$$

We prove that we got correct solutions

$$\begin{aligned} \mathbf{U}t &= \frac{1}{x}2x + \frac{1}{y}(-2y) = 0, \\ \mathbf{U}s &= \frac{1}{2x}2x = 1. \end{aligned}$$

Problem 3.7.5 *Find the orbits of the* **conformal transformation group** *on $\mathbb{R}^2$ where the coordinates of the infinitesimal generator are given by*

$$\xi(x, y) = y^2 - x^2, \quad \eta(x, y) = -2xy.$$

Solution Let us write an equation for the orbits

$$\frac{dx}{y^2 - x^2} = -\frac{dy}{2xy}. \tag{3.7.30}$$

Using the substitution $z = \frac{y}{x}$ we obtain

$$\begin{aligned} \frac{dx}{z^2 - 1} &= -\frac{xdz + zdx}{2z}, \\ \frac{dx}{x} &= \frac{1 - z^2}{z(1 + z^2)}dz, \\ x &= c\frac{z}{z^2 + 1}. \end{aligned}$$

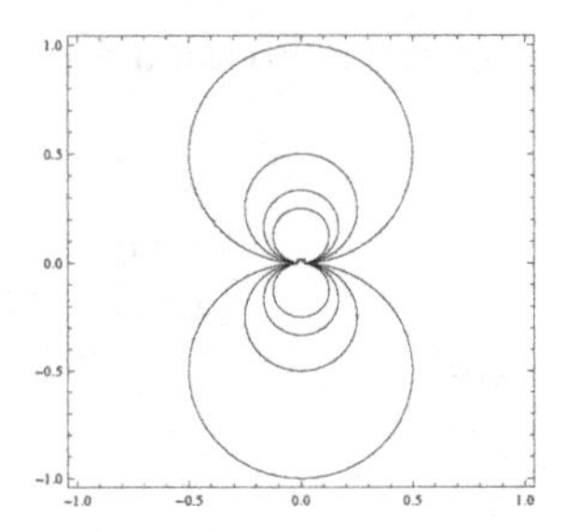

Figure 3.3: The sketch of some orbits of the conformal transformation group.

Substituting back the expression for z we can conclude that the orbits of the conformal transformation group are given by

$$\frac{y}{x^2+y^2} = \text{const.},$$

the point $(0,0)$ is invariant under transformations of this group. The plot of some orbits is given in Figure 3.3.

3.8 Exercises

The five exercises in this section should help the reader become more familiar with the important ideas of orbit, invariant points, invariant curves and invariant family of curves.

1. Find invariant curves and the family of curves which are invariant under the action of the helical transformation group on $\mathbb{R}^3$

$$\begin{aligned} x_\epsilon &= x\cos(\epsilon) - y\sin(\epsilon), \\ y_\epsilon &= x\sin(\epsilon) + y\cos(\epsilon), \\ z_\epsilon &= z + m\epsilon, \quad \epsilon, m \in \mathbb{R}. \end{aligned}$$

2. Find the invariant points, invariant curves and the invariant family of curves which are invariant under the action of the projective group of transformation on $\mathbb{R}^2$ with

$$\xi = xy, \quad \eta = y^2.$$

3. Find the invariant points, invariant curves and the family of curves which are invariant under the action of the following projective trans-

formation group on $\mathbb{R}^2$

$$\begin{aligned} x_\epsilon &= \frac{x}{1-\epsilon y}, \\ y_\epsilon &= \frac{y}{1-\epsilon y}, \qquad \epsilon \in \mathbb{R}. \end{aligned}$$

4. Find the invariant points, invariant curves and the invariant family of curves on $\mathbb{R}^2$ under the action of the point transformation group

$$\begin{aligned} x_\epsilon &= \ln\left(e^x + \epsilon\right), \\ y_\epsilon &= y e^\epsilon, \qquad \epsilon \in \mathbb{R}. \end{aligned}$$

5. Find the invariant point, invariant curves and the invariant family of curves on $\mathbb{R}^2$ which are invariant under the action of the Lorentz group

$$\begin{aligned} x_\epsilon &= x\cosh(\epsilon) + y\sinh(\epsilon), \\ y_\epsilon &= x\sinh(\epsilon) + y\cosh(\epsilon), \quad \epsilon \in \mathbb{R}. \end{aligned}$$

Chapter 4

First order ODEs

One of the main reasons why Sophus Lie wanted to develop a new approach to the theory of point transformations was that he was completely dissatisfied with the theory of differential equations in his time. The majority of textbooks looked more like cookbooks where one could find instructions explaining how to solve this or that equation step by step using different substitutions. Sophus Lie decided to look for the meaning of all these substitutions and understand why they even exist. His approach was very successful and Lie group theory is now a well established theory that provides a proper basis for many methods in the theory of ODEs and PDEs. Despite that if you look at a typical textbook for beginners today you will see the same situation as more than a hundred years ago: textbooks are just filled with recipes explaining how to use this or that substitution. The student has to learn different types of equations and corresponding substitutions by heart without any intuition about their interaction and origin. In this chapter we try to address this problem. We take the most simple type of ODEs – the first order ODEs and examine in great detail the geometrical side of the ODE theory and a connection to the Lie group theory. With this simple and transparent example we can demonstrate almost all practical tips and tricks given by the Lie group methods in application to differential equations. The practical problems described in detail in the Section 4.6 will help the students feel confident with problems of such type even if they have had only a very short introduction to the theory of ODEs before.

4.1 Geometrical image of first order ODEs

In the case of a first order ODE we have only one independent variable x, one dependent variable y and their first derivative denoted by $\frac{dy}{dx}$ or y'. The most general form of a first order ordinary differential equation is

$$F\left(x, y, \frac{dy}{dx}\right) = F(x, y, y') = 0,$$

where the given function F is a well defined function in a region of $\mathbb{R}^3$ with coordinates (x, y, y'). This form is called an implicit form of a first order ODE. We say that we have a partial solution of this equation if we found a differentiable function $y = y(x)$ such that the above equation is satisfied in the region where it is defined. A partial solution is defined up to one arbitrary constant. This integration constant parametrizes the family of partial solutions. We say that the general solution of this differential equation is a one-parameter family of curves (called also the family of integral curves of the equation) in $\mathbb{R}^2$ or in some region $\mathcal{M}$ of $\mathbb{R}^2$.

If the first order ODE is given, or can be put in the form

$$y' = f(x, y) \tag{4.1.1}$$

we say that we have a first order ODE in an explicit form. In this chapter we will mostly work with first order ODEs in an explicit form. For simplicity we suppose that the function $f(x, y)$ is a continuous function in some open subset $\mathcal{M}$ of $\mathbb{R}^2$.

We notice that the Equation (4.1.1) defines a slope at every point (x, y) of the region $\mathcal{M}$. We call it a slope field or a direction field for a differential equation. In all points where $f(x, y)$ is well defined every solution to the differential equation must follow the direction field, running tangent to every little piece of slope and touching it.

For each fixed point (x_0, y_0) in $\mathcal{M}$ where the function $f(x, y)$ and its partial derivative $\frac{\partial y}{\partial x}$ are continuous functions we have the possibility to find the unique integral curve that passes through this point (x_0, y_0). If we fix one point (x_0, y_0) in the region $\mathcal{M}$ we choose a particular solution by selecting one of the curves from the family of integral curves. It means we solve a Cauchy problem or an initial value problem for this equation

$$y' = f(x, y), \quad y(x_0) = y_0. \tag{4.1.2}$$

Now we look for the one-parameter group of point transformations for which this family of integral curves is invariant. If we find this group $G(\mathbb{R}^2)$ with the infinitesimal generator $\mathbf{U}$, we say that the equation has the symmetry $G(\mathbb{R}^2)$ (or in the infinitesimal form the symmetry generator $\mathbf{U}$) or the differential equation admits the group $G(\mathbb{R}^2)$. To illustrate the action of the group, let us look at another geometrical interpretation of the Equation (4.1.1).

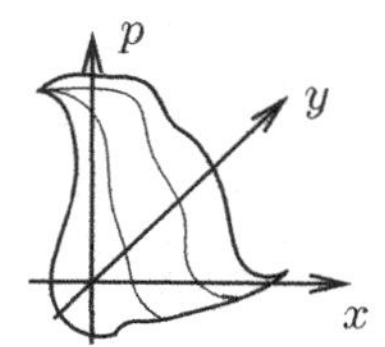

Figure 4.1: The frame of a first order ODE $y' = p = f(x, y)$.

We denote $p = y'$, then we get $p = f(x, y)$ it is an equation for a surface in $\mathbb{R}^3$, Figure 4.1.

This surface is called the frame of the Equation (4.1.1). It is evident that, if an equation admits a symmetry group, then the action of the group $G(\mathbb{R}^2)$ provides invariant transformations of the frame.

Let us take a very simple example to demonstrate these relations.

Example 4.1.1 We take an equation in the form $y' = f(x)$.

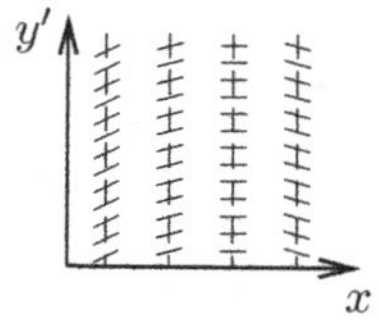

Figure 4.2: The direction field of an ODE of the type $y' = f(x)$.

The direction field has the following property: along each straight line $x = \text{const.}$ the slope $y' = p$ has the same value for all values y, Figure 4.2. Let us look at the corresponding frame $p = f(x)$. We notice that the position of any point on the surface does not depend on the value of y. We can obtain this surface if we take a curve in the plane (x, p) which satisfies the equation $p = f(x)$ and move it parallel to the y-axis. Because of that we can suppose that the equation $y' = f(x)$ must admit the translation group in the y-direction, Figure 4.3.

On the other hand we know that the translation generator has the form $\mathbf{U} = \frac{\partial}{\partial y}$. We look at all possible invariant families of curves for the one-parameter group $G(\mathbb{R}^2)$ of point transformations with this generator. The equation for the invariant family of curves has the form

$$\left(\xi(x, y)\frac{\partial}{\partial x} + \eta(x, y)\frac{\partial}{\partial y}\right)\Psi = 1.$$

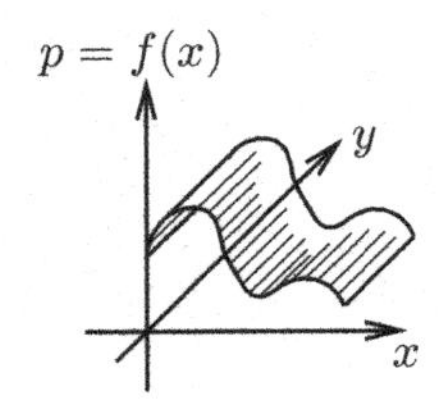

Figure 4.3: The frame is translation invariant in the y-direction, it is a surfase of a cylindric type.

If we solve this equation we obtain a solution in the form $\Psi(x, y) = C$, where $C =$ constant is a parameter which marks the curves in the family. Because of the translation generator $\xi(x, y) = 0$ and $\eta(x, y) = 1$, to find the invariant family of curves we have to solve

$$\frac{\partial}{\partial y}\Psi = 1.$$

The general solution of this equation has the form $\Psi = y - F(x)$. Thus we obtain for the invariant family of curves the following expression

$$y - F(x) = \text{const.}$$

When we differentiate this expression, we obtain the equation

$$dy - F'(x)dx = 0 \;\Rightarrow\; \frac{dy}{dx} = F'(x) = f(x).$$

It means we have found the most general first order ODE that admits the translation group in the y-direction

$$y' = f(x).$$

Its integral curves build the invariant family of curves for the translation group.

In fact let us integrate the equation $dy = f(x)dx$ in a classical way. We obtain a one-parameter family of integral curves

$$y = \int_{x_0}^{x} f(x)\, dx = \int_{x_0}^{x} F'(x)\, dx = F(x) - F(x_0) = F(x) + \text{const.}$$
$$\Rightarrow\; y - F(x) = \text{const.}$$

It is the same family of curves as calculated before. These calculations demonstrate how knowledge of the symmetry group can be used to find a solution of the equation and in which relation an ODE and a symmetry group are.

The knowledge of the admitted symmetry group helps us to make a statement about the form of the open subset $\mathcal{M}$ where the equation possesses a unique solution. We demonstrate this approach with the same simple example.

Let $f(x)$ be a continuous function over the interval $x \in [a, b]$. We suppose that we are able to find an integral curve of $y' = f(x)$ for $x \in [a, b]$. Then the action of the group is equivalent to the translation of a one integral curve parallel to the y-axis.

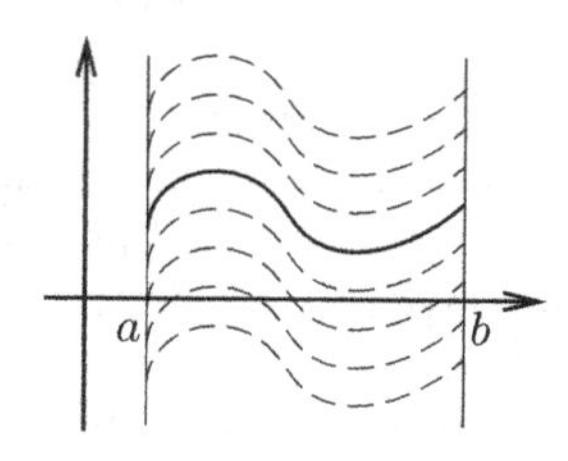

Figure 4.4: A natural region of definition for an ODE of the form $y' = f(x)$.

Because the function $f(x)$ is continuous in $[a, b]$ and the translation group is also a continuous one-parameter group of point transformations we cover the strip $[a, b] \times \mathbb{R}$ with integral curves without intersections or gaps. We obtain a strip $D = [a, b] \times \mathbb{R}$ where exactly one integral curve passes through each point. It means that in each point (x_0, y_0) of this region we can find a unique solution for the initial value problem (4.1.2). We have seen that if we are able to find an integral curve of $y' = f(x)$ for $x \in [a, b]$ then in the whole strip $D = [a, b] \times \mathbb{R}$ the initial value problem (4.1.2) has a unique solution for any point $(x_0, y_0) \in D$. We have found that the natural domain where the equation of the type $y' = f(x)$ is defined is a strip $D = [a, b] \times \mathbb{R}$, see Figure 4.4.

If a solution of $y' = f(x)$ is well defined in every point $x \in (c, b]$ and has a singularity in a point c then every integral curve of this equation in the strip $D = (c, b] \times \mathbb{R}$ will have the same singularity in $x = c$. It means that all integral curves in the strip $D = (c, b] \times \mathbb{R}$ are not well defined on the straight line $x = c$. This straight line is an orbit of the translation group admitted by the differential equation. We found that a natural region of definition for a first order ODE which admits a one-parameter continuous group of point transformations lies between two orbits of the group. Singularities of this first order ODE can lie just on orbits. A first order ODE can possess a so-called singular solution, i.e. a solution in each point of which the uniqueness of solutions is violated. If this equation admits a one-parameter continuous group of point transformations then all such singular solutions coincide with orbits of the group.

We obtain a typical example of such behavior if the function $f(x)$ is a discontinuous function in the point $c \in (a, b)$ and $f(x) \to \pm\infty$ by $x \to c \pm 0$. It means that the slope y' tends to infinity along this line. $\diamond$

4.2 A linear first order differential operator $\mathcal{A}$ related to a first order ODE

To describe the action of a one-parameter group $G(\mathbb{R}^2)$ of point transformations locally we introduced an infinitesimal generator $\mathbf{U}$. In many applications it is much easier to work with this linear differential operator than with a group transformation given in a closed form. Our goal is to find a linear first order differential operator which represents a first order ODE and is more convenient for further investigations than the equation itself.

Let us consider a first order differential equation $\frac{dy}{dx} = f(x, y)$ now, which has the family of integral curves $\Phi(x, y) = \text{const}$. For any $f(x, y)$ it is possible to rewrite this equation in the following form

$$\frac{dy}{dx} = f(x, y) = \frac{\mathcal{Y}(x, y)}{\mathcal{X}(x, y)}, \tag{4.2.3}$$

for instance, if we take $f(x, y) = \mathcal{Y}(x, y)$ and $\mathcal{X}(x, y) = 1$. Then we can rewrite the differential equation in the form

$$\mathcal{X}(x, y)dy - \mathcal{Y}(x, y)dx = 0,$$

or formally rewrite (4.2.3) as

$$\frac{dy}{\mathcal{Y}(x, y)} = \frac{dx}{\mathcal{X}(x, y)}. \tag{4.2.4}$$

We can look at this equation as a characteristic equation of a first order PDE with a function $u(x, y)$

$$\mathcal{X}(x, y)\frac{\partial u(x, y)}{\partial x} + \mathcal{Y}(x, y)\frac{\partial u(x, y)}{\partial y} = \mathcal{A}u(x, y) = 0$$

where we introduced a linear differential operator $\mathcal{A}$

$$\mathcal{A} = \mathcal{X}\frac{\partial}{\partial x} + \mathcal{Y}\frac{\partial}{\partial y}. \tag{4.2.5}$$

We see that the form of the linear operator $\mathcal{A}$ is not unique, because we can choose the functions $\mathcal{X}(x,y), \mathcal{Y}(x,y)$ in arbitrary many different ways. On the other hand we can obtain a linear operator $\mathcal{A}$ (4.2.5) for any first order ODE (4.2.3) and for a nonlinear one as well. In fact the one-parameter family of integral curves of (4.2.3) is exactly the one-parameter family of characteristics to PDE described by the operator $\mathcal{A}$ independently of the form in which we choose this operator.

Now, when a local action of the group of point transformations is represented with a linear first order differential operator $\mathbf{U}$ and an action of the first order ODE – with a linear first order differential operator $\mathcal{A}$, both actions are locally represented by the objects with the same structure. This idea belongs to Sophus Lie and is used in modern analysis very often. He decided that working with simple linear objects instead of complicated nonlinear ones can be very fruitful. We will see now how we can use this idea in the theory of first order ODEs.

If the described family of integral curves of Equation (4.2.3) admits the symmetry group $G(\mathbb{R}^2)$ with the infinitesimal generator $\mathbf{U} = \xi(x,y)\frac{\partial}{\partial x} + \eta(x,y)\frac{\partial}{\partial y}$, then this family of integral curves will be invariant under the action of the group, i.e.

$$\mathbf{U}\Phi(x,y) = \xi(x,y)\Phi_x + \eta(x,y)\Phi_y = 1.$$

On the other hand we have $\mathcal{A}\Phi(x,y) = 0$ because $\Phi(x,y) = C = \text{const.}$ are integral curves of the equation. Therefore the solution of the PDE is given in the form $u(x,y) = F(\Phi(x,y))$, where F is any differentiable function. So, we obtain

$$\mathcal{A}u(x,y) = \mathcal{A}(F(\Phi(x,y))) = \frac{dF}{d\Phi} \cdot \mathcal{A}\Phi(x,y) = 0.$$

We can see that if (4.2.3) admits a symmetry group $G(\mathbb{R}^2)$ with an infinitesimal generator $\mathbf{U}$ then in solutions of this equation the relations $\mathbf{U}\Phi(x,y) = 1$ and $\mathcal{A}\Phi(x,y) = 0$ are fulfilled.

Up to now we have discussed the relations between a first ODE and a symmetry group admitted by this equation using simple examples. For further study we need a definition which explains exactly what it means when a symmetry group is admitted by a differential equation. Later we will look at high order ODEs and also at systems of ODEs and introduce a more general definition instead of the one that we need for this chapter. Here we will introduce a general definition for the symmetry group of a system of differential equations similar to the definition introduced in [59]. We suppose that $x \in \mathbb{R}^p$ and $y \in \mathbb{R}^q$ and denote the set of independent and dependent variables defined on an open subset $\mathcal{M} \in \mathbb{R}^{(p+q)}$ and G is a point transformations group on $\mathcal{M}$.

Definition 4.2.1 (Symmetry group of a system of differential equations) *Let* $\mathbf{S}$ *be a system of differential equations. A symmetry group of the system* $\mathbf{S}$ *is a local group of transformations* G*, acting on an open subset* $\mathcal{M}$ *of the space of independent and dependent variables of the system, with the property that whenever* $y = \Phi(x)$ *is a solution of* $\mathbf{S}$ *and whenever* $g_\varepsilon(\Phi(x))$ *is defined for* $g_\varepsilon \in G$*, then* $y_\varepsilon = g_\varepsilon(\Phi(x))$ *is also a solution of the system.*

We say as well: $\mathbf{S}$ *is invariant under the group* G*, or: a group* G *is admitted by* $\mathbf{S}$*, or:* G *is a symmetry group of* $\mathbf{S}$*.*

4.3 Solution of a first order ODE using an integrating factor

Now we have a following situation: we have a differential equation $y' = f(x, y)$ and a one-parameter group which leaves the family of integral curves of this equation invariant. How can we use this knowledge to integrate the given equation?

Sophus Lie was the first to suggest methods of solution of an ODE using its symmetry group. We formulate the main idea of the first method in the next theorem.

Theorem 4.3.1 (Lie's integrating factor, Lie, 1874) *A first order ordinary differential equation*

$$\mathcal{X}(x, y)dy - \mathcal{Y}(x, y)dx = 0 \tag{4.3.6}$$

admits a one-parameter group $G(\mathbb{R}^2)$ *with an infinitesimal generator*

$$\mathbf{U} = \xi(x, y)\frac{\partial}{\partial x} + \eta(x, y)\frac{\partial}{\partial y}$$

if and only if the function

$$\mu(x, y) = \frac{1}{\mathcal{X}\eta - \mathcal{Y}\xi} = -\frac{1}{\begin{vmatrix} \xi & \eta \\ \mathcal{X} & \mathcal{Y} \end{vmatrix}} \tag{4.3.7}$$

is an integrating factor for the Equation (4.3.6), provided that $\mathcal{X}\eta - \mathcal{Y}\xi \neq 0$*.*

Proof. Let us denote the family of integral curves of Equation (4.3.6) as $\Phi(x, y) = C, \ \ C = \text{const}$. If this equation admits the symmetry group $G(\mathbb{R}^2)$

with $\mathbf{U} = \xi\frac{\partial}{\partial x} + \eta\frac{\partial}{\partial y}$, then the family $\Phi(x,y) = C$ is invariant under the action of the group, i.e. $\mathbf{U}\Phi = \xi\Phi_x + \eta\Phi_y = 1$. We look at Equation (4.3.6) as the characteristic equation for a first order PDE and obtain relations

$$\mathcal{X}dy - \mathcal{Y}dx = 0 \Leftrightarrow \frac{dy}{\mathcal{Y}} = \frac{dx}{\mathcal{X}} \Leftrightarrow \mathcal{X}\frac{\partial\Phi}{\partial x} + \mathcal{Y}\frac{\partial\Phi}{\partial y} = 0,$$

where $\Phi(x,y) = C$ is the first integral of this equation.
Correspondingly we obtain a system of equations

$$\begin{aligned} \mathcal{X}\frac{\partial\Phi}{\partial x} + \mathcal{Y}\frac{\partial\Phi}{\partial y} &= 0, \\ \xi\frac{\partial\Phi}{\partial x} + \eta\frac{\partial\Phi}{\partial y} &= 1. \end{aligned} \tag{4.3.8}$$

If we multiply the first equation of (4.3.8) with ξ and the second one with $\mathcal{X}$ and subtract the first equation from the second one we obtain

$$(\mathcal{X}\eta - \xi\mathcal{Y})\frac{\partial\Phi}{\partial y} = \mathcal{X} \Rightarrow \frac{\partial\Phi}{\partial y} = \frac{\mathcal{X}}{\mathcal{X}\eta - \mathcal{Y}\xi}.$$

Respectively, if we multiply the second equation of (4.3.8) with $\mathcal{Y}$ and the first one with η and subtract the second equation from the first one, we obtain

$$(\eta\mathcal{X} - \mathcal{Y}\xi)\frac{\partial\Phi}{\partial x} = -\mathcal{Y} \Rightarrow \frac{\partial\Phi}{\partial x} = -\frac{\mathcal{Y}}{\mathcal{X}\eta - \mathcal{Y}\xi}.$$

Thus we get the exact differential of the function Φ

$$d\Phi = \frac{\partial\Phi}{\partial x}dx + \frac{\partial\Phi}{\partial y}dy = -\frac{\mathcal{Y}}{\mathcal{X}\eta - \mathcal{Y}\xi}dx + \frac{\mathcal{X}}{\mathcal{X}\eta - \mathcal{Y}\xi}dy$$

or

$$d\Phi = \frac{1}{\mathcal{X}\eta - \mathcal{Y}\xi}(\mathcal{X}dy - \mathcal{Y}dx) = \mu(x,y)(\mathcal{X}\eta - \mathcal{Y}\xi). \tag{4.3.9}$$

Eventually, after an integration we obtain from (4.3.9) the family of integral curves

$$\Phi(x,y) = \int \frac{\mathcal{X}dy - \mathcal{Y}dx}{\mathcal{X}\eta - \mathcal{Y}\xi} = \text{const.}$$

□

Remark 4.3.1 *The function $\mu(x,y)$ is called either Lie's integrating factor or Euler's multiplier.*

The theorem above shows that after multiplication with the factor $\mu(x, y)$ the equation will be equivalent to an exact differential equation, which can be integrated easily.

Let us look at one example of an application of this method.

Example 4.3.1 We want to solve the equation

$$\frac{dy}{dx} = \frac{y}{x + F(y)}, \tag{4.3.10}$$

where F is an arbitrary continuous function of y. We know that this equation admits the one-parameter group of Galilean transformations

$$x_\varepsilon = x + \varepsilon y, \qquad y_\varepsilon = y.$$

The infinitesimal generator of this group is equal to

$$\mathbf{U} = y\frac{\partial}{\partial x},$$

in other words $\xi(x, y) = y$ and $\eta(x, y) = 0$. We transform Equation (4.3.10) to

$$(x + F(y))\, dy - y dx = 0, \quad \text{where } \mathcal{X} = x + F(y),\ \mathcal{Y} = y.$$

Then the integrating factor is equal to

$$\mu(x, y) = \frac{1}{(x + F(y)) \cdot 0 - y \cdot y} = -\frac{1}{y^2}.$$

Therefore, the exact differential equation is written in the form

$$-\frac{x + F(y)}{y^2} dy + \frac{1}{y} dx = 0,$$

i.e. $\Phi_y = -\frac{x+F(y)}{y^2}$, $\Phi_x = \frac{1}{y}$. Respectively the relation $\Phi_{xy} = \Phi_{yx}$ must be true. We prove this relation now and obtain $\Phi_{yx} = -\frac{1}{y^2}$ and $\Phi_{xy} = -\frac{1}{y^2}$ and see that it is really an exact differential equation. To get the family of integral curves we integrate Φ_x with respect to x and obtain

$$\Phi(x, y) = \frac{x}{y} + \phi(y),$$

where $\phi(y)$ is an arbitrary function of y. Then we insert $\Phi(x, y)$ in the equation for Φ_y to define the unknown function $\phi(y)$

$$-\frac{x}{y^2} + \phi'(y) = -\frac{x}{y^2} - \frac{F(y)}{y^2} \;\Rightarrow\; \phi(y) = -\int \frac{F(y)}{y^2}\, dy.$$

Thus the solution of Equation (4.3.10) has the form

$$\Phi(x, y) = \frac{x}{y} - \int \frac{F(y)}{y^2}\, dy = C = \text{const.}$$

We can take the infinitesimal generator $\mathbf{U}$ of the Galilean group of point transformations and prove if this family of integral curves is invariant under the action of this group. In fact

$$\mathbf{U}\Phi(x, y) = y\frac{\partial}{\partial x}\left(\frac{x}{y} - \int \frac{F(y)}{y^2}\, dy\right) = 1.$$

We see that with the help of the integrating factor we can find the general solution of a first order ODE, even if this solution is given in the implicit form $\Phi(x, y) = C$. ◇

4.4 Solution of a first order ODE using canonical variables of the admitted symmetry

The second way to use the known symmetry group is connected directly with the structure of the group. It is the method of canonical variables. We have seen that for each group there exist canonical variables which reduce the infinitesimal generator to its normal form $\mathbf{U} = \frac{\partial}{\partial s}$, i.e. to the generator of a translation group. But the change of variables (x, y) to (s, t) will also change our equation. We recognized earlier that the frame of the equation, the surface $p = f(x, y)$, can be used to find the symmetry group. This frame will also be changed after substitution. We know that the action of the one-parameter group corresponds to the moving points on the frame. If our group is transformed to a translation group now, then in the new variables the frame is a general cylindric surface and we can easily integrate the equation.

Let us look at how this integrating procedure works in the general case. We have the infinitesimal generator

$$\mathbf{U} = \xi(x, y)\frac{\partial}{\partial x} + \eta(x, y)\frac{\partial}{\partial y}$$

and are able to find canonical variables which solve the system of equations

$$\xi(x, y)t_x + \eta(x, y)t_y = 0,$$
$$\xi(x, y)s_x + \eta(x, y)s_y = 1.$$

In these new variables the action of the infinitesimal operator is reduced to translation only, i.e. $\mathbf{U} = \frac{\partial}{\partial s}$. The old and new variables are connected by $s = s(x, y)$ and $t = t(x, y)$. Thus, we have

$$ds = \frac{\partial s}{\partial x}dx + \frac{\partial s}{\partial y}dy, \; dt = \frac{\partial t}{\partial x}dx + \frac{\partial t}{\partial y}dy,$$

and because $\frac{dy}{dx} = f(x, y) = f(x(s,t), y(s,t))$ we obtain

$$\frac{ds}{dt} = \frac{\frac{\partial s}{\partial x}dx + \frac{\partial s}{\partial y}dy}{\frac{\partial t}{\partial x}dx + \frac{\partial t}{\partial y}dy} = \frac{\frac{\partial s}{\partial x} + \frac{\partial s}{\partial y}\frac{dy}{dx}}{\frac{\partial t}{\partial x} + \frac{\partial t}{\partial y}\frac{dy}{dx}} = \frac{\frac{\partial s}{\partial x} + f(x,y)\frac{\partial s}{\partial y}}{\frac{\partial t}{\partial x} + f(x,y)\frac{\partial t}{\partial y}} = F(s,t).$$

As we have learned in Section 4.1 the most general first order ODE which is invariant under the action of the translation group has the form $s' = F(t)$. (In Section 4.1 we have the infinitesimal generator in the form $\mathbf{U} = \frac{\partial}{\partial y}$ and we have seen in Example 4.1.1 that the most general first order ODE that is invariant under the action of this translation group has the form $y' = f(x)$.) This means that we get $\frac{ds}{dt} = F(t)$ and the solution is given by

$$s = \int F(t)\, dt + C, \quad C = \text{const}.$$

Example 4.4.1 We take the same equation as in the previous example

$$\frac{dy}{dx} = \frac{y}{x + F(y)}, \quad \mathbf{U} = y\frac{\partial}{\partial x}.$$

Then the system of equations for the canonical variables is

$$ys_x = 1, \quad yt_x = 0,$$

and we can choose as new variables

$$s = \frac{x}{y}, \qquad t = y.$$

With the new variables the equation takes the form

$$\frac{ds}{dt} = \frac{\frac{\partial s}{\partial x} + f(x(s,t), y(s,t))\frac{\partial s}{\partial y}}{\frac{\partial t}{\partial x} + f(x(s,t), y(s,t))\frac{\partial t}{\partial y}} = \frac{\frac{1}{t} + \frac{t}{st+F(t)}\left(-\frac{s}{t}\right)}{\frac{t}{st+F(t)}}$$

$$= \frac{1 - \frac{st}{st+F(t)}}{\frac{t^2}{st+F(t)}} = \frac{st + F(t) - st}{t^2} = \frac{F(t)}{t^2}.$$

Respectively we obtain the family of integral curves in the form

$$s = \int \frac{F(t)}{t^2}\, dt + C \text{ or } \frac{x}{y} = \int \frac{F(y)}{y^2}\, dy + C, \quad C \in \mathbb{R}.$$

We got the solution of the equation in the same form as in the previous section. ◇

4.5 Relation between the infinitesimal generator U and the operator $\mathcal{A}$

Up to now we have used the following method to prove whether a given equation admits a symmetry or not: we studied the properties of the family of integral curves. If this family is invariant under the action of the group, then the equation admits the symmetry. There exists also a local method to prove if the given equation admits the symmetry or not.

We have seen that to the ordinary differential equation $\mathcal{X}dy - \mathcal{Y}dx = 0$ corresponds the partial differential equation $\mathcal{X}\frac{\partial \Phi}{\partial x} + \mathcal{Y}\frac{\partial \Phi}{\partial y} = 0$, it means to each first order ODE we can find the linear operator $\mathcal{A} = \mathcal{X}\frac{\partial}{\partial x} + \mathcal{Y}\frac{\partial}{\partial y}$. We can investigate the relations between the operators $\mathbf{U}$ and $\mathcal{A}$.

If we have a family of integral curves, then $\mathcal{A}\Phi = 0$ as was shown before, because they are solutions of this equation. It also means $\mathbf{U}(\mathcal{A}\Phi) = 0$. If we act with the operator $\mathbf{U}$ first, then as we have seen before $\mathbf{U}\Phi = F(\Phi)$ and $\mathcal{A}(\mathbf{U}\Phi) = \mathcal{A}(F(\Phi)) = F' \cdot \mathcal{A}\Phi = 0$. Our goal now is to avoid the unknown function $F(\Phi)$ and get a convenient condition to prove if the given group is a symmetry group of the equation or not. We have two linear operators $\mathbf{U}$ and $\mathcal{A}$ and we can use this linearity to formulate the condition we are looking for. Let us study the commutator $[\cdot,\cdot]$ of these operators

$$[\mathbf{U}, \mathcal{A}] = \mathbf{U}\mathcal{A} - \mathcal{A}\mathbf{U}. \tag{4.5.11}$$

We have proved that $\mathbf{U}\mathcal{A}\Phi = \mathcal{A}\mathbf{U}\Phi = 0$ if $\Phi =$ const. describes the family of integral curves of the operator $\mathcal{A}$. Now we can calculate the structure of the commutator

$$\begin{aligned}
[\mathbf{U}, \mathcal{A}] &= \left(\xi\frac{\partial}{\partial x} + \eta\frac{\partial}{\partial y}\right)\left(\mathcal{X}\frac{\partial}{\partial x} + \mathcal{Y}\frac{\partial}{\partial y}\right) - \left(\mathcal{X}\frac{\partial}{\partial x} + \mathcal{Y}\frac{\partial}{\partial y}\right)\left(\xi\frac{\partial}{\partial x} + \eta\frac{\partial}{\partial y}\right)\\
&= \xi\frac{\partial \mathcal{X}}{\partial x}\frac{\partial}{\partial x} + \xi\mathcal{X}\frac{\partial^2}{\partial x^2} + \xi\frac{\partial \mathcal{Y}}{\partial x}\frac{\partial}{\partial y} + \xi\mathcal{Y}\frac{\partial^2}{\partial x\partial y} + \eta\frac{\partial \mathcal{X}}{\partial y}\frac{\partial}{\partial x} + \eta\mathcal{X}\frac{\partial^2}{\partial x\partial y}\\
&\quad + \eta\frac{\partial \mathcal{Y}}{\partial y}\frac{\partial}{\partial y} + \eta\mathcal{Y}\frac{\partial^2}{\partial y^2} - \mathcal{X}\frac{\partial \xi}{\partial x}\frac{\partial}{\partial x} - \mathcal{X}\xi\frac{\partial^2}{\partial x^2} - \mathcal{X}\frac{\partial \eta}{\partial x}\frac{\partial}{\partial y} - \mathcal{X}\eta\frac{\partial^2}{\partial x\partial y}\\
&\quad - \mathcal{Y}\frac{\partial \xi}{\partial y}\frac{\partial}{\partial x} - \mathcal{Y}\xi\frac{\partial^2}{\partial x\partial y} - \mathcal{Y}\frac{\partial \eta}{\partial y}\frac{\partial}{\partial y} - \mathcal{Y}\eta\frac{\partial^2}{\partial y^2}\\
&= \left(\xi\frac{\partial \mathcal{X}}{\partial x} + \eta\frac{\partial \mathcal{X}}{\partial y} - \mathcal{X}\frac{\partial \xi}{\partial x} - \mathcal{Y}\frac{\partial \xi}{\partial y}\right)\frac{\partial}{\partial x}\\
&\quad + \left(\xi\frac{\partial \mathcal{Y}}{\partial x} + \eta\frac{\partial \mathcal{Y}}{\partial y} - \mathcal{X}\frac{\partial \eta}{\partial x} - \mathcal{Y}\frac{\partial \eta}{\partial y}\right)\frac{\partial}{\partial y}\\
&= (\mathbf{U}\mathcal{X} - \mathcal{A}\xi)\frac{\partial}{\partial x} + (\mathbf{U}\mathcal{Y} - \mathcal{A}\eta)\frac{\partial}{\partial y}.
\end{aligned}$$

It is also a linear partial differential operator and we know that this operator has the same set of solutions – the same family of integral curves as the operator $\mathcal{A}$. As a consequence, these two operators are not independent, but proportional to each other in every point (x, y), i.e.

$$[\mathbf{U}, \mathcal{A}] = \lambda(x, y)\mathcal{A}, \quad \text{or} \quad [\mathbf{U}, \mathcal{A}] = 0 \ (\mathrm{mod}\mathcal{A}). \tag{4.5.12}$$

Both equations $\mathcal{A}\Phi = 0$ and $[\mathbf{U}, \mathcal{A}]\Phi = 0$ are equivalent. The Equation (4.5.12) is called a determining equation. In an ideal case we can use this equation to find a symmetry group of a first order ODE and then solve this ODE using the symmetry properties. But in the case of the first ODEs the determining equation is equivalent to the first order ODE and we have to solve a problem which has the same difficulty level as before. The reason for this situation is that any first order ODE has infinitely many symmetries and there is no algorithmic way to find any one of them.

4.6 Problems with solutions

In this section we provide twenty-two problems with detailed solutions. It allows us to explain the relationships between the first order ODE and the admitted symmetry in every detail.

If the symmetry group is known we can use it to solve the given first order ODE using an integrating factor or a method of canonical variables. The first method of solution using an integrating factor is explained in Problems 4.6.2-4.6.4, 4.6.17, 4.6.21 and 4.6.22. The second method showing how one can solve a first order ODE using canonical variables is demonstrated in Problems 4.6.1 and 4.6.16. These two methods are equivalent as long as we have a first order ODE. Yet the method of canonical variables can be extended to more complicated cases of high order ODEs or high order PDEs.

To understand better the connection between the symmetry group and a differential equation we show how to find the most general form of a first order ODE which admits the given symmetry in Problems 4.6.5-4.6.8 and 4.6.10.

In Problems 4.6.9, 4.6.11 and 4.6.12 we verify if the given equation admits the provided symmetry group.

Students are often not very familiar with the theory of first order PDEs, so in Problems 4.6.13-4.6.15 we discuss how to solve a Cauchy problem for such equations.

Problems 4.6.18-4.6.22 demonstrate how to find a first order ODE if its one-parameter family of integral curves is given.

Problem 4.6.1 *Solve the equation*

$$2x^4y\frac{\mathrm{d}y}{\mathrm{d}x} + 4x^3y^2 + 2x = 0$$

using the known infinitesimal symmetry

$$\mathbf{U} = x\frac{\partial}{\partial x} - y\frac{\partial}{\partial y}.$$

Solution We obtain the coefficients $\xi(x, y) = x$, $\eta(x, y) = -y$ from the infinitesimal generator $\mathbf{U}$. The original equation can be rewritten in the following way

$$\frac{dy}{dx} = -\frac{4x^3y^2 + 2x}{2x^4y} = -\frac{2x^2y^2 + 1}{x^3y}, \quad \textit{for} \quad x \neq 0, y \neq 0.$$

It means the original equation is equivalent to the system of equations

$$\frac{dx}{dy} = -\frac{x^3y}{2x^2y^2 + 1}, \quad x = 0.$$

($y = 0$ do not solve the original equation). The canonical variables for the operator $\mathbf{U}$ are first integrals of the system of equations

$$\mathbf{U}t = 0, \quad \mathbf{U}s = 1.$$

The variable t is defined by the characteristic system of equations

$$\frac{dx}{x} = \frac{dy}{-y} = \frac{dt}{0}.$$

Its solution takes the form

$$\ln y = -\ln x + \text{const.} \quad \textit{or} \quad t = F(xy).$$

We take the simplest form for the variable t, i.e. $t = xy$ with $dt = xdy + ydx$. From the characteristic system of equations for the variable s follows

$$\begin{aligned} \frac{dx}{x} &= \frac{dy}{-y} = \frac{ds}{1}, \\ s &= \ln x, \\ ds &= \frac{1}{x}dx. \end{aligned}$$

The ratio of both differentials ds and dt is given by

$$\frac{ds}{dt} = \frac{\frac{1}{x}dx}{xdy + ydx} = \frac{\frac{1}{x}\frac{dx}{dy}}{x + y\frac{dx}{dy}} = \frac{\frac{1}{x}\frac{-x^3y}{2x^2y^2+1}}{x + y\frac{-x^3y}{2x^2y^2+1}} = \frac{-x^2y}{2x^2y^2+1}\frac{2x^2y^2+1}{2x^3y^2+x-x^3y^2},$$

$$\frac{ds}{dt} = \frac{-xy}{x^2y^2+1} = \frac{-t}{t^2+1}.$$

It represents the original equation in new coordinates s, t. After integration we obtain

$$s = -\frac{1}{2}\ln(1+t^2) + \text{const.}$$

We rewrite the solution in old variables x, y

$$\ln x + \frac{1}{2}\ln(1+x^2y^2) = \text{const.}$$

or

$$C = x\sqrt{1+x^2y^2}, \quad C = \text{const.}$$

Raising to the second power we get

$$y^2 = \frac{c - x^2}{x^4}.$$

Correspondingly the complete set of solutions of the original equation is given by the set

$$y^2 = \frac{C - x^2}{x^4}, \quad x = 0$$

(if we keep the first solution in the form $C = x\sqrt{1+x^2y^2}$ then the solution $x = 0$ is included in this one-parameter family of solutions for $C = 0$, otherwise we have to note it separately).

Problem 4.6.2 *Find the general solution of the differential equation*

$$\frac{\mathrm{d}y}{\mathrm{d}x} = xy(1 + \ln y + x^2) \tag{4.6.13}$$

using Lie's integrating factor and the symmetry group with the infinitesimal generator equal to

$$\mathbf{U} = \frac{1}{2x}\frac{\partial}{\partial x} - y\frac{\partial}{\partial y}.$$

Solution The coefficients in $\mathbf{U}$ are $\xi(x,y) = \frac{1}{2x}$, $\eta(x,y) = -y$. The original Equation (4.6.13) we rewrite in the following way

$$\frac{dy}{dx} = \frac{\mathcal{Y}}{\mathcal{X}} = \frac{x(1+x^2+\ln y)}{1/y}.$$

The integrating factor is equal to

$$\mu(x,y) = \frac{1}{\mathcal{X}\eta - \mathcal{Y}\xi} = \frac{1}{\frac{1}{y}(-y) - \frac{1}{2x}x(1+x^2+\ln y)} = \frac{-2}{3+\ln y + x^2}.$$

Therefore the related exact differential equation is

$$\frac{-2}{y(3+x^2+\ln y)}dy - x(1+x^2+\ln y)\frac{-2}{3+x^2+\ln y}dx = 0,$$

i.e. the auxiliary function $\Phi(x,y)$ can be defined from

$$\Phi_x = \frac{2x(1+x^2+\ln y)}{3+x^2+\ln y} = \frac{2x(3+x^2+\ln y) - 4x}{3+x^2+\ln y} = 2x - \frac{4x}{3+x^2+\ln y},$$

and

$$\Phi_y = \frac{-2}{y(3+x^2+\ln y)}.$$

We know that $\Phi_{xy} = \Phi_{xy}$ must hold (it is easy to prove by differentiation). Using this knowledge we find the one-parameter family of integral curves $\Phi(x,y) = \text{const}$.
We take Φ_x and integrate it with respect to x

$$\Phi(x,y) = \int \left(2x - \frac{4x}{3+x^2+\ln y}\right) dx = x^2 - 2\ln(3+x^2+\ln y) + F(y).$$

Then we insert $\Phi(x,y)$ in the equation for Φ_y to define the unknown function $F(y)$

$$\Phi_y = \frac{1}{y}\frac{-2}{3+x^2+\ln y} + F'(y) = \frac{-2}{y(3+x^2+\ln y)} \quad \Rightarrow \quad F(y) = \text{const}.$$

Thus the solution of the given equation has the form

$$\Phi = x^2 - 2\ln(3+x^2+\ln y) = C = \text{const}.$$

or

$$\ln y = C\, e^{x^2/2} - x^2 - 3,$$

where C is an integration constant.

Problem 4.6.3 *Find the integrating factor for the differential equation*

$$xy\prime = \frac{y}{\ln x + y}, \tag{4.6.14}$$

that admits the symmetry

$$\mathbf{U} = xy\frac{\partial}{\partial x}$$

and solve the equation.

Solution Let us rewrite the differential equation in the form

$$-ydx + x(\ln x + y)dy = 0.$$

Then the corresponding operator $\mathcal{A} = \mathcal{X}(x,y)\frac{\partial}{\partial x} + \mathcal{Y}(x,y)\frac{\partial}{\partial y}$ takes the form

$$\begin{aligned} \mathcal{X}(x,y) &= -y, \\ \mathcal{Y}(x,y) &= x(\ln x + y). \end{aligned}$$

Since $\mathbf{U} = xy\frac{\partial}{\partial x}$ we have

$$\begin{aligned} \xi(x,y) &= xy, \\ \eta(x,y) &= 0. \end{aligned}$$

According to Lie theorem an integrating factor can be calculated in the following way:

$$\mu(x,y) = \frac{1}{\begin{vmatrix} \xi & \eta \\ \mathcal{X} & \mathcal{Y} \end{vmatrix}} = \frac{1}{xy^2}.$$

Multiplying the Equation (4.6.14) to it we obtain

$$\frac{dx}{xy} + \frac{\ln x + y}{y^2}dy = 0.$$

Integrating the system

$$\begin{aligned} \frac{\partial \Phi}{\partial x} &= -\frac{1}{xy}, \\ \frac{\partial \Phi}{\partial y} &= \frac{\ln x + y}{y^2} + \frac{1}{y}, \end{aligned}$$

we obtain the following solution of the original Equation (4.6.14)

$$y + C(\ln x + y\ln y) = 0.$$

Problem 4.6.4 *Find the integrating factor for the differential equation*

$$y\prime = y \sin x, \tag{4.6.15}$$

that admits the symmetry

$$\mathbf{U} = y\frac{\partial}{\partial y} \tag{4.6.16}$$

and solve it.

Solution Let us rewrite the differential equation in the form

$$-\sin x dx + \frac{1}{y}dy = 0.$$

We take the coordinates of the corresponding operator $\mathcal{A}$ as

$$\begin{aligned} \mathcal{X} &= -\sin x, \\ \mathcal{Y} &= \frac{1}{y}. \end{aligned}$$

Since $\mathbf{U} = y\frac{\partial}{\partial y}$ we have

$$\begin{aligned} \xi(x, y) &= 0, \\ \eta(x, y) &= y. \end{aligned}$$

According to Lie theorem an integrating factor can be calculated in the following way

$$\mu(x, y) = \frac{1}{\begin{vmatrix} \xi & \eta \\ \mathcal{X} & \mathcal{Y} \end{vmatrix}} = 1.$$

This means that our equation already has a form of an exact differential, and we can integrate the system

$$\begin{aligned} \frac{\partial \Phi}{\partial x} &= -\sin x, \\ \frac{\partial \Phi}{\partial y} &= \frac{1}{y}. \end{aligned}$$

We obtain the following solution of the original equation

$$y = Ce^{-\cos x}.$$

Problem 4.6.5 *Find the most general form of a first order ODE which admits the symmetry described by the infinitesimal generator*

$$\mathbf{U} = l\frac{\partial}{\partial x} - k\frac{\partial}{\partial y}.$$

Solution The Lie equation

$$\frac{dx}{l} = -\frac{dy}{k}$$

for this infinitesimal generator gives the orbits of the group in the form

$$kx + ly = C, \quad C \in \mathbb{R}.$$

Now we find family of integral curves of the ODE which we looked for

$$\begin{aligned} \frac{dx}{l} &= d\Phi \Leftrightarrow \\ \Phi(x, y) &= \frac{x}{l} + \phi(kx + ly) = C, \end{aligned}$$

where function ϕ is arbitrary.

Differentiating the last equation gives us the most general form of an ODE with this family of the integral curves

$$\left(\frac{x}{l} + k\phi\prime(kx + ly)\right) dx + (l\phi\prime(kx + ly))dy = 0.$$

We represent this equation in explicit form

$$\frac{dx}{dy} = -\frac{1}{l^2\phi\prime(kx + ly)} + \frac{k}{l}.$$

Since ϕ is an arbitrary function of its argument, we can rewrite the equation in the form

$$\frac{dy}{dx} = F(kx + ly),$$

where F is an arbitrary function.

Problem 4.6.6 *Find the most general form of a first order ODE which admits the symmetry represented by the infinitesimal generator*

$$\mathbf{U} = x\frac{\partial}{\partial x} + \frac{\partial}{\partial y}.$$

Solution The first of the Lie equations for this infinitesimal generator

$$\frac{dx}{x} = \frac{dy}{y}$$

gives us the group's orbits in the form

$$\ln|x| - y = C.$$

Now we can find a family of the integral curves which corresponds to the first order ODE to find

$$\Phi(x,y) = y + \phi(\ln|x| - y) = C,$$

where the function ϕ is an arbitrary continuously differentiable function of its argument.

Differentiating the last equation leads to the most general form of an ODE with such integral curves

$$\frac{\phi\prime(\ln|x| - y)}{x}dx + (1 - \phi\prime(\ln|x| - y))dy = 0,$$

or in the explicit form

$$\frac{dy}{dx} = \frac{\phi\prime(\ln|x| - y)}{x(1 - \phi\prime(\ln|x| - y))}.$$

Since ϕ is an arbitrary function, we can rewrite it in a more convenient form

$$xy\prime = f(\ln|x| - y) = F(xe^{-y}),$$

where $f = e^F$ and F is an arbitrary continuous function as well as f.

Problem 4.6.7 *Find the most general form of a first order ODE which admits the symmetry with the infinitesimal generator*

$$\mathbf{U} = \frac{\partial}{\partial x} + y\frac{\partial}{\partial y}.$$

Solution Exchanging x and y in the previous example, we obtain the ODE in the form

$$y\phi\prime(ye^{-x})dx + (1 - \phi\prime(ye^{-x}))dy = 0,$$

or in an explicit form

$$\frac{dy}{dx} = \frac{y\phi\prime(ye^{-x})}{-1 + \phi\prime(ye^{-x})}.$$

Since ϕ is an arbitrary function, we can rewrite it in the form

$$y\prime = yF(ye^{-x}),$$

where F is an arbitrary continuous function.

Problem 4.6.8 *Find the most general form of a first order ODE with symmetry provided by the infinitesimal generator*

$$\mathbf{U} = x\frac{\partial}{\partial y}.$$

Solution The first Lie equation

$$\frac{dy}{x} = \frac{dx}{0} = \frac{d\Phi}{1}$$

gives us integral curves in the form

$$\Phi(x, y) = \frac{y}{x} + \phi(x) = C,$$

where the function ϕ is an arbitrary continuously differentiable function.

After differentiation of the last equation we obtain the most general form of a first order ODE with such integral curves

$$\left(-\frac{y}{x^2} + \phi\prime(x)\right) dx + \frac{dy}{x} = 0,$$

or in the explicit form

$$x\frac{dy}{dx} = y - \phi\prime(x)x^2.$$

Since ϕ is an arbitrary function, we can rewrite it in a more convenient form

$$xy\prime = y + F(x)$$

where F is an arbitrary continuous function.

Problem 4.6.9 *Verify that the differential equation*

$$\frac{\mathrm{d}y}{\mathrm{d}x} = (\ln y - e^x)\exp(x + e^x)$$

admits the symmetry group

$$x_\epsilon = \ln(e^x + \epsilon), \quad y_\epsilon = e^\epsilon y.$$

Solution First we find the infinitesimal generator of the given transformation group. It easy to see that

$$\xi(x, y) = \left.\frac{dx_\epsilon}{\epsilon}\right|_{\epsilon=0} = \left.\frac{1}{e^x + \epsilon}\right|_{\epsilon=0} = e^{-x},$$
$$\eta(x, y) = \left.\frac{dy_\epsilon}{\epsilon}\right|_{\epsilon=0} = \left. e^\epsilon y\right|_{\epsilon=0} = y.$$

The infinitesimal generator $\mathbf{U}$ takes the following form

$$\mathbf{U} = e^{-x}\frac{\partial}{\partial x} + y\frac{\partial}{\partial y}.$$

Rewriting the main equation as

$$dy - (\ln y - e^x)e^{x+e^x}dx = 0$$

and taking

$$\mathcal{X} = 1,\ \mathcal{Y} = (\ln y - e^x)e^{x+e^x},$$

we obtain the operator $\mathcal{A}$ which is associated with this ODE

$$\mathcal{A} = \frac{\partial}{\partial x} + (\ln y - e^x)e^{x+e^x}\frac{\partial}{\partial y}.$$

To check whether the differential equation admits the group or not, we should verify that there exists a function $\lambda(x, y)$ such that the relation holds

$$[\mathbf{U}, \mathcal{A}] = \lambda(x, y)\mathcal{A}.$$

We insert the expressions for $\mathcal{A}, \mathbf{U}$ and calculate

$$\begin{aligned}
&[\mathbf{U}, \mathcal{A}] = \mathbf{U}\mathcal{A} - \mathcal{A}\mathbf{U} = \\
&e^{-x}(\ln ye^{x+e^x}(1+e^x) - e^{2x+e^x}(2+e^x))\frac{\partial}{\partial y} + \\
&y\frac{1}{y}e^{x+e^x}\frac{\partial}{\partial y} - (-e^{-x}\frac{\partial}{\partial x} + (\ln y - e^x)e^{x+e^x}\frac{\partial}{\partial y}) = \\
&e^{-x}\frac{\partial}{\partial x} + (\ln ye^{e^x}(1+e^x) - e^{x+e^x}(2+e^x) + e^{x+e^x} - (\ln y - e^x)e^{x+e^x})\frac{\partial}{\partial y} = \\
&e^{-x}\left(\frac{\partial}{\partial x} + e^{x+e^x}\big(\ln y(1+e^x) - e^x(2+e^x) + e^x - (\ln y - e^x)e^x\big)\frac{\partial}{\partial y}\right) = \\
&e^{-x}\left(\frac{\partial}{\partial x} + e^{x+e^x}\big(\ln y - e^x\big)\frac{\partial}{\partial y}\right) = e^{-x}\mathcal{A}.
\end{aligned}$$

It means that there exists the function $\lambda(x, y) = e^{-x}$. We proved that the commutator of the operators $\mathcal{A}, \mathbf{U}$ is proportional in each point (x, y) to the operator $\mathcal{A}$ and consequently the corresponding equation admits the symmetry described by the infinitesimal generator $\mathbf{U}$.

Problem 4.6.10 *Let a one-parameter group of point transformations be given by*

$$\mathbf{U} = x^3\frac{\partial}{\partial x} + y^3\frac{\partial}{\partial y}.$$

Find the most general ODE of the first order which admits this group of point transformations.

Solution At first let us write down the equation for the group orbits,

$$\frac{dx}{x^3} = \frac{dy}{y^3},$$

from which we can easily obtain the solution

$$-\frac{1}{2x^2} = -\frac{1}{2y^2} + C.$$

In order to find the most general form of the first order ODE which admits this group, we introduce the function $\Phi(x, y)$ that denotes the family of integral curves. We obtain $\Phi(x, y)$ as a solution of the equation

$$\frac{dx}{x^3} = d\Phi,$$

$$\Phi(x,y) = -\frac{1}{2x^2} + C = -\frac{1}{2x^2} + \varphi\left(\frac{1}{2y^2} - \frac{1}{2x^2}\right),$$

where φ is an arbitrary function of its argument. After differentiating the formula for $\Phi(x, y)$ we obtain the equation

$$\frac{1}{x^3}dx + \varphi'\left(\frac{1}{2y^2} - \frac{1}{2x^2}\right)\frac{1}{x^3}dx - \varphi'\left(\frac{1}{2y^2} - \frac{1}{2x^2}\right)\frac{1}{y^3}dy = 0,$$

or while φ' is an arbitrary continuously differentiable function. We rewrite the equation using a new notation F for an arbitrary continuous function

$$\left(1 + \varphi'\left(\frac{1}{2y^2} - \frac{1}{2x^2}\right)\right)\frac{1}{x^3}dx = \varphi'\left(\frac{1}{2y^2} - \frac{1}{2x^2}\right)\frac{1}{y^3}dy,$$

$$\frac{dy}{dx} = \frac{y^3}{x^3}F\left(\frac{1}{y^2} - \frac{1}{x^2}\right),$$

which is the most general form of equation that admits the given group of transformations.

Problem 4.6.11 *Let the group of scaling transformations be given by*

$$\mathbf{U} = x\frac{\partial}{\partial x} + 2y\frac{\partial}{\partial y}.$$

The first order ordinary differential equation has the form

$$xy' = y(\ln y - 2\ln x).$$

Prove that this equation admits this scaling symmetry group (use the determining equation).

Solution In order to use the determining equation let us find the operator $\mathcal{A}$ that corresponds to the given differential equation. We rewrite the equation in the following form

$$\frac{x}{y}dy = \ln\left(\frac{y}{x^2}\right)dx,$$

and find the expression for $\mathcal{A}$

$$\mathcal{A} = \frac{x}{y}\frac{\partial}{\partial x} + \ln\frac{y}{x^2}\frac{\partial}{\partial y}.$$

To use the determining equation we first find the commutator of $\mathbf{U}$ and $\mathcal{A}$ as follows

$$[\mathbf{U}, \mathcal{A}] = (\mathbf{U}X - \mathcal{A}\xi)\frac{\partial}{\partial x} + (\mathbf{U}Y - \mathcal{A}\eta)\frac{\partial}{\partial y},$$

or

$$\begin{aligned}[\mathbf{U}, \mathcal{A}] = & \left(\left(x\frac{\partial}{\partial x} + 2y\frac{\partial}{\partial y}\right)\frac{y}{x} - \frac{x}{y}\right)\frac{\partial}{\partial x} \\ & + \left(\left(x\frac{\partial}{\partial x} + 2y\frac{\partial}{\partial y}\right)\ln\frac{y}{x^2} - \left(\frac{x}{y}\frac{\partial}{\partial x} + \ln\frac{y}{x^2}\frac{\partial}{\partial y}\right)2y\right)\frac{\partial}{\partial y}.\end{aligned}$$

Thus, we obtain

$$[\mathbf{U}, \mathcal{A}] = -\frac{2x}{y}\frac{\partial}{\partial x} - 2\ln\frac{y}{x^2}\frac{\partial}{\partial y} = \lambda\mathcal{A}, \tag{4.6.17}$$

where $\lambda = -2$. We have proved that the equation admits the given symmetry group.

Problem 4.6.12 *Prove that the relation*

$$[\mathbf{U}, \mathcal{A}] = \lambda(x, y)\mathcal{A},$$

holds if $\mathbf{U}$ *is an infinitesimal generator of a group of point transformations set by a formula*

$$\mathbf{U} = \frac{\partial}{\partial x} + \frac{y}{x}\frac{\partial}{\partial y},$$

and $\mathcal{A}$ is a differential operator corresponding to the equation

$$y' = \frac{y}{x} + \frac{y^2}{x}, \quad x \neq 0.$$

Solution We rewrite the ODE in the form

$$\frac{dx}{x} = \frac{dy}{y+y^2},$$

then we obtain an expression for the differential operator $\mathcal{A}$

$$\mathcal{A} = x\frac{\partial}{\partial x} + (y+y^2)\frac{\partial}{\partial y}, \quad \mathcal{X} = x, \quad \mathcal{Y} = y + y^2.$$

Using the formula

$$[\mathbf{U}, \mathcal{A}] = (\mathbf{U}\mathcal{X} - \mathcal{A}\xi)\frac{\partial}{\partial x} + (\mathbf{U}\mathcal{Y} - \mathcal{A}\eta)\frac{\partial}{\partial y}$$

we calculate the commutator of $\mathcal{A}$ and the given infinitesimal generator $\mathbf{U}$

$$[\mathbf{U}, \mathcal{A}] = (1-0)\frac{\partial}{\partial x} + (\frac{y}{x} + \frac{2y^2}{x} + \frac{yx}{x^2} - \frac{y+y^2}{x})\frac{\partial}{\partial y} = \frac{\partial}{\partial x} + \frac{y+y^2}{x}\frac{\partial}{\partial y}.$$

We have proved that

$$[\mathbf{U}, \mathcal{A}] = \lambda(x,y)\mathcal{A} \tag{4.6.18}$$

where $\lambda(x,y) = \frac{1}{x}$. We have proved also that the equation admits the given symmetry group.

Problem 4.6.13 *Solve the Cauchy problem for the equation*

$$(x^2+y^2)\frac{\partial u}{\partial x} + 2xy\frac{\partial u}{\partial y} = xu, \quad u(x,y)|_{y=1} = x. \tag{4.6.19}$$

Solution We denote the differential operator of this first order PDE as $\mathcal{B}$. Then the differential equation can be rewritten as follows

$$\mathcal{B}u = xu.$$

In order to solve it let us find the canonical variables s and t for the given operator $\mathcal{B}$ using the system of equations

$$\mathcal{B}s = 1, \quad \mathcal{B}t = 0.$$

We look at the first equation of this system

$$(x^2+y^2)\frac{\partial s}{\partial x}+2xy\frac{\partial s}{\partial y}=1.$$

A solution can be found immediately, it is equal to

$$s=\frac{1}{x+y}.$$

The second equation for the canonical coordinate t has the form

$$(x^2+y^2)\frac{\partial t}{\partial x}+2xy\frac{\partial t}{\partial y}=0.$$

Introducing the polar coordinates and separating the variables we obtain the following equation

$$\frac{dr}{r}=\frac{\cos\varphi(2\sin^2\varphi+1)}{\sin\varphi(2\cos^2\varphi-1)}d\varphi,$$

after integration we obtain the solution

$$r=c\frac{2\sin\varphi}{\cos^2\varphi-\sin^2\varphi}.$$

Coming back to the coordinates x and y we obtain a simple expression for the second canonical coordinate t

$$t=\frac{x^2-y^2}{2y}.$$

Now we introduce canonical variables correspondingly as

$$s=\frac{1}{x+y},\quad t=\frac{x^2-y^2}{2y}. \tag{4.6.20}$$

Using these variables we can express y and x as

$$y=\frac{1}{2ts^2+2s)},\quad x=\frac{2ts+1}{2ts^2+2s}.$$

Hence Equation (4.6.19) can now be rewritten as follows

$$\frac{\partial u}{\partial s}=\frac{2ts+1}{2ts^2+2s}u.$$

It is easy to solve the equation and obtain the solution

$$\begin{aligned} u &= C(t)\sqrt{ts^2+s} \\ &= C\left(\frac{x^2-y^2}{2y}\right)\sqrt{\frac{x^2-y^2}{2y}\frac{1}{(x+y)^2}+\frac{1}{x+y}} = C\left(\frac{x^2-y^2}{2y}\right)\sqrt{\frac{1}{2y}}. \end{aligned}$$

Here $C(t)$ is actually an arbitrary function of the argument t to be defined using the initial conditions. Taking into consideration the initial conditions given we obtain the solution of the Cauchy problem

$$u(x,y) = \frac{\sqrt{x^2-y^2+y}}{y}.$$

Problem 4.6.14 *Solve the Cauchy problem for the equation*

$$y\frac{\partial u}{\partial x} - x\frac{\partial u}{\partial y} = x+y, \quad u(x,y)|_{x=0} = y.$$

Solution For this PDE we can write a corresponding characteristic system

$$\frac{dx}{y} = -\frac{dy}{x} = -\frac{du}{x+y}.$$

From the first equation we obtain the general solution of the homogeneous equation that we denote as $u_0(x,y)$

$$\begin{aligned} \frac{dx}{y} &= -\frac{dy}{x} \Rightarrow xdx = -ydy \\ \Rightarrow x^2 &= -y^2 + C \Rightarrow u_0(x,y) = \phi(x^2+y^2), \end{aligned}$$

where ϕ is an arbitrary function of its argument.
Now we can examine the second equation

$$-\frac{dy}{x} = -\frac{du}{x+y}.$$

If we express x through y and ϕ, insert this expression into the given equation and integrate it we obtain the general solution of the non-homogeneous PDE

$$\frac{x+\sqrt{\phi^2-x^2}}{\sqrt{\phi^2-x^2}} = du,$$

after transformation

$$-\sqrt{\phi^2-x^2}+x+\phi = u(x,y),$$

we obtain the expression for the general solution of non-homogeneous PDE

$$u(x,y) = x - y + \phi(x^2 + y^2),$$

where ϕ is an arbitrary function of its argument.

Taking into consideration the given initial conditions we obtain the corresponding form of ϕ and solve the Cauchy problem

$$u(x,y) = x - y + 2\sqrt{x^2 + y^2}.$$

Problem 4.6.15 *Solve the Cauchy problem for the equation*

$$(x^2y - x)\frac{\partial u}{\partial x} - xy^2\frac{\partial u}{\partial y} = u, \quad u(x,y)|_{y=1} = x.$$

using **Mathematica**.

Solution Here we provide the solution procedure of a Cauchy problem for a PDE with **Mathematica**. We show a simulated script of the program with comments.

```
Clear[pde, solution];
pde = - x y^2 D[u[x, y], y] + (-x + x^2 y) D[u[x, y], x]
      - u[x, y] == 0;
solution = DSolve[{pde, u[x, 1] == x}, u,
           {x, y}] // PowerExpand // FullSimplify
 {{u -> Function[{x,y}, -(((x y - Log[y])
            ProductLog[-E^(-x y) y (x y - Log[y])])/(x y))]}}
```

The command `FullSimplify` has not worked properly. We simplify the expression and define the solution now as follows.

```
solution = {{u -> Function[{x, y}, (xy - Log[y])^2/(xy)]}};
```

Let us check that this function solves the equation indeed.

```
pde/ . solution//FullSimplify
{True}
```

We prove that the initial condition is also satisfied.

```
s[x_,y_]=u[x,y]/. solution[[1]]; s[x,1]
x
```

It means that the solution of the Cauchy problem is given by

$$u(x,y) = \frac{(xy - \ln y)^2}{xy}.$$

Problem 4.6.16 *The differential equation*

$$y' = -\frac{y + 2x}{x}$$

admits the symmetry group with the infinitesimal operator defined by

$$\xi(x, y) = \frac{1}{y + 2x}, \qquad \eta(x, y) = 0.$$

Use this symmetry and solve the differential equation (use the method of canonical variables).

Solution The canonical variables s and t can be found from the following equations

$$\mathbf{U}s = 1, \quad \mathbf{U}t = 0,$$

or explicitly

$$\frac{1}{y + 2x}\frac{\partial s}{\partial x} = 1, \qquad \frac{1}{y + 2x}\frac{\partial t}{\partial x} = 0.$$

Thus we immediately obtain the canonical variables in the form

$$t = y, \quad s = xy + x^2.$$

Now let us find the differentials of s and t

$$dt = dy, \quad ds = ydx + xdy + 2xdx.$$

Knowing the explicit formulas for the differentials we calculate the derivative of s with respect to t and substitute the expression for y' from the given differential equation

$$\frac{ds}{dt} = \frac{(2x + y)dx + xdy}{dy} = \frac{(2x + y) + x\frac{dy}{dx}}{\frac{dy}{dx}} = \frac{2x + y - y - 2x}{-\frac{y+2x}{x}} = 0.$$

This means that $s =$ const. and the solution of the original differential equation is

$$yx + x^2 = \text{const.}$$

Problem 4.6.17 *The differential equation*

$$y' = 1 + \frac{\tan(x - y)}{x}$$

admits the symmetry group

$$\mathbf{U} = \frac{1}{x\cos(x-y)}\frac{\partial}{\partial y}.$$

Use this symmetry and solve the differential equation (use the Lie's integrating factor).

Solution Let us calculate Lie's integrating factor

$$\mu(x,y) = \frac{x\cos(x-y)}{x} = \cos(x-y).$$

We rewrite the original equation in the following form

$$xdy - (x + \tan(x-y))dx = 0,$$

and multiply it on the obtained integrating factor $\mu(x,y)$

$$x\cos(x-y)dy - (x\cos(x-y) + \sin(x-y))dx = 0.$$

We know that this expression is an exact differential. That means that there is such a function $\Phi(x,y)$ that $d\Phi(x,y) = \Phi_x(x,y)dx + \Phi_y(x,y)dy = 0$, where

$$\Phi_x(x,y) = -x\cos(x-y) - \sin(x-y), \quad \Phi_y(x,y) = x\cos(x-y).$$

Before finding $\Phi(x,y)$ we check whether our calculations were correct. We calculate $\Phi_{xy}(x,y)$ and $\Phi_{yx}(x,y)$. If we have found Lie's factor correctly these two functions should coincide. Let us check it

$$\Phi_{yx}(x,y) = \cos(x-y) - x\sin(x-y),$$

$$\Phi_{xy}(x,y) = -x\sin(x-y) + \cos(x-y),$$

so, it means that $\Phi_{yx}(x,y) = \Phi_{xy}(x,y)$. Integrating the equations for the partial derivatives of $\Phi(x,y)$ we finally obtain $\Phi(x,y) = -x\sin(x-y)$ and the general solution of the original equation is

$$\Phi(x,y) = -x\sin(x-y) = \text{const}.$$

Problem 4.6.18 *Let a family of curves be set by a system of equations*

$$x^2 - 2y\ln x = Cy, \quad y = 0, \quad x = 0, \quad C = \text{const}. \tag{4.6.21}$$

Find a first order ODE that possesses the given family of curves as a family of integral curves.

Solution Let us find the expression for C

$$C = \frac{x^2}{y} - 2\ln x, \quad y \neq 0.$$

Calculating a total differential of both sides of the equation we obtain

$$\frac{2xydx - x^2dy}{y^2} - 2\frac{dx}{x} = 0.$$

Separating the terms with dx from the terms with dy we obtain

$$\frac{2x^2ydx - 2y^2}{xy^2}dx = \frac{x^3}{xy^2}dy.$$

We should remember that $y = 0$ and $x = 0$ also solve our equation, while the equation we have obtained does not have these solutions. Therefore we should multiply both sides of the equation with xy^2 and only this new equation

$$(2x^2y - 2y^2)dx = x^3dy \tag{4.6.22}$$

will be the correct answer to the problem.

Problem 4.6.19 *Let a family of curves be set by the formulas*

$$x = Cy + ye^x, \quad y = 0, \quad x = 0, \quad C = \text{const.}$$

Find a first order ODE that possesses the given family of curves as a family of integral curves.

Solution Let us find the expression for the constant C

$$\frac{x}{y} - e^x = C, y \neq 0. \tag{4.6.23}$$

Calculating a total differential of both sides of the equation we obtain

$$\frac{ydx - xdy}{y^2} - e^x dx = 0,$$

separating the terms with dx from the terms with dy we obtain

$$\left(\frac{1}{y} - e^x\right)dx = \frac{x}{y^2}dy.$$

We should remember that $y = 0$ and $x = 0$ also solve our equation, while the equation we have obtained does not have the solution $y = 0$. Therefore we should multiply both sides of the equation with y^2 and only this new equation

$$y(1 - ye^x)dx = xdy \tag{4.6.24}$$

will be the correct answer to the problem.

Problem 4.6.20 *Let a one-parameter family of curves be set by a formula*

$$x^2 = (C + y)e^y, \quad C = \text{const.}$$

Find a first order ODE that has the given family of curves as the family of integral curves.

Solution Let us find the expression for the constant C

$$C = x^2e^{-y} - y.$$

Calculating a total differential of both sides of the equation we obtain

$$2xe^{-y} - x^2e^{-y}dy - dy = 0,$$

separating the terms with dx from the terms with dy we obtain the solution of the given problem

$$2xe^{-y}dx = (x^2e^{-y} + 1)dy. \tag{4.6.25}$$

Problem 4.6.21 *Let a family of curves be set by a formula*

$$y^3 - x^3 - xy + x = C, \quad C = \text{const.}$$

Find a first order ODE for which this family describes the family of integral curves.

Solution Let us calculate a total differential of both sides of the equation

$$3y^2dy - 3x^2dx - ydx - xdy + dx = 0.$$

Separating the terms with dx from the terms with dy we obtain the first order ODE in the form

$$(3x^2 + y - 1)dx = (3y^2 - x)dy. \tag{4.6.26}$$

Problem 4.6.22 *Let a one-parameter family of curves be given by formulas*

$$x = Cy - y^2, \quad y = 0, \quad C = \text{const.}$$

Find a first order ODE for which this one-parameter family describes the family of integral curves.

Solution If $y \neq 0$ we obtain

$$C = \frac{x + y^2}{y}.$$

Calculating a total differential for both sides of the equation we obtain

$$\frac{ydx + 2y^2dy - xdy - y^2dy}{y^2} = 0,$$

separating the terms with dx from the terms with dy we obtain a first order ODE

$$\frac{1}{y}dx = \left(\frac{x}{y^2} - 1\right)dy.$$

We remember that $y = 0$ also solves our equation, while the equation we have obtained does not have this solution. Therefore we should multiply both sides of the equation with y^2 to avoid division by 0 and only this new equation

$$ydx = \left(x - y^2\right)dy$$

will be the correct answer to the problem.

4.7 Exercises

Here we provide a few exercises for our reader to try independently. All these exercises can be solved with the methods described in this chapter and have the same difficulty level as the problems in the previous section. All the practical tips and tricks demonstrated in this chapter are very helpful for the material discussed later.

1. The first order differential equation

$$\frac{\mathrm{d}y}{\mathrm{d}x} = \frac{y}{x - f(y)g(\frac{y}{x})},$$

where $f(\cdot), g(\cdot)$ are smooth functions of their arguments, admits a symmetry group with an infinitesimal generator

$$\mathbf{U} = \frac{xy}{f(y)}\frac{\partial}{\partial x} + \frac{y^2}{f(y)}\frac{\partial}{\partial y}.$$

Solve this equation using the symmetry group.

2. Find the solution of the differential equation

$$\frac{\mathrm{d}y}{\mathrm{d}x} = \frac{y}{x + e^y}$$

using the symmetry group with the infinitesimal generator equal to

$$\mathbf{U} = y\frac{\partial}{\partial x}.$$

3. Solve the equation

$$y\frac{\partial u}{\partial x} - x\frac{\partial u}{\partial y} = x^2.$$

4. Solve the following equations

   (a) $2y\frac{\partial u}{\partial x} + 3x^2\frac{\partial u}{\partial y} = 0$,

   (b) $x\frac{\partial u}{\partial x} + y\frac{\partial u}{\partial y} = au, \quad a = \text{const.}, \ a \neq 0$,

   (c) and the Cauchy problem for the equation

$$x\frac{\partial u}{\partial x} + (xz + y)\frac{\partial u}{\partial y} + z\frac{\partial u}{\partial z} = 0, \quad u(x, y, z)|_{x+y-z=0} = 1 - x$$

   using **Mathematica** [67].

   (d) Davis [24] states that the function $\mu = (x^2+y^2)^{-1}$ is an integrating factor for the equation

$$(y + xF(x^2 + y^2))\mathrm{d}x - (x - yF(x^2 + y^2))\mathrm{d}y = 0,$$

   where F is an arbitrary function. Prove this statement and that the equation is the most general equation of the first order, which is invariant under the rotation group on the plane.

Chapter 5

Prolongation procedure

In the previous chapters we dealt only with one-parameter groups of point transformations and applied our knowledge to the simplest type of differential equations - first order ordinary differential equations. This chapter addresses more complicated equations, for instance, second or higher order differential equations.

At first we discuss how we can prolong the action of a one-parameter group of point transformations on the first and higher derivatives of a dependent variable. We explain it at an intuitive level using geometric properties of these objects. We define a prolongation of an infinitesimal generator and obtain a linear operator in a higher dimensional space. Then in Section 5.2 we discuss the invariance of differential expressions under the action of a one-parameter group of point transformations. Using this idea we can define the determining equations which play a crucial role in the applications of Lie group theory to differential equations. In Section 5.3 we provide nineteen problems with solutions which allow one to deepen the knowledge in the methods described before. In Section 5.4 we give three exercises for the reader to try independently.

5.1 Differential invariants and prolongation of one-parameter groups of point transformations

In the case of the first order differential equation

$$y' = f(x, y)$$

we introduced the frame of the equation: $p = f(x, y)$. Then the action of the symmetry group corresponds to the movement of a point on this surface. We can always calculate $(x_\varepsilon, y_\varepsilon, p_\varepsilon)$ for every point (x, y, p). Why is it possible? The one-parameter group maps only one point to another point. We have to ask ourselves, why we can be sure that the slope $p = y'$ is also mapped to $p_\varepsilon = y'_\varepsilon$ and why this map is one-to-one. The reason is that we deal with a continuous group. We have proved before that each smooth curve is mapped to another one. But each curve defines a unique slope in each point.

Let us calculate this slope after the action of the one-parameter group. We have the simple formula $\frac{dy}{dx}$ to obtain the slope for every point (x, y) of the curve. Therefore, we should calculate $\frac{dy_\varepsilon}{dx_\varepsilon}$ in the image point $(x_\varepsilon, y_\varepsilon)$.

We know that locally the action of the transformation group can be represented as the Taylor series

$$\begin{aligned} x_\varepsilon &= x + \xi(x, y) \cdot \varepsilon + \mathcal{O}(\varepsilon^2), \\ y_\varepsilon &= y + \eta(x, y) \cdot \varepsilon + \mathcal{O}(\varepsilon^2). \end{aligned}$$

Thus, locally, along the image curve we obtain

$$\begin{aligned} dx_\varepsilon &= dx + d\xi(x, y) \cdot \varepsilon + \mathcal{O}(\varepsilon^2), \\ dy_\varepsilon &= dy + d\eta(x, y) \cdot \varepsilon + \mathcal{O}(\varepsilon^2). \end{aligned}$$

Correspondingly, the fraction of these differentials is the new slope in the image point

$$\frac{dy_\varepsilon}{dx_\varepsilon} = \frac{dy + d\eta(x, y) \cdot \varepsilon + \mathcal{O}(\varepsilon^2)}{dx + d\xi(x, y) \cdot \varepsilon + \mathcal{O}(\varepsilon^2)}.$$

We calculate the linear part of this function expanding Taylor series again. First we should recall the simple formula for the linear part of the expression $(1+z)^n$ for small values of z. We have $(1+z)^n = 1+nz+\mathcal{O}(z^2)$, for $z \ll 1$. In our case for the expression in the denominator we have $n = -1$. We divide the numerator and denominator of the fraction above by dx and then use this expansion

$$\begin{aligned} \frac{dy_\varepsilon}{dx_\varepsilon} &= \frac{\frac{dy}{dx} + \frac{d\eta}{dx} \cdot \varepsilon + \mathcal{O}(\varepsilon^2)}{1 + \frac{d\xi}{dx} \cdot \varepsilon + \mathcal{O}(\varepsilon^2)} \\ &= \left(\frac{dy}{dx} + \frac{d\eta}{dx} \cdot \varepsilon + \mathcal{O}(\varepsilon^2)\right)\left(1 - \frac{d\xi}{dx} \cdot \varepsilon + \mathcal{O}(\varepsilon^2)\right) \\ &= \frac{dy}{dx} + \varepsilon \cdot \left(\frac{d\eta}{dx} - \frac{d\xi}{dx} \cdot \frac{dy}{dx}\right) + \mathcal{O}(\varepsilon^2). \end{aligned}$$

The linear part of the image slope has the form

$$\frac{dy_\varepsilon}{dx_\varepsilon} = \frac{dy}{dx} + \varepsilon \cdot \left(\frac{d\eta}{dx} - \frac{d\xi}{dx} \cdot \frac{dy}{dx}\right).$$

We denote

$$\eta^{(1)}\left(x, y, y^{(1)}\right) = \frac{d\eta}{dx} - \frac{d\xi}{dx} \cdot \frac{dy}{dx} = \frac{d\eta}{dx} - \frac{d\xi}{dx} \cdot y^{(1)},$$

where the function $\eta^{(1)}\left(x, y, y^{(1)}\right)$ depends on three variables - x, y and on the value of the slope in the source point $y^{(1)}$ which is often denoted just as y'. Now we have a similar representation for the change of the slope

$$y_\varepsilon^{(1)} = y^{(1)} + \eta^{(1)} \cdot \varepsilon + \mathcal{O}(\varepsilon^2) \tag{5.1.1}$$

as for the change of the coordinates x and y. We can use this procedure infinitely many times and calculate the change of each derivative in the point $(x_\varepsilon, y_\varepsilon)$ in the same way

$$y_\varepsilon^{(2)} = y^{(2)} + \eta^{(2)} \cdot \varepsilon + \mathcal{O}(\varepsilon^2),$$

where

$$\eta^{(2)}\left(x, y, y^{(1)}, y^{(2)}\right) = \frac{d\eta^{(1)}}{dx} - \frac{d\xi}{dx} \cdot \frac{d^2y}{dx^2} = \frac{d\eta^{(1)}}{dx} - y^{(2)} \frac{d\xi}{dx}.$$

Because $\frac{d^2 y_\varepsilon}{dx_\varepsilon^2} = \frac{d}{dx_\varepsilon}\left(\frac{dy_\varepsilon}{dx_\varepsilon}\right)$, we calculate the change of the slope, or in other words the slope of the slope, and so on.

Hence, we get a recursive formula that describes the action of the group on each derivative of the function $y(x)$ in the image point $(x_\varepsilon, y_\varepsilon)$.

$$y_\varepsilon^{(n)} = y^{(n)} + \varepsilon \cdot \eta^{(n)} + \cdots, \tag{5.1.2}$$

where

$$\eta^{(n)}\left(x, y, y^{(1)}, \ldots, y^{(n-1)}\right) = \frac{d\eta^{(n-1)}}{dx} - y^{(n)} \frac{d\xi}{dx}. \tag{5.1.3}$$

At first glance the formula for $\eta^{(n)}$ looks very short and convenient to use but that is not so. Let us look at the function $\eta^{(1)}\left(x, y, y^{(1)}\right)$ in detail. We should calculate the total directional derivative, i.e.

$$\begin{aligned}
\eta^{(1)}\left(x, y, y^{(1)}\right) &= \frac{d\eta}{dx} - y^{(1)} \frac{d\xi}{dx} \\
&= \frac{\partial \eta}{\partial x}\frac{dx}{dx} + \frac{\partial \eta}{\partial y}\frac{dy}{dx} - \frac{dy}{dx}\left(\frac{\partial \xi}{\partial x} + \frac{\partial \xi}{\partial y}\frac{dy}{dx}\right) \\
&= \frac{\partial \eta}{\partial x} + \left(\frac{\partial \eta}{\partial y} - \frac{\partial \xi}{\partial x}\right)\frac{dy}{dx} - \frac{\partial \xi}{\partial y}\left(\frac{dy}{dx}\right)^2 \\
&= \frac{\partial \eta}{\partial x} + \left(\frac{\partial \eta}{\partial y} - \frac{\partial \xi}{\partial x}\right) y^{(1)} - \frac{\partial \xi}{\partial y}\left(y^{(1)}\right)^2.
\end{aligned} \tag{5.1.4}$$

Now, if we insert this expression into the formula for $\eta^{(2)}$ we get a dozen terms to calculate instead of the four terms for $\eta^{(1)}$. But the recursive formula (5.1.3), written using total derivatives, is easier to remember than the formula where all partial derivatives are presented.

The function $\eta^{(1)}\left(x, y, y^{(1)}\right)$ defines the slope $y_\varepsilon^{(1)}$ as a function of the coordinates (x, y) and $y^{(1)}$ only. Geometrically, this means that if any two curves are tangent to each other in the source point (x, y), then their images are tangent in the image point $(x_\varepsilon, y_\varepsilon)$ as well. It also means that the point transformation $(x, y) \to (x_\varepsilon, y_\varepsilon)$, which was defined in some subset of $\mathbb{R}^2$, is now extended to a point transformation $(x, y, y^{(1)}) \to (x_\varepsilon, y_\varepsilon, y_\varepsilon^{(1)})$ in some subset of $\mathbb{R}^3$, or correspondingly in our frame notation: $(x, y, p) \to (x_\varepsilon, y_\varepsilon, p_\varepsilon)$, with the rule

$$\begin{aligned} x_\varepsilon &= X_\varepsilon(x, y) \\ y_\varepsilon &= Y_\varepsilon(x, y) \\ y_\varepsilon^{(1)} &= Y_\varepsilon^{(1)}(x, y, y^{(1)}), \quad p_\varepsilon^{(1)} = Y_\varepsilon^{(1)}(x, y, y^{(1)}). \end{aligned} \tag{5.1.5}$$

This new point transformation is called the first extended point transformation or the first prolongation of the group action.

Example 5.1.1 Let the original point transformation be a scaling transformation

$$\begin{aligned} x_\varepsilon &= \mathrm{e}^\varepsilon x, \\ y_\varepsilon &= \mathrm{e}^\varepsilon y. \end{aligned}$$

Then we obtain for the first derivative

$$\frac{dy_\varepsilon}{dx_\varepsilon} = \frac{\frac{\partial y_\varepsilon}{\partial x} + \frac{\partial y_\varepsilon}{\partial y}\frac{dy}{dx}}{\frac{\partial x_\varepsilon}{\partial x} + \frac{\partial x_\varepsilon}{\partial y}\frac{dy}{dx}} = \frac{\mathrm{e}^\varepsilon \frac{dy}{dx}}{\mathrm{e}^\varepsilon} = \frac{dy}{dx}.$$

If we use the infinitesimal generator

$$\mathbf{U} = x\frac{\partial}{\partial x} + y\frac{\partial}{\partial y}$$

of this group we get for $\eta^{(1)}\left(x, y, y^{(1)}\right)$ the expression

$$\begin{aligned} \eta^{(1)}\left(x, y, y^{(1)}\right) &= \frac{d\eta}{dx} - \frac{dy}{dx}\frac{d\xi}{dx} = \frac{dy}{dx} - \frac{dy}{dx}\frac{dx}{dx} \\ &= \frac{\partial y}{\partial x} + \frac{\partial y}{\partial y}\frac{dy}{dx} - \frac{dy}{dx}\left(\frac{\partial x}{\partial x} + \frac{\partial x}{\partial y}\frac{dy}{dx}\right) = 0, \end{aligned}$$

i.e. $\eta^{(1)}\left(x, y, y^{(1)}\right) = 0$, and up to the linear part we obtain

$$\frac{dy_\varepsilon}{dx_\varepsilon} = \frac{dy}{dx} + 0 \cdot \varepsilon.$$

So the slope does not change under the action of this group. $\diamond$

Let our point transformations form a one-parameter group in the space $\mathbb{R}^2$. Is the group property also preserved after the first prolongation?

We prove now that the point transformations $(x, y, p) \to (x_\varepsilon, y_\varepsilon, p_\varepsilon)$ in $\mathbb{R}^3$ form a group as well. The prolongation procedure immediately gives us the possibility to prove that the identity transformation belongs to the set. The second step is to show that the superposition of two successive point transformations is equivalent to the action of the one transformation in the set

$$(x, y) \xrightarrow{g_1} (x_1, y_1) \xrightarrow{g_2} (x_2, y_2) \quad \Leftrightarrow \quad (x, y) \xrightarrow{g_3} (x_2, y_2).$$

The group property for the original set of transformations is given by

$$\begin{aligned}
&g_{\varepsilon_1} : (x, y) \to (x_{\varepsilon_1}, y_{\varepsilon_1}),\\
&g_{\tilde{\varepsilon}_2} : (x_{\varepsilon_1}, y_{\varepsilon_1}) \to (x_{\varepsilon_2}, y_{\varepsilon_2}),\\
&g_{\varepsilon_1+\tilde{\varepsilon}_2} : (x, y) \to (x_{\varepsilon_2}, y_{\varepsilon_2}), \quad \varepsilon_2 = \varepsilon_1 + \tilde{\varepsilon}_2.
\end{aligned}$$

Now we extend each of these transformations on first derivatives (slopes)

$$\begin{aligned}
y^{(1)}_{\varepsilon_1} &= \frac{dy_{\varepsilon_1}}{dx_{\varepsilon_1}} = \frac{dy}{dx} + \varepsilon_1 \eta^{(1)}(x, y) + \mathcal{O}({\varepsilon_1}^2),\\
y^{(1)}_{\tilde{\varepsilon}_2} &= \frac{d(y_{\varepsilon_1})_{\tilde{\varepsilon}_2}}{d(x_{\varepsilon_1})_{\tilde{\varepsilon}_2}} = \frac{dy_{\varepsilon_1}}{dx_{\varepsilon_1}} + \tilde{\varepsilon}_2 \eta^{(1)}(x_{\varepsilon_1}, y_{\varepsilon_1}) + \mathcal{O}({\tilde{\varepsilon}_2}^2)\\
&= \frac{dy}{dx} + \varepsilon_1 \eta^{(1)}(x, y) + \tilde{\varepsilon}_2 \eta^{(1)}(x_{\varepsilon_1}, y_{\varepsilon_1}) + \mathcal{O}({\tilde{\varepsilon}_2}^2)\\
&= \frac{dy}{dx} + (\varepsilon_1 + \tilde{\varepsilon}_2)\eta^{(1)}(x, y) + \mathcal{O}(\min({\varepsilon_1}^2, \varepsilon_1\tilde{\varepsilon}_2, {\tilde{\varepsilon}_2}^2)),
\end{aligned}$$

because $\eta^{(1)}(x_{\varepsilon_1}, y_{\varepsilon_1}) = \eta^{(1)}(x+\varepsilon_1\xi+\mathcal{O}({\varepsilon_1}^2), y+\varepsilon_1\eta+\mathcal{O}({\varepsilon_1}^2)) = \eta^{(1)}(x, y)+\ldots$. On the other hand we have the same structure if we use $y^{(1)}_{\varepsilon_1+\tilde{\varepsilon}_2}$, so we proved the second group property. In a similar way we show that each transformation of the prolonged group also possesses the inverse transformation. Thus, we proved Theorem 5.1.1.

Theorem 5.1.1 *The action of a one-parameter group $G : \mathbb{R}^2 \to \mathbb{R}^2$ (which maps $(x, y) \to (x_\varepsilon, y_\varepsilon)$) induces the action of the first prolonged group $G^{(1)}$: $\mathbb{R}^3 \to \mathbb{R}^3$ (which maps $(x, y, p) \to (x_\varepsilon, y_\varepsilon, p_\varepsilon)$).*

In the space $\mathbb{R}^2$ we describe locally the action of the group with the help of the infinitesimal generator

$$\mathbf{U} = \xi(x,y)\frac{\partial}{\partial x} + \eta(x,y)\frac{\partial}{\partial y}$$

(which is also called a differential operator or a vector field). The functions $\xi(x,y)$ and $\eta(x,y)$ are the components of a vector field in the two-dimensional tangent space and indicate the rate of change in the corresponding direction. Now we prolong the group $G(\mathbb{R}^2)$ to the group of point transformations in $\mathbb{R}^3$. The local action of the group should take into account the change in the additional direction $y' = p$. We have calculated this action before, $y_\varepsilon^{(1)} = y^{(1)} + \varepsilon\eta^{(1)} + \mathcal{O}(\varepsilon^2)$, so we know the linear part of this change. This means, the corresponding infinitesimal generator of the first prolonged group has the form

$$\mathrm{pr}^{(1)}\mathbf{U} = \mathbf{U}^{(1)} = \xi(x,y)\frac{\partial}{\partial x} + \eta(x,y)\frac{\partial}{\partial y} + \eta^{(1)}\left(x,y,y^{(1)}\right)\frac{\partial}{\partial y'}.$$

We will use the first derivative y' as a new variable in this formula. We notice that $\eta^{(1)} = \frac{d\eta}{dx} - y'\frac{d\xi}{dx}$ depends only on the components $\xi(x,y)$ and $\eta(x,y)$ of the infinitesimal operator $\mathbf{U}$. Accordingly, we can consider the operator $\mathbf{U}^{(1)}$ as the prolongation or extension of the operator $\mathbf{U}$.

Example 5.1.2 The infinitesimal generator of the rotation group in $\mathbb{R}^2$ has the form

$$\mathbf{U} = -y\frac{\partial}{\partial x} + x\frac{\partial}{\partial y}.$$

We calculate the infinitesimal generator of the first prolonged group. Because $\xi(x,y) = -y$ and $\eta(x,y) = x$, we obtain

$$\begin{aligned} \xi_x &= 0, & \eta_x &= 1, \\ \xi_y &= -1, & \eta_y &= 0, \end{aligned}$$

and therefore

$$\begin{aligned} \eta^{(1)} &= \eta_x + \eta_y\frac{dy}{dx} - \frac{dy}{dx}\left(\xi_x + \xi_y \cdot y'\right) = 1 - y'\left(0 - y'\right) \\ &= 1 + (y')^2. \end{aligned}$$

Thus, the infinitesimal generator of the first prolonged group is

$$\mathrm{pr}^{(1)}\mathbf{U} = \mathbf{U}^{(1)} = -y\frac{\partial}{\partial x} + x\frac{\partial}{\partial y} + \left(1 + (y')^2\right)\frac{\partial}{\partial y'}.$$

◇

Example 5.1.3 The infinitesimal generator of the translation group is given by

$$\mathbf{U} = \frac{\partial}{\partial x}.$$

Because of

$$\begin{aligned} \xi(x, y) &= 1, & \eta(x, y) &= 0, \\ \xi_x &= \xi_y = 0, & \eta_x &= \eta_y = 0, \end{aligned}$$

we see that $\eta^{(1)}\left(x, y, y^{(1)}\right) = 0$ and we have

$$\mathbf{U}^{(1)} = \frac{\partial}{\partial x} = \mathbf{U}.$$

◇

This simple example proves Corollary 5.1.1.

Corollary 5.1.1 *The infinitesimal generator in the normal form preserves this form after the prolongation procedure.*

Having calculated the change of the first derivative $\frac{dy_\varepsilon}{dx_\varepsilon}$ we have proved that the change of the n-th derivative $\frac{d^n y_\varepsilon}{dx_\varepsilon^n}$ can be calculated in a similar way. For this change we got the recursive formula

$$y_\varepsilon^{(n)} = \frac{d^n y_\varepsilon}{dx_\varepsilon^n} = \frac{d^n y}{dx^n} + \varepsilon \eta^{(n)}(x, y, \ldots, y^{(n)}) + \mathcal{O}(\varepsilon^2),$$

where

$$\eta^{(n)}\left(x, y, y^{(1)}, \ldots, y^{(n)}\right) = \frac{d\eta^{(n-1)}}{dx} - \frac{d\xi}{dx} y^{(n)}.$$

We have also proved that the first prolongation of a group $G(\mathbb{R}^2)$ is a group $G^{(1)} : (x, y, p) \to (x_\varepsilon, y_\varepsilon, p_\varepsilon)$ of point transformations in $\mathbb{R}^3$.

Now if we use the second derivative $\frac{d^2 y}{dx^2}$ as a new variable $y^{(2)} = r$ we construct the second prolongation. The action of the group will be prolonged to point transformations $(x, y, p, r) \to (x_\varepsilon, y_\varepsilon, p_\varepsilon, r_\varepsilon)$ in $\mathbb{R}^4$. We can easily prove that the second prolonged group $G^{(2)}$ is a one-parameter continuous group in this space. We can represent the local action of this group by the infinitesimal generator as well

$$\mathrm{pr}^{(2)}\mathbf{U} = \mathbf{U}^{(2)} = \xi(x, y)\frac{\partial}{\partial x} + \eta(x, y)\frac{\partial}{\partial y} + \eta^{(1)}(x, y, y')\frac{\partial}{\partial y'} + \eta^{(2)}(x, y, y', y'')\frac{\partial}{\partial y''},$$

where $\mathbf{U}^{(2)}$ describes the first prolongation of $\mathbf{U}^{(1)}$. Both coefficients $\eta^{(1)}$ and $\eta^{(2)}$ are completely defined by the coefficients $\xi(x, y)$ and $\eta(x, y)$. It also means that the second prolongation of the operator $\mathbf{U}$ can be considered as just the prolongation of the original local action on the new variables $y' = p$ and $y'' = r$. The prolongation procedure can be used infinitely many times up to any order. The local action of the prolonged group will be described by the formula

$$\mathrm{pr}^{(n)}\mathbf{U} = \mathbf{U}^{(n)} = \xi\frac{\partial}{\partial x} + \eta\frac{\partial}{\partial y} + \eta^{(1)}\frac{\partial}{\partial y'} + \eta^{(2)}\frac{\partial}{\partial y''} + \ldots + \eta^{(n)}\frac{\partial}{\partial y^{(n)}}. \quad (5.1.6)$$

By the prolongation procedure we use all higher derivatives as new variables.

Example 5.1.4 Let us take the translation group

$$\mathbf{U} = \frac{\partial}{\partial x}, \quad \xi(x,y) = 1,\ \eta(x,y) = 0,\ \eta^{(1)} = 0.$$

Find the n-th prolongation $\mathbf{U}^{(n)}$ of this generator.
After the prolongation formula

$$\begin{aligned}\eta^{(2)} &= \frac{d\eta^{(1)}}{dx} - y''\frac{d\xi}{dx} = 0 + y'' \cdot 0 \\ \eta^{(3)} &= \frac{d\eta^{(2)}}{dx} - y'''\frac{d\xi}{dx} = 0\end{aligned}$$

$$\Rightarrow \quad \mathbf{U}^{(n)} = \frac{\partial}{\partial x} = \mathbf{U}.$$

◇

This example proves Corollary 5.1.2.

Corollary 5.1.2 *The infinitesimal generator of the translation group preserves its form after any prolongation.*

Example 5.1.5 A scaling group has the generator

$$\mathbf{U} = x\frac{\partial}{\partial x} + y\frac{\partial}{\partial y}.$$

Find the n-th prolongation.

After the prolongation formula we get

$$
\begin{aligned}
\eta^{(1)} &= \frac{d\eta}{dx} - y'\frac{d\xi}{dx} = \frac{d}{dx}(y) - y'\frac{d}{dx}(x) = \frac{dy}{dx} - \frac{dy}{dx} = 0,\\
\mathbf{U}^{(1)} &= x\frac{\partial}{\partial x} + y\frac{\partial}{\partial y} = \mathbf{U},\\
\eta^{(2)} &= \frac{d\eta^{(1)}}{dx} - y''\frac{d\xi}{dx} = \frac{d}{dx}(0) - y''\frac{d}{dx}(x) = -y'',\\
\mathbf{U}^{(2)} &= x\frac{\partial}{\partial x} + y\frac{\partial}{\partial y} - y''\frac{\partial}{\partial y''},\\
\eta^{(3)} &= \frac{d}{dx}(-y'') - y'''\frac{d\xi}{dx} = -2y''',\\
\mathbf{U}^{(3)} &= x\frac{\partial}{\partial x} + y\frac{\partial}{\partial y} - y''\frac{\partial}{\partial y''} - 2y'''\frac{\partial}{\partial y'''},\\
&\vdots\\
\mathbf{U}^{(n)} &= x\frac{\partial}{\partial x} + y\frac{\partial}{\partial y} - y''\frac{\partial}{\partial y''} - 2y'''\frac{\partial}{\partial y'''} - \ldots - (n-1)y^{(n)}\frac{\partial}{\partial y^{(n)}}.
\end{aligned}
$$

$\diamond$

Let us summarize the results of this section. We started with the one-parameter continuous group $G(\mathbb{R}^2)$ of point transformations in the base space $\mathbb{R}^2 = \mathbb{R}\times\mathbb{R} \ni (x, y)$. We proved that the action of the group $G(\mathbb{R}^2)$ induced the transformations of all derivatives $y^{(k)} \to y_\varepsilon^{(k)}$ in a natural way. Following this, we can introduce the vector space $J^{(k)} = \mathbb{R}\times\mathbb{R}\times\mathbb{R}^k$ with coordinates labeled $(x, y, y', \ldots, y^{(k)})$. A smooth transformation in the base space $\mathbb{R}^2$ will uniquely induce the transformations in each $J^{(k)}$-space, where $J^{(k)}$ is called the k-th order jet bundle.

We notice even more: we have defined the arbitrary smooth transformation only for the variables x and y in the base space $\mathbb{R}^2$ and immediately obtain a unique transformation rule for any derivative $y^{(k)}$ due to the differential calculus. The new point transformation $J^{(k)} \to J^{(k)}$ has the infinitesimal generator $\mathbf{U}^{(k)}$. The arbitrary change of base variables x and y is defined by both functions $\xi(x, y)$ and $\eta(x, y)$ whereas the change of any other variables $y^{(l)}$, $l = 1, \ldots, k$ is uniquely prescribed by the same functions $\xi(x, y)$ and $\eta(x, y)$ and their derivatives.

Let us take an arbitrary space $\mathbb{R}^{k+2}$, where we can define a general point transformation

$$(u_1, \ldots, u_{k+2}) \to (u_{1,\varepsilon}, \ldots, u_{k+2,\varepsilon}).$$

In this case we define the change of each variable independently and the change of a part of variables does not imply the change of others. This feature distinguishes the general space $\mathbb{R}^{k+2}$ from a k-th order jet bundle $J^{(k)}$. In the latter case the base space is always two-dimensional (so far as we speak about one-parameter group of point transformations in $\mathbb{R}^2$)

$$\begin{aligned} x_\varepsilon &= X_\varepsilon(x, y), \\ y_\varepsilon &= Y_\varepsilon(x, y), \end{aligned}$$

and thus, the change of all derivatives is prescribed uniquely by the point transformation in the base space.

The idea to generalize the point transformations was introduced by the Swedish mathematician Bäcklund. He was influenced by the work of Sophus Lie and proved in which case we get a self-contained generalization of point transformations. His idea was to use the base space of dimension $2 + n$,

$$\begin{aligned} x_\varepsilon &= X_\varepsilon(x, y, y', \ldots, y^{(n)}), \\ y_\varepsilon &= Y_\varepsilon(x, y, y', \ldots, y^{(n)}), \\ &\vdots \\ y_\varepsilon^{(n)} &= Y_\varepsilon^{(n)}(x, y, y', \ldots, y^{(n)}), \end{aligned}$$

i.e. to define arbitrarily the transformations of variables x, y and derivatives up to the order n. Following his idea we should use the common differentiation rules and uniquely define the change of all higher derivatives $l \geq n + 1$ and get a jet bundle with a $(n+2)$-dimensional base space. But this procedure does not contradict itself only if $n = 1$. It means that the base space can be 2- or 3-dimensional. It means, for instance, that if we define arbitrarily smooth transformation that describes the change of the first and second derivative then changing all higher derivatives we either obtain a contradiction (i.e. such transformation does not exist at all) or the change can be reduced to a transformation with 2- or 3-dimensional base space. Consequently, the maximal extension of the finite-dimensional base space is only possible up to $n = 1$, i.e. we can change x, y, y' arbitrarily. The transformation of the type

$$\begin{aligned} x_\varepsilon &= X_\varepsilon(x, y, y'), \\ y_\varepsilon &= Y_\varepsilon(x, y, y'), \\ y'_\varepsilon &= Y'_\varepsilon(x, y, y') \end{aligned}$$

is called the *contact transformation*. The class of differential equations that admit a one-parameter group of contact transformations is poorer than the class of the equations that admit a one-parameter group of the point transformations. The general procedure of construction and usage of contact transformations is very similar to the procedure for point transformations and can be found in [40].

Another possible generalization is to use the infinite-dimensional base space and prescribe change of each variable and each derivative up to the arbitrary order. Such transformations are called Lie-Bäcklund transformations, but they are not the subject of this book.

Let us instead show how we can use the prolonged group to solve or simplify differential equations.

5.2 Differential invariants of n-th order

First we recall that we defined the invariant functions on $\mathbb{R}^2$ under the action of the group $G(\mathbb{R}^2)$ in the following way

$$u(x_\varepsilon, y_\varepsilon) = u(x, y) \quad \Leftrightarrow \quad \mathbf{U}u(x, y) = 0.$$

Then we used the characteristic Equation (it is also a part of Lie equations)

$$\frac{dx}{\xi(x, y)} = \frac{dy}{\eta(x, y)}$$

and calculated the invariant $u = \Phi(x, y)$, which defined the orbits of $G(\mathbb{R}^2)$. Then we prolonged the action of $G(\mathbb{R}^2)$ to obtain the n-th prolongation $G^{(n)} = \mathrm{pr}^{(n)}G$ on $J^{(n)}$ (on the n-th order jet bundle).

Now we can ask which functions are invariant under the action of this group $\mathrm{pr}^{(n)}G$. We, for example, can also take a first prolongation $\mathrm{pr}^{(1)}G$ of a one-parameter group $G(\mathbb{R}^2)$ of point transformations on $\mathbb{R}^2$ and discuss how a first order ODE which admits the group $G(\mathbb{R}^2)$ is related to $\mathrm{pr}^{(1)}G$.

Example 5.2.1 Let us look at a first order differential equation once more

$$y' = \frac{\mathcal{Y}(x, y)}{\mathcal{X}(x, y)}.$$

If we denote $p = y'$, the equation $\mathcal{X}p - \mathcal{Y} = 0$ describes a surface in $\mathbb{R}^3$ with coordinates (x, y, p). We know that an action of the symmetry group $G(\mathbb{R}^2)$ admitted by this equation is equivalent to a motion of the corresponding point on the surface. Further we proved that $\mathbf{U}$ and $\mathcal{A}$ satisfy the equation

$$[\mathbf{U}, \mathcal{A}] = \lambda(x, y)\mathcal{A}.$$

Now let us take the first prolongation of $\mathbf{U}$, i.e. $\mathbf{U}^{(1)} = \mathrm{pr}^{(1)}\mathbf{U}$ and ask ourselves which function will be invariant under the action of $G^{(1)}$.

We should look for functions that satisfy

$$\mathbf{U}^{(1)}v(x,y,y') = 0,$$

where the infinitesimal generator of $G^{(1)}$ has the form

$$\mathbf{U}^{(1)} = \xi(x,y)\frac{\partial}{\partial x} + \eta(x,y)\frac{\partial}{\partial y} + \eta^{(1)}(x,y,y')\frac{\partial}{\partial y'}.$$

If we can define the invariant function $u(x,y,y') \equiv u(x,y,p)$ or the invariant surface $u(x,y,p)$, then we have an invariant surface in the bundle $J^{(1)}$ (with coordinates (x,y,p)), or in other words, we have to do with the differential equation which admits this group. Therefore, to find $v(x,y,y')$ we have to solve

$$\mathbf{U}^{(1)}v(x,y,y') = \xi\frac{\partial v}{\partial x} + \eta\frac{\partial v}{\partial y} + \left(\frac{\partial \eta}{\partial x} + y'\frac{\partial \eta}{\partial y} - y'\frac{\partial \xi}{\partial x} - (y')^2\frac{\partial \xi}{\partial y}\right)\frac{\partial v}{\partial y'} = 0. \tag{5.2.7}$$

This equation should be satisfied just on the surface $p = f(x,y)$, where $y' = \frac{\mathcal{Y}(x,y)}{\mathcal{X}(x,y)}$ or respectively $v(x,y,y') = \mathcal{X}y' - \mathcal{Y} \Rightarrow \frac{\partial u}{\partial y'} = \mathcal{X}(x,y)$. We insert those last formulas in the Equation (5.2.7) and obtain

$$\xi\frac{\partial \mathcal{X}}{\partial x}y' - \xi\frac{\partial \mathcal{Y}}{\partial x} + \eta\frac{\partial \mathcal{X}}{\partial y}y' - \eta\frac{\partial \mathcal{Y}}{\partial y} + \frac{\partial \eta}{\partial x}\mathcal{X} + y'\frac{\partial \eta}{\partial y}\mathcal{X} - y'\frac{\partial \xi}{\partial x}\mathcal{X} - (y')^2\frac{\partial \xi}{\partial y}\mathcal{X} = 0.$$

If we insert $y' = \frac{\mathcal{Y}}{\mathcal{X}}$ to be on the frame of the equation and multiply the result to $\mathcal{X}$ we get

$$\xi\frac{\partial \mathcal{X}}{\partial x}\mathcal{Y} - \xi\frac{\partial \mathcal{Y}}{\partial x}\mathcal{X} + \eta\frac{\partial \mathcal{X}}{\partial y}\mathcal{Y} - \eta\frac{\partial \mathcal{Y}}{\partial y}\mathcal{X} + \frac{\partial \eta}{\partial x}\mathcal{X}^2 + \frac{\partial \eta}{\partial y}\mathcal{Y}\mathcal{X} - \frac{\partial \xi}{\partial x}\mathcal{X}\mathcal{Y} - \mathcal{Y}^2\frac{\partial \xi}{\partial y} = 0.$$

This expression can be rearranged to

$$\mathcal{Y}\mathbf{U}\mathcal{X} - \mathcal{X}\mathbf{U}\mathcal{Y} - \mathcal{X}\mathcal{A}\eta - \mathcal{Y}\mathcal{A}\xi = \mathcal{Y}\left(\mathbf{U}\mathcal{X} - \mathcal{A}\xi\right) - \mathcal{X}\left(\mathbf{U}\mathcal{Y} - \mathcal{A}\eta\right) = 0,$$

and after multiplaying both sides of the equation with $\frac{1}{\mathcal{X}\mathcal{Y}}$ we have

$$\frac{\mathbf{U}\mathcal{X} - \mathcal{A}\xi}{\mathcal{X}} = \frac{\mathbf{U}\mathcal{Y} - \mathcal{A}\eta}{\mathcal{Y}} = \lambda(x,y),$$

or

$$[\mathbf{U}, \mathcal{A}] = \lambda(x,y)\mathcal{A}.$$

Thus, we got the former condition for the symmetry group to be admitted by the first order ODE $y' = \frac{\mathcal{Y}(x,y)}{\mathcal{X}(x,y)}$.

We have proved that the conditions

$$[\mathbf{U}, \mathcal{A}] = \lambda(x,y)\mathcal{A}, \qquad \mathbf{U}^{(1)}u(x,y,y')\big|_{y'=\frac{\mathcal{Y}}{\mathcal{X}}} = 0$$

are equivalent and both mean that the first order ODE $y' = \frac{\mathcal{Y}(x,y)}{\mathcal{X}(x,y)}$ admits the symmetry group $G(\mathbb{R}^2)$. $\diamond$

Let us now summarize the latest results. We have seen in previous chapters that $u = u(x,y)$ is invariant under the action of the group $G(\mathbb{R}^2)$ if $\mathbf{U}u = 0$ and the equation $y' = \frac{\mathcal{Y}}{\mathcal{X}}$ admits $G(\mathbb{R}^2)$ if $[\mathbf{U}, \mathcal{A}] = \lambda(x,y)\mathcal{A}$.

Now we have proved that the expression $v = v(x,y,y')$ is invariant under the action of the group $G^{(1)}$ if $\mathbf{U}^{(1)}v(x,y,y') = 0$. If it is evaluated on the frame $y' = f(x,y)$, i.e. if $\mathbf{U}^{(1)}v(x,y,y')\big|_{y'=f(x,y)} = 0$ then this condition is equivalent to $[\mathbf{U}, \mathcal{A}] = \lambda(x,y)\mathcal{A}$ and means that $y' = f(x,y)$ admits the symmetry group $G(\mathbb{R}^2)$.

Therefore, we can say that the equation $\mathbf{U}u(x,y) = 0$ defines invariant functions of $G(\mathbb{R}^2)$ on the underlying space $\mathcal{M} \subset \mathbb{R}^2$ (also called base space) and the corresponding equation $\mathbf{U}^{(1)}v(x,y,y') = 0$ defines invariant "functions" $v(x,y,y')$ of the group $G^{(1)}$ on the jet bundle $J^{(1)}$, i.e. these "functions" contain derivatives up to the first order. We introduce now for the jet bundle $J^{(1)}$ the following notation $J^{(1)} = \mathcal{M} \times V^{(1)}$, where we denote with $\mathcal{M}$ the base space and with $V^{(1)}$ the space of the first derivatives (in the studied case a one-dimensional space with coordinate y' also denoted as p). Now we can give a definition of an invariant expression which possesses a first derivative of dependent variable.

Definition 5.2.1 (Differential invariant) *The invariant expression $v(x,y,y')$ calculated as a solution of*

$$\mathbf{U}^{(1)}v(x,y,y') = 0 \tag{5.2.8}$$

is called a differential invariant of the first order of the group $G^{(1)}$ (i.e. it contains a first derivative).

Using both invariants, i.e. the invariant function $u(x,y)$ which is invariant under the action of the group $G(\mathbb{R}^2)$ and the first differential invariant $v(x,y,y')$ which is invariant under the action of the first prolonged group $G^{(1)}$ we can find a first order ODE which admits the group $G(\mathbb{R}^2)$. We obtain the most general differential equation of the first order which admits this symmetry group $G(\mathbb{R}^2)$ if we take an arbitrary function $F(\cdot)$ from $u(x,y)$ and set it equal to $v(x,y,y')$

$$v(x,y,y') = F(u(x,y)), \quad \text{or} \quad \mathrm{inv}_{G^{(1)}} = F(\mathrm{inv}_G).$$

As we proved before the conditions

$$\mathbf{U}^{(1)}u(x,y,y')\big|_{y'=\frac{y}{x}} = 0 \quad \Leftrightarrow \quad [\mathbf{U},\mathcal{A}] = \lambda(x,y)\mathcal{A}. \tag{5.2.9}$$

are equivalent for the first order differential equations and each means that the equation admits the symmetry group $G(\mathbb{R}^2)$, so we checked once again that the first ODE $v(x,y,y') = F(u(x,y))$ admits the one-parameter group $G(\mathbb{R}^2)$. We explain these ideas step-by-step in the next Example.

Example 5.2.2 We consider the one-parameter group of Galilean transformations on the plane

$$x_\varepsilon = x + \varepsilon y,$$
$$y_\varepsilon = y.$$

Thus, we have $\mathbf{U} = y\frac{\partial}{\partial x}$ and $y = \text{const.} = u(x,y)$ is an invariant function of $G(\mathbb{R}^2)$. Earlier we proved that the equation

$$y' = \frac{y}{x + F(y)}$$

admits this group. Let us calculate the first differential invariants of Galilean transformations group now. For the group G we obtain

$$G: \quad \mathbf{U}u(x,y) = 0 \;\Rightarrow\; y\frac{\partial u}{\partial x} = 0 \;\Rightarrow\; u(x,y) = y \text{ or } u(x,y) = F(y).$$

The first prolongation $G^{(1)}$ has the infinitesimal generator in the form

$$G^{(1)}: \quad \mathbf{U}^{(1)} = y\frac{\partial}{\partial x} + \eta^{(1)}\frac{\partial}{\partial y'}; \quad \eta^{(1)} = \frac{d\eta}{dx} - y'\frac{d\xi}{dx} = -(y')^2.$$

This means that in order to find the first differential invariant we have to solve

$$\mathbf{U}^{(1)}u(x,y,y') = 0 \text{ or } y\frac{\partial y}{\partial x} - (y')^2\frac{\partial y}{\partial y'} = 0.$$

The characteristic system for this PDE is simple in this case:

$$\frac{dx}{y} = -\frac{dy'}{(y')^2},$$

and because y is an invariant (if we are on the frame), we obtain

$$\frac{x}{y} = \frac{1}{y'} + c \;\Rightarrow\; \widetilde{v(x,y,y')} = \frac{x}{y} - \frac{1}{y'}.$$

An invariant is not uniquely defined, if we multiply an invariant with another one we still have an invariant expression. We multiply now $\widetilde{v(x,y,y')}$ which is a first differential invariant of the first prolonged Galilean group with invariant function $u(x,y) = y$ then we obtain a first differential invariant in the form

$$v(x,y,y') = i_1(x,y,y') = y\widetilde{v(x,y,y')} = x - \frac{y}{y'},$$

where $v(x,y,y')$ (also often denoted as $i_1(x,y,y')$) is the first differential invariant.

The equation $v(x,y,y') = F(u(x,y))$ or correspondingly $x - \frac{y}{y'} = F(y)$ also describes the most general first order differential equation which admits the Galilean group $G(\mathbb{R}^2)$ of point transformations and it coincides with the equation $y' = \frac{y}{x+F(y)}$ studied before. ◇

Remark 5.2.1 *The first differential invariant is often denoted as $v(x,y,y')$ since the first publications by Sophus Lie. We will also use the notation $i_1(x,y,y')$ to emphasize that it is the first differential invariant in a series of higher order differential invariants which we introduce now.*

In exactly the same way as we introduced the first differential invariant we can calculate all higher order differential invariants i_n, i.e. after the rule

$$\mathbf{U}^{(n)} i_n(x,y,\ldots,y^{(n)}) = 0. \tag{5.2.10}$$

Using this definition we obtain the method obtaining the most general equation of n-th order $y^{(n)} = f(x,y,\ldots,y^{(n-1)})$ which admits this group $G(\mathbb{R}^2)$. We set

$$i_n = \Phi(u, i_1, i_2, \ldots, i_{n-1}),$$

where Φ is an arbitrary function of its arguments and obtain the most general n-th order ODE which admits the one-parameter group $G(\mathbb{R}^2)$ of point transformations.

Now, let us study the case of a second order differential equation which admits the group of Galilean transformations. First we look at the last example again and calculate the second order differential invariant.

Example 5.2.3 Let us calculate the second differential invariant of the Galilean group of transformations on $\mathbb{R}^2$ and find the most general second order differential equation which admits this group.

For the Galilean group we have the infinitesimal generator $\mathbf{U} = y\frac{\partial}{\partial x}$ and as calculated before the two invariants of the group G and $G^{(1)}$

$$\begin{aligned} u(x,y) &= y, \\ i_1(x,y,y') &= \frac{x}{y} - \frac{1}{y'}. \end{aligned}$$

Respectively we obtain the first order differential equation invariant under the action of the Galilean group in the form $y' = \frac{y}{x+F(y)}$.

Since we know $\eta^{(1)}(x,y,y') = -(y')^2$, we get for the second prolongation

$$\eta^{(2)}(x,y,y',y'') = \frac{d\eta^{(1)}}{dx} - y''\frac{d\xi}{dx} = -2y'y'' - y''y' = -3y'y'',$$

and thus the second prolongation of the infinitesimal operator $\mathbf{U}$ takes the form

$$\mathbf{U}^{(2)} = y\frac{\partial}{\partial x} - (y')^2\frac{\partial}{\partial y'} - 3y'y''\frac{\partial}{\partial y''}.$$

To find the differential invariant of the second order $i_2(x,y,y',y'')$, we have to solve a first order partial differential equation

$$\mathbf{U}^{(2)}i_2(x,y,y',y'') = y\frac{\partial i_2}{\partial x} - (y')^2\frac{\partial i_2}{\partial y'} - 3y'y''\frac{\partial i_2}{\partial y''} = 0.$$

The characteristic system for this equation will be

$$\frac{dx}{y} = \frac{dy'}{-(y')^2} = \frac{dy''}{-3y'y''}.$$

The first equation from this system we solved in the example before. Let us consider the second equation

$$\frac{dy'}{y'} = \frac{dy''}{3y''} \Rightarrow 3\ln y' = \ln y'' - \ln i_2 \Rightarrow i_2(x,y,y',y'') = \frac{y''}{(y')^3}.$$

We obtain the most general ordinary differential equation of the second order which is invariant under the action of the Galilean group of transformations if we set

$$\begin{aligned} i_2 &= \Phi(u, i_1) = \Phi(u,v), \\ \frac{y''}{(y')^3} &= \Phi\left(y, x - \frac{y}{y'}\right) \\ \Rightarrow y'' &= (y')^3\Phi\left(y, x - \frac{y}{y'}\right). \end{aligned}$$

◇

We obtained expressions of differential invariants of the n-th order using the prolongation of the operator $\mathbf{U}$ up to the order n first. Then we looked for invariant expressions for this prolonged operator. This procedure can take a lot of time in some cases. The following theorem helps us to avoid a lot of calculations and get an expression for a higher order differential invariant just using the first two invariants.

Theorem 5.2.1 (Sophus Lie) *All higher order differential invariants can be found merely by differentiation, provided that two first invariants $u(x, y)$ and $v(x, y, y')$ (also referred to as $i_1(x, y, y')$) are known.*

To clarify this procedure we provide an example.

Example 5.2.4 Let us calculate $i_2(x, y, y', y'')$ for the Galilean group of transformations using Theorem 5.2.1. Again we examine the Galilean group. We have $u(x, y) = y$ and $v(x, y, y') = i_1(x, y, y') = \frac{x}{y} - \frac{1}{y'}$. From the last example we know that we should get for the invariant $i_2(x, y, y', y'')$ the expression $i_2 = \frac{y''}{(y')^3}$. To get this expression we differentiate the first differential invariant with respect to x

$$\left(v - \frac{x}{y} + \frac{1}{y'}\right)_x = -\frac{1}{y} + \frac{xy'}{y^2} - \frac{y''}{(y')^2} = 0.$$

Because $\frac{x}{y} = \frac{1}{y'} + v$ we insert in the above equation the expression for $\frac{x}{y}$ and obtain

$$-\frac{1}{y} + \frac{y'}{y} \cdot \frac{x}{y} - \frac{y''}{(y')^2} = -\frac{1}{y} + \frac{1}{y} + \frac{y'}{y}v - \frac{y''}{(y')^2} = \frac{y'}{y}v - \frac{y''}{(y')^2} = 0.$$

Now this expression contains just invariants and it leads to the invariant expressions

$$\frac{v(x, y, y')}{y} = \frac{v(x, y, y')}{u(x, y)} = \frac{y''}{(y')^3}.$$

Because $\frac{v(x,y,y')}{u(x,y)}$ is an invariant then the right part is invariant too. We have found a new invariant $i_2 = \frac{y''}{(y')^3}$ different from previous invariants. It is a second differential invariant because it contains the second derivative. $\diamond$

We summarize relations between differential invariants of a one-parameter group $G(\mathbb{R}^2)$ of point transformations and equations which admit this group in Table 5.1. In this section we discussed how to use n-th order differential invariants to get a differential equation which admits the corresponding one-parameter group of point transformations. Geometrically we presented each

Table 5.1: In the first column of the table we provide the group and its prolongations. In the second column the corresponding infinitesimal generator and its prolongations are displayed. In the third column the invariants and differential invariants are presented. In the fourth column the corresponding invariant functions or differential equations are displayed. In the last column the type of the invariant expression in the third column is noticed.

G	$\mathbf{U}u = 0$	$u = u(x, y)$	$\Phi(u)$	invariant functions
$G^{(1)}$	$\mathbf{U}^{(1)}i_1 = 0$	$i_1 = v = v(x, y, y')$	$i_1 = v = \Phi(u)$	invariant first order ODE
$G^{(2)}$	$\mathbf{U}^{(2)}i_2 = 0$	$i_2 = i_2(x, y, y', y'')$	$i_2 = \Phi(u, v)$ $= \Phi(u, i_1)$	invariant second order ODE
$G^{(n)}$	$\mathbf{U}^{(n)}i_n = 0$	$i_n =$ $i_n(x, y, y', \ldots, y^{(n)})$	$i_n = \Phi(u, i_1,$ $\ldots, i_{n-1})$	invariant n-th order ODE

n-th order ODE as a frame in the jet bundle $J^{(n)}$. We proved that an n-th order ODE admits a symmetry group $G(\mathbb{R}^2)$ if and only if the action of the prolonged group $G^{(n)}$ can be locally presented as a motion of a point on this frame. Now we can try to find an admitted symmetry group for a given ODE using this idea. We demand that the prolonged group $G^{(n)}$ leaves the point on the frame. We have seen before that this condition on two functions $\xi(x, y)$ and $\eta(x, y)$ in the case of a first order ODE is as difficult as the equation itself. A first order ODE have infinitely many symmetries and because of that we cannot use any one of them. In the case of n-th order ODE we obtain an over-determined system of equations on the functions $\xi(x, y)$, $\eta(x, y)$ and their derivatives which defines the infinitesimal generator $\mathbf{U}^{(n)}$ of the admitted group $G^{(n)}$. This system can have no, one or many solutions. If we are able to find the functions $\xi(x, y)$ and $\eta(x, y)$ then we obtain the infinitesimal operator (or many of them) of the admitted symmetry group and the problem is solved. We introduce now the definition of the system of equations, which define for us the admitted symmetry group.

Definition 5.2.2 (Determining equations) *The system of equations*

$$\mathbf{U}^{(n)}\Delta(x, y, y', \ldots, y^{(n)})\big|_{\Delta(x,y,y',\ldots,y^{(n)})=0} = 0 \qquad (5.2.11)$$

determines all infinitesimal generators $\mathbf{U}^{(n)}$ *of the group* $G^{(n)}$ *admitted by the* n*-th order differential equation*

$$\Delta(x, y, y', \ldots, y^{(n)}) = 0$$

and therefore, it is known as a determining system of equations for the symmetry group $G(\mathbb{R}^2)$.

By solving determining equations we can obtain many infinitesimal generators $\mathbf{U}_i^{(n)}, \quad i = 1, 2, \ldots k$ and correspondingly $\mathbf{U}_i, \quad i = 1, 2, \ldots k$. Each of them defines a one-parameter symmetry group of point transformations admitted by the given equation $\Delta(x, y, y', \ldots, y^{(n)}) = 0$. We will see a bit later that all these generators $\mathbf{U}_i, \quad i = 1, 2, \ldots k$ build a linear space and an algebra, which is called Lie algebra. The dimension of the underlying linear space of linear operators defines the dimension of the algebra. We will discuss all properties of such algebras in the next chapter. Here let us just make a hint that the set of admitted operators is not arbitrary, it has an algebraic structure and because of that is called an algebra.

We have seen before that a first order ODE has infinitely many symmetries but we do not have any algorithmic way to find at least one of them. For the higher order ODEs the situation is quite different. Some of the ODEs do not admit any one-parameter symmetry group of point transformations while others can have many of them. How many symmetries can a higher order ODE admit? The answer can be formulated in the following theorem.

Theorem 5.2.2 *[53],[33]. Let us consider an ordinary differential equation of* n*-th order*

$$y^{(n)} = f(x, y, y', \ldots, y^{(n-1)}).$$

If $n = 1$*, an infinitesimal generator of a symmetry algebra contains an arbitrary function, i.e. first order differential equations admit an infinite symmetry algebra.*

If $n \geq 3$*, then an ordinary differential equation of order* n *admits at maximum* $(n+4)$ *infinitesimal generators, i.e. symmetries (or possesses a* $(n+4)$*-dimensional Lie algebra) and this maximum is reached by the equation* $y^{(n)} = 0$.

If $n = 2$*, then a second order ordinary differential equation admits at most* 8 *symmetries (i.e. the* 8*-dimensional Lie algebra) and this maximum is reached by* $y'' = 0$,

An equation of the second order can have 0, 1, 2, 3 *or* 8 *symmetries.*

This theorem gives us a short overview how many symmetries we can expect from an ODE. In Chapter 7 we will discuss how to use the found symmetries to solve or to simplify the given ODE.

5.3 Problems with solutions

In this section we give nineteen problems that should help the reader to systemize and apply in practice the main ideas of this chapter: a prolongation procedure, differential invariants and a relation between a high order ODE and the corresponding symmetry group.

In Problems 5.3.1, 5.3.2 and 5.3.10 we discuss the prolongation procedure for different one-parameter groups of point transformations. We verify if the given equation admits the provided symmetry group in Problems 5.3.3, 5.3.6, 5.3.14, 5.3.15 and 5.3.18. We find the most general form of an ODE which admits one or two given symmetries in the Problems 5.3.4, 5.3.5, 5.3.8, 5.3.12 and 5.3.13. Using the prolongation procedure we calculate differential invariants in the Problems 5.3.7 and 5.3.11. We calculate higher order differential invariants using differentiation in accordance with Lie theorem in Problems 5.3.9, 5.3.16 and 5.3.17. In Problem 5.3.19 we provide a solution of a Cauchy problem for a first order PDE.

Problem 5.3.1 *Find the first prolongation of the infinitesimal generator of the conformal transformation group*

$$\mathbf{U} = (y^2 - x^2)\frac{\partial}{\partial x} - 2xy\frac{\partial}{\partial y}.$$

Solution The first prolongation of the infinitesimal generator $\mathbf{U}$ has the following form

$$\mathbf{U}^{(1)} = \xi(x, y)\frac{\partial}{\partial x} + \eta(x, y)\frac{\partial}{\partial y} + \eta^{(1)}\frac{\partial}{\partial y'}.$$

Functions $\xi(x, y)$ and $\eta(x, y)$ define the infinitesimal generator $\mathbf{U}$ and have the form

$$\xi(x, y) = (y^2 - x^2), \ \eta(x, y) = -2xy.$$

Therefore

$$\begin{aligned}\eta^{(1)}(x, y, y') &= \eta_x + \eta_y y' - y'(\xi_x + \xi_y y') = -2y + (-2x)y' - y'(-2x + 2yy') \\ &= -2y - 2y(y')^2.\end{aligned}$$

Thus, the infinitesimal generator of the first prolonged conformal transformation group is equal to

$$\mathbf{U}^{(1)} = (y^2 - x^2)\frac{\partial}{\partial x} - 2xy\frac{\partial}{\partial y} - (2y + 2y(y')^2)\frac{\partial}{\partial y'}.$$

Problem 5.3.2 *Find the second prolongation of the following infinitesimal generator*

$$\mathbf{U} = (x-y)\frac{\partial}{\partial x} + (x+y)\frac{\partial}{\partial y}.$$

Solution Because in this case

$$\xi(x,y) = x - y,\ \eta(x,y) = x + y,$$

we obtain

$$\xi_x = \eta_x = \eta_y = 1,\ \xi_y = -1.$$

Therefore

$$\eta^{(1)}(x,y,y') = \eta_x + \eta_y y' - y'(\xi_x + \xi_y y') = 1 + y' - y'(1-y') = 1 + (y')^2,$$

and

$$\eta^{(2)}(x,y,y',y'') = \eta_x^{(1)} + \eta_y^{(1)} y' - y''(\xi_x + \xi_y y') = 2y'y'' + 0 - y''(1-y') = (3y'-1)y''.$$

It follows that the second prolongation of the generator $\mathbf{U}$ has the form

$$\mathbf{U}^{(2)} = (x-y)\frac{\partial}{\partial x} + (x+y)\frac{\partial}{\partial y} + (1+(y')^2)\frac{\partial}{\partial y'} + (3y'-1)y''\frac{\partial}{\partial y''}.$$

Problem 5.3.3 *Prove that the first order ODE*

$$y' = -2xy\ln(x^2 + \ln y)$$

admits the symmetry group with the infinitesimal operator

$$\mathbf{U} = \frac{1}{2x}\frac{\partial}{\partial x} - y\frac{\partial}{\partial y}.$$

Use the first prolongation of this operator and the determining equations.

Solution With the formula (5.1.4) we calculate the first prolongation of the given infinitesimal generator

$$\begin{aligned}
\eta^{(1)}(x,y,y') &= \frac{d\eta}{dx} - y'\frac{d\xi}{dx} = -y' + \frac{y'}{2x^2}, \\
\mathbf{U}^{(1)} &= \frac{1}{2x}\frac{\partial}{\partial x} - y\frac{\partial}{\partial y} + y'\left(\frac{1}{2x^2} - 1\right)\frac{\partial}{\partial y'}.
\end{aligned}$$

Let us denote $\Delta(x,y,y') = y' + 2xy\ln(x^2+\ln y)$ then the given equation can be rewritten in the form $\Delta(x,y,y') = 0$. Now we find how the prolonged infinitesimal generator $\mathbf{U}^{(1)}$ acts on the frame of the given equation

$$\left(\frac{1}{2x}\frac{\partial}{\partial x} - y\frac{\partial}{\partial y} + y'\left(\frac{1}{2x^2}-1\right)\frac{\partial}{\partial y'}\right)(y' + 2xy\ln(x^2+\ln y))\Big|_{\Delta(x,y,y')=0}$$

$$= \left(\frac{y}{x}\ln(x^2+\ln y) + \frac{2xy}{x^2+\ln y} - 2xy\ln(x^2+\ln y)\right.$$

$$\left. - \frac{2xy}{x^2+\ln y} + y'\left(\frac{1}{2x^2}-1\right)\right)\Big|_{\Delta(x,y,y')=0}$$

$$= \left(\frac{1}{2x^2}\left(2xy\ln(x^2+\ln y) + y'\right) - y'\left(2xy\ln(x^2+\ln y) + y'\right)\right)\Big|_{\Delta(x,y,y')=0}$$

$$= \left(\frac{1}{2x^2}-1\right)\left(2xy\ln(x^2+\ln y) + y'\right)\Big|_{\Delta(x,y,y')=0}.$$

Due to the fact that we are on the solution manifold $\Delta(x,y,y') = 0$ the second factor in the last line is equal to zero which means that the whole product is equal to zero. This fact finishes the proof.

Problem 5.3.4 *Find the most general form of a first order differential equation which admits the following symmetry*

$$\mathbf{U} = x^2\frac{\partial}{\partial x} + xy\frac{\partial}{\partial y}.$$

Solution Let us calculate the invariant function of the group with the infinitesimal generator $\mathbf{U}$

$$\mathbf{U}\ u(x,y) = 0 \Rightarrow x^2\frac{\partial u}{\partial x} + xy\frac{\partial u}{\partial y} = 0 \Rightarrow \frac{dx}{x^2} = \frac{dy}{xy} = \frac{du}{0}.$$

From the characteristic system it follows that

$$\text{const.} = u(x,y) = \frac{y}{x}$$

is an invariant function of the group $G(\mathbb{R}^2)$ generated by $\mathbf{U}$.
Because $\xi(x,y) = x^2$, $\eta(x,y) = xy$, we obtain for the first prolongation

$$\eta^{(1)}(x,y,y') = \eta_x + \eta_y y' - y'(\xi_x + \xi_y y') = y + xy' - y'(2x + 0\cdot y') = y - xy'.$$

Therefore the first prolongation of the generator $\mathbf{U}$ has the form

$$\mathbf{U}^{(1)} = x^2\frac{\partial}{\partial x} + xy\frac{\partial}{\partial y} + (y - xy')\frac{\partial}{\partial y'}.$$

The characteristic system of the equation which allows us to find the first order differential invariant $v(x, y, y')$ is

$$\frac{dx}{x^2} = \frac{dy}{xy} = \frac{dy'}{(y - xy')} = \frac{du}{0}.$$

Let us take the following pair of equations first

$$\frac{dx}{x^2} = \frac{dy'}{(y - xy')}.$$

Since $y = ux$, we get

$$\frac{dx}{x} = \frac{dy'}{(u - y')} \quad \Rightarrow \quad x(u - y') = v(x, y, y') = \text{const}.$$

We obtain the most general differential equation of the first order which admits the one-parameter group $G(\mathbb{R}^2)$ with the given infinitesimal generator if we put $v(x, y, y') = F(u(x, y))$. After the transformation this equation takes the form

$$y' = \frac{y - F\left(\frac{y}{x}\right)}{x}.$$

Problem 5.3.5 *Find the most general equation of the second order which admits the group with the infinitesimal generator*

$$\mathbf{U} = \frac{\partial}{\partial x} + y\frac{\partial}{\partial y}.$$

Solution The coefficients of the infinitesimal generator $\mathbf{U}$ are $\xi(x, y) = 1$, $\eta(x, y) = y$, it means that $\xi_x = \xi_y = \eta_x = 0$, $\eta_y = 1$. Using the expressions, we calculate $\eta^{(1)}(x, y, y') = \eta_x + \eta_y y' - y'(\xi_x + \xi_y y') = y'$ and $\eta^{(2)}(x, y, y', y'') = \eta_x^{(1)} + \eta_y^{(1)} y' - y''(\xi_x + \xi_y y') = y''$. Thus the second prolongation of the generator $\mathbf{U}$ is given by

$$\mathbf{U}^{(2)} = \frac{\partial}{\partial x} + y\frac{\partial}{\partial y} + y'\frac{\partial}{\partial y'} + y''\frac{\partial}{\partial y''}.$$

If we look for the differential invariants of the second order we have to solve an equation of the type

$$\mathbf{U}^{(2)} i_2(x, y, y', y'') = \frac{\partial i_2}{\partial x} + y\frac{\partial i_2}{\partial y} + y'\frac{\partial i_2}{\partial y'} + y''\frac{\partial i_2}{\partial y''} = 0.$$

The corresponding characteristic system has the form

$$\frac{dx}{1} = \frac{dy}{(y)} = \frac{dy'}{(y')} = \frac{dy''}{(y'')}.$$

Solving this system, we obtain the following invariants

$$u(x,y) = ye^{-x}, \ v(x,y,y') = i_1(x,y,y') = \frac{y'}{y}, \ i_2(x,y,y',y'') = \frac{y''}{y'}.$$

The most general equation of the second order which admits the group G has the form $i_2 = F(u,v)$, where F is the arbitrary function of its arguments, in other words

$$y'' = y'F\left(ye^{-x}, \frac{y'}{y}\right).$$

Problem 5.3.6 *Prove (using the second prolongation of* $\mathbf{U}$*) that the second order differential equation*

$$y'' = x^3 + (y - xy')^2$$

admits the symmetry group with the following infinitesimal generator

$$\mathbf{U} = x\frac{\partial}{\partial y}.$$

Solution Functions $\xi(x,y)$ and $\eta(x,y)$ are the coefficients of the infinitesimal generator $\mathbf{U}$ and have the form

$$\xi(x,y) = 0, \ \eta(x,y) = x.$$

Therefore

$$\eta^{(1)}(x,y,y') = \eta_x + \eta_y y' - y'(\xi_x + \xi_y y') = 1,$$

and the first prolongation of the infinitesimal generator $\mathbf{U}$ consequently is

$$\mathbf{U}^{(1)} = x\frac{\partial}{\partial y} + \frac{\partial}{\partial y'}.$$

For the second prolongation we have

$$\eta^{(2)}(x,y,y',y'') = \eta_x^{(1)} + \eta_y^{(1)} y' - y''(\xi_x + \xi_y y') = 0,$$

and correspondingly the second prolongation of the generator $\mathbf{U}$ has the same form as $\mathbf{U}^{(1)}$, i.e.

$$\mathbf{U}^{(2)} = x\frac{\partial}{\partial y} + \frac{\partial}{\partial y'}.$$

To prove that the second order differential equation $\Delta(x,y,y',y'') = 0$ admits the symmetry group with the infinitesimal generator $\mathbf{U}$, we should verify

$$\mathbf{U}^{(2)}\Delta|_{\Delta=0} = 0, \qquad \Delta(x,y,y',y'') = y'' - x^3 - (y - xy')^2.$$

$$x\left(\frac{\partial}{\partial y} + \frac{\partial}{\partial y'}\right)(y'' - x^3 - (y-xy')^2) = -2(y-xy')x - 2(y-xy')(-x) = 0.$$

So we proved that the symmetry defined by $\mathbf{U}$ is admitted by this equation.

Problem 5.3.7 *Find the differential invariants of the third order for the group of the non-uniform dilation*

$$x_\epsilon = xe^\epsilon, \quad y_\epsilon = ye^{k\epsilon}.$$

Solution The coefficients of the infinitesimal generator $\mathbf{U}$ have the form

$$\begin{aligned}\xi(x,y) &= \frac{\partial x_\epsilon}{\partial \epsilon}\bigg|_{\epsilon=0} = xe^\epsilon\bigg|_{\epsilon=0} = x,\\ \eta(x,y) &= \frac{\partial y_\epsilon}{\partial \epsilon}\bigg|_{\epsilon=0} = kye^{k\epsilon}\bigg|_{\epsilon=0} = ky.\end{aligned}$$

The infinitesimal generator is equal to

$$\mathbf{U} = x\frac{\partial}{\partial x} + ky\frac{\partial}{\partial y}.$$

Therefore to obtain the third prolongation we calculate

$$\begin{aligned}\eta^{(1)}(x,y,y') &= \eta_x + \eta_y y' - y'(\xi_x + \xi_y y') = ky' - y' = y'(k-1),\\ \eta^{(2)}(x,y,y',y'') &= \eta_x^{(1)} + \eta_y^{(1)} y' - y''(\xi_x + \xi_y y')\\ &= (k-1)y'' - y'' = (k-2)y'',\\ \eta^{(3)}(x,y,y',y'',y''') &= \eta_x^{(2)} + \eta_y^{(2)} y' - y''(\xi_x + \xi_y y')\\ &= (k-2)y''' - y''' = (k-3)y'''.\end{aligned}$$

The third prolongation of the infinitesimal generator $\mathbf{U}$ has the following form

$$\mathbf{U}^{(3)} = x\frac{\partial}{\partial x} + ky\frac{\partial}{\partial y} + (k-1)y'\frac{\partial}{\partial y'} + (k-2)y''\frac{\partial}{\partial y''} + (k-3)y'''\frac{\partial}{\partial y'''}.$$

The characteristic system of equations for this infinitesimal generator takes the form

$$\frac{dx}{x} = \frac{dy}{ky} = \frac{dy'}{(k-1)y'} = \frac{dy''}{(k-2)y''} = \frac{dy'''}{(k-3)y'''}.$$

By solving the first equation, i.e.

$$\frac{dx}{x} = \frac{dy}{ky},$$

we obtain the invariant function $u(x,y) = \frac{x^k}{y}$. The first differential invariant $i_1(x,y,y') = v(x,y,y')$ could be obtained from the following equation

$$\frac{dx}{x} = \frac{dy'}{(k-1)y'} \quad \Rightarrow \quad v(x,y,y') = \frac{x^{(k-1)}}{y'}.$$

Analogously, we obtain for the second and third differential invariants the following expressions

$$i_2(x, y, y', y'') = \frac{x^{(k-1)}}{(k-2)y''}, \; i_3(x, y, y', y'', y''') = \frac{x^{(k-1)}}{(k-3)y'''}.$$

Problem 5.3.8 *Find the most general form of the second order differential equation which admits two one-parameter symmetry groups with generators*

$$\mathbf{U}_1 = \frac{\partial}{\partial y}, \quad \mathbf{U}_2 = x\frac{\partial}{\partial x} + y\frac{\partial}{\partial y}.$$

Solution The prolongations of the operators $\mathbf{U}_1$ and $\mathbf{U}_2$ have the form

$$\mathbf{U}_1^{(2)} = \frac{\partial}{\partial y}, \; \mathbf{U}_2^{(2)} = x\frac{\partial}{\partial x} + y\frac{\partial}{\partial y} - y''\frac{\partial}{\partial y''}.$$

Let us find the invariants of the infinitesimal generator $\mathbf{U}_2^{(2)}$. We solve the characteristic system of equations

$$\frac{dx}{x} = \frac{dy}{y} = \frac{dy'}{0} = \frac{dy''}{-y''}.$$

Solving that system of equations, we get the following invariants

$$u(x, y) = \frac{y}{x}, \; v(x, y, y') = y', \; i_2(x, y, y', y'') = y''x.$$

We obtain the most general differential equation of the second order which admits the Lie group $G_2(\mathbb{R}^2)$ with the infinitesimal generator $\mathbf{U}_2$ if we set

$$i_2 = f(u, v), \; \Rightarrow \; y''x = F\left(\frac{y}{x}, y'\right).$$

We apply the generator $\mathbf{U}_1^{(2)}$ to the equation to exclude from this family of equations all of them that are not invariant under the action of the group $G_1(\mathbb{R}^2)$ with the generator $\mathbf{U}_1$. We obtain

$$\mathbf{U}_1^{(2)}\left(y''x - F\left(\frac{y}{x}, y'\right)\right) = \frac{1}{x}\frac{\partial}{\partial z}F(z, y') = 0, \quad z = \frac{y}{x}$$

consequently the function $F(z, y')$ does not depend on the first argument. We get the most general differential equation of the second order which admits both groups with generators $\mathbf{U}_1$ and $\mathbf{U}_2$ in the form

$$y''x = F(y').$$

Problem 5.3.9 *The infinitesimal generator has the form*

$$\mathbf{U} = xy\frac{\partial}{\partial x}.$$

Find the second differential invariant i_2 using the invariant differentiation.

Solution At first let us calculate the first prolongation of the given infinitesimal generator as follows

$$\begin{aligned}
\eta^{(1)}(x, y, y') &= \frac{d\eta}{dx} - y'\frac{d\xi}{dx} = -y'\left(\frac{ydx + xdy}{dx}\right) = -yy' - xy'^2, \\
\mathbf{U}^{(1)} &= xy\frac{\partial}{\partial x} - (yy' + xy'^2)\frac{\partial}{\partial y'}.
\end{aligned}$$

As usual we should solve the system of characteristic equations

$$\frac{dx}{xy} = \frac{dy}{0} = -\frac{dy'}{(yy' + xy'^2}$$

If we look at the second term then it is really easy to notice that the orbits are given by the formula $u(x, y) = y$. To find the next first integral of the system we take the first and the last term of the system and solve the following equation

$$-(yy' + xy'^2)dx = xydy'.$$

It can be integrated after the introduction of a new variable $t = xy'$

$$-(yy' + ty')(dt - \frac{t}{y'}dy) = tydy',$$

or after the transformation

$$(y'y + y't)dt = t^2dy'.$$

Thus,

$$\left(\frac{y}{t^2} + \frac{1}{t}\right)dt = \frac{dy'}{y'},$$

and we obtain the first integral in the form

$$\frac{y}{t} + \ln t = \ln y' + C.$$

Substituting $t = xy'$ we obtain the first integral, i.e. the first differential invariant $v(x, y, y') = i_1(x, y, y')$

$$v(x, y, y') = \ln x - \frac{y}{xy'}.$$

The second differential invariant i_2 according to the Lie theorem can be found by differentiation

$$\frac{dv}{du} = \frac{d(\ln x - \frac{y}{xy'})}{dy} = \left(\frac{1}{xy'} - \frac{1}{xy'} + \frac{y}{x^2y'^2} + \frac{yy''}{xy'^3}\right)$$
$$= y\left(\frac{1}{x^2y'^2} + \frac{y''}{xy'^3}\right) = u\left(\frac{1}{x^2y'^2} + \frac{y''}{xy'^3}\right).$$

Thus, we obtain the second differential invariant for this group of point transformations

$$i_2(x, y, y'y'') = \left(\frac{1}{x^2y'^2} + \frac{y''}{xy'^3}\right). \tag{5.3.12}$$

Problem 5.3.10 *Find the first prolongation of the operator*

$$\mathbf{U} = y^n \exp\left[(1-n)\int P(x)\mathrm{d}x\right]\frac{\partial}{\partial y}.$$

Find the first differential invariant of this transformation group. Find the form of the most general differential equation of the first order which admits this symmetry.

Solution For the first prolongation of the infinitesimal generator $\mathbf{U}$ we have

$$\mathrm{pr}^{(1)}\mathbf{U} = \mathbf{U}^{(1)} = y^n \exp\left[(1-n)\int P(x)\mathrm{d}x\right]\frac{\partial}{\partial y} + \eta^{(1)}\frac{\partial}{\partial y'},$$

where

$$\eta^{(1)}(x, y, y') = \frac{d\eta}{dx} - y'\frac{d\xi}{dx}$$
$$= ny^{(n-1)}y' \exp\left[(1-n)\int P(x)dx\right] + y^n(1-n)\exp\left[(1-n)\int P(x)dx\right]$$
$$= y^{(n-1)} \exp\left[(1-n)\int P(x)dx\right](ny' + (1-n)P(x)y),$$

and characteristic system of equations for invariants of $\mathbf{U}^{(1)}$ reduces to

$$u(x, y) = x = \text{const.}$$
$$\frac{dy}{y} = \frac{dy'}{ny' + P(x)(1-n)y}.$$

The first equation gives us the form of orbits and the second we rewrite in the form

$$dy\left(n\frac{y'}{y} + P(x)(1-n)\right) = dy'.$$

Note, that since x is invariant, we can integrate it explicitly. Let us change the variables so that $\psi = \frac{y'}{y}$. Then we have

$$dy' = d(\psi y) = yd\psi + \psi dy,$$

and the equation transforms to

$$dy(n\psi + P(x)(1-n)) = yd\psi + \psi dy.$$

After separation of variables we get

$$\frac{dy}{y} = \frac{d\psi}{(n-1)(\psi - P(x))}.$$

We integrate the equation and obtain the first differential invariant

$$i_1(x, y, y') = v(x, y, y') = \frac{y^{n-1}}{\psi - P(x)} = \frac{y^n}{y' - P(x)y}.$$

The most general differential equation of the first order which admits the symmetry in question can be immediately written in the form (recalling the equation for orbits that is presented by $x = \text{const.}$)

$$\frac{y^n}{y' - P(x)y} = F(x)$$

or

$$y' = P(x)y + y^n F(x),$$

where F is an arbitrary function. It is the so-called Bernoulli equation.

Problem 5.3.11 *Find the differential invariants of the 1st, 2nd and 3rd order for the group with the infinitesimal generator*

$$\mathbf{U} = x^2 \frac{\partial}{\partial y}.$$

Solution The orbits, i.e. the invariant functions can be immediately found from the form of the infinitesimal generator

$$u(x, y) = F(x),$$

where F is an arbitrary function of its argument.
Knowing the first prolongation $\mathbf{U}^{(1)}$ (that we can find using the prolongation formula)

$$\mathbf{U}^{(1)} = x^2 \frac{\partial}{\partial y} + 2x \frac{\partial}{\partial y'},$$

we can calculate the invariant of the first order from the equation

$$x\frac{\partial v}{\partial y} + 2\frac{\partial v}{\partial y'} = 0,$$

using the characteristic system of equations

$$\frac{dy}{x} = \frac{dy'}{2}.$$

Then we get the differential invariant of the first order

$$v(x, y, y') = \frac{2y}{x} - y'.$$

Invariants of the 2nd and the 3rd order according to the Lie theorem can be found merely by differentiation using the first two invariants $u(x, y)$ and $v(x, y, y')$

$$0 = \frac{d}{dx}\left(\frac{2y}{x} - y'\right) = \frac{2y'}{x} - \frac{2y}{x^2} - y'' \Rightarrow$$
$$2uv = u^2y'' - 2y \quad \Rightarrow \quad i_2(x, y, y', y'') = x^2y'' - 2y.$$

Now we can also calculate i_3

$$0 = \frac{d}{dx}\left(x^2y'' - 2y\right) = 2xy'' + x^2y''' - 2y'.$$

We know that

$$y' = \frac{2y}{x} - v,$$
$$y'' = \frac{i_2 - 2y}{x^2}.$$

Substituting these formulas into the obtained equation we get

$$2u\frac{i_2 - 2y}{u^2} + u^2y''' - \frac{4y}{u} + 2v = 0, \Rightarrow$$
$$y''' = \frac{2v}{u^2} - \frac{2i_2}{u^3} \quad \Rightarrow \quad i_3(x, y, y', y'', y''') = y'''.$$

Problem 5.3.12 *Find the most general form of the second order differential equation which admits two-parameter symmetry group with generators*

$$\mathbf{U}_1 = x\frac{\partial}{\partial x} + y\frac{\partial}{\partial y}, \quad \mathbf{U}_2 = x^2\frac{\partial}{\partial x} + y^2\frac{\partial}{\partial y}.$$

Solution First, let us calculate the first and the second differential invariants for $\mathbf{U}_1$ and $\mathbf{U}_2$. For the prolongation of $\mathbf{U}_1$, we have

$$\begin{aligned}
\mathrm{pr}^{(1)}\mathbf{U}_1 &= x\frac{\partial}{\partial x} + y\frac{\partial}{\partial y} + \eta^{(1)}\frac{\partial}{\partial y'} \\
&= x\frac{\partial}{\partial x} + y\frac{\partial}{\partial y} + (y' - y')\frac{\partial}{\partial y'} \\
&= x\frac{\partial}{\partial x} + y\frac{\partial}{\partial y}.
\end{aligned}$$

Thus, for orbits and the first differential invariant, we obtain the reduced characteristic system of equations

$$\frac{dx}{x} = \frac{dy}{y}, \quad y' = \text{const.},$$

and we immediately get the invariants $u_{(1)}(x, y)$ and $v_{(1)}(x, y, y')$ for the operator $\mathbf{U}_1$,

$$u_{(1)}(x, y) = \frac{x}{y}, \quad i_{1,(1)}(x, y, y') = v_{(1)}(x, y, y') = y'.$$

To obtain the second differential invariant $i_{2,(1)}$ for the operator $\mathbf{U}_1$, we can simply differentiate $v_{(1)} = i_{1,(1)}(x, y, y')$ over $u_{(1)}(x, y)$ according to the Lie theorem

$$\frac{dy'}{d\left(\frac{x}{y}\right)} = \frac{y^2 dy'}{ydx - xdy} = \frac{yy''}{1 - \frac{x}{y}dy} = \frac{yy''}{1 - uv}.$$

Therefore $i_{2,(1)} = yy''$, and the most general form of the second order ODE that admits $\mathbf{U}_1$ is

$$yy'' = F\left(\frac{x}{y}, y'\right).$$

For the second symmetry, generated by $\mathbf{U}_2$, we have

$$\begin{aligned}
\mathrm{pr}^{(1)}\mathbf{U}_2 &= x^2\frac{\partial}{\partial x} + y^2\frac{\partial}{\partial y} + \eta^{(1)}\frac{\partial}{\partial y'} \\
&= x^2\frac{\partial}{\partial x} + y^2\frac{\partial}{\partial y} + (2yy' - 2xy')\frac{\partial}{\partial y'} \\
&= x\frac{\partial}{\partial x} + y\frac{\partial}{\partial y} + 2y'(y - x)\frac{\partial}{\partial y'}.
\end{aligned}$$

To obtain orbits $u_{(2)}(x,y)$ and the first differential invariant $i_{1,(2)}(x,y,y')$ for the operator $\mathbf{U}_2$, we have to integrate the following characteristic system of equations

$$\frac{dx}{x^2}=\frac{dy}{y^2}=\frac{dy'}{2y'(y-x)}.$$

Therefore we have

$$u_{(2)}(x,y)=\frac{1}{x}-\frac{1}{y}$$

for orbits. Having $x=\frac{y}{uy+1}$ we substitute this expression into the second equation of the characteristic system of equations

$$\frac{dy}{y^2}=\frac{dy'}{2y'\left(\frac{uy^2+y-y}{uy+1}\right)}=\frac{(uy+1)dy'}{2uy'y^2}.$$

Thus,

$$\frac{dy'}{y'}=\frac{2udy}{uy+1}\Leftrightarrow 2\ln(uy+1)=C\ln y',$$

or

$$i_{1,(2)}(x,y,y')=\frac{(uy+1)^2}{y'}=\frac{y^2}{x^2y'}.$$

Differentiating with respect to $u_{(2)}(x,y)$, we simply derive the second differential invariant $i_{2,(2)}(x,y,y')$ for the operator $\mathbf{U}_2$

$$\frac{d\left(\frac{y^2}{x^2y'}\right)}{d\left(\frac{1}{x}-\frac{1}{y}\right)}=\frac{2yx^2y'^2-2xy^2y'-x^2y^2y''}{-x^2y'^2+\frac{y^2}{x^2y'}}$$
$$=\frac{\frac{2yy'}{i_{1,(2)}}-2xy'-x^2y''}{y\left(i_{1,(2)}-\frac{1}{i_{1,(2)}}\right)}.$$

We can simplify this formula and the most general form of the second order ODE that admits $\mathbf{U}_2$ will be

$$\left(\frac{y}{x}-1\right)\left(2\left(\frac{1}{i_{1,(2)}(x,y,y')}+1\right)+\frac{xy''}{y'}\right)=\mathcal{F}(u_{(2)}(x,y),i_{1,(2)}(x,y,y')).$$

Also, according to the admitted symmetry $\mathbf{U}_1$ our differential equation must have a form $yy''=F(\frac{x}{y},y')$ where F is an arbitrary function. Comparing both forms of equations we can derive the dependence between functions $\mathcal{F}$

and F

$$\frac{\mathcal{F}(\frac{1}{x}-\frac{1}{y},\frac{y^2}{x^2y'})}{(\frac{y}{x}-1)}-2\left(\frac{1}{i_{1,(2)}}+1\right)=\frac{xy''}{y'}$$
$$=F\left(\frac{x}{y},y'\right)\frac{x}{yy'}=f\left(\frac{x}{y},y'\right),$$

where f is an arbitrary function of its argument. Obviously, we can exclude the dependence on the orbits of the second operator because having fixed $\frac{x}{y}$ and y' the right side of the equation becomes fixed while on the left side $\frac{1}{x}-\frac{1}{y}$ can vary. This means the function $\mathcal{F}$ must depend only on the second argument.
So, finally we obtain the most general form of the second order ODE that admits both symmetries in question

$$y''=\left(\frac{1}{y'}-\frac{x}{yy'}\right)\mathcal{K}\left(\frac{y^2}{x^2y'}\right),$$

where $\mathcal{K}$ is an arbitrary function of its argument.

Problem 5.3.13 *Let a one-parameter group of point transformations be generated by*

$$\mathbf{U}=x^3\frac{\partial}{\partial x}+y^3\frac{\partial}{\partial y}. \tag{5.3.13}$$

Find the most general ordinary differential equation of the first order which admits this group of transformations (find the first prolongation of $\mathbf{U}$ *and calculate the corresponding invariants).*

Solution Using the formula for the first prolongation of $\mathbf{U}$

$$\mathrm{pr}^{(1)}\mathbf{U}=x^3\frac{\partial}{\partial x}+y^3\frac{\partial}{\partial y}+\eta^{(1)}\frac{\partial}{\partial y'},$$

where

$$\eta^{(1)}(x,y,y')=\frac{\partial\eta}{\partial x}-y'\frac{\partial\xi}{\partial y},$$

we obtain

$$\mathrm{pr}^{(1)}\mathbf{U}=x^3\frac{\partial}{\partial x}+y^3\frac{\partial}{\partial y}+(3y^2y'-3x^2y')\frac{\partial}{\partial y'}.$$

Now, to get invariants we need to integrate the characteristic system of equations

$$\frac{dx}{x^3}=\frac{dy}{y^3}=\frac{dy'}{3y'(y^2-x^2)}.$$

Obviously, the first equation gives us the formula for orbits, obtained before

$$\frac{1}{x^2} - \frac{1}{y^2} = u(x, y),$$

which is our first invariant. Expressing x and substituting it into the second equation, we transform the equation to

$$\frac{dy}{y^3} = \frac{dy'}{3y'y^2(\frac{Cy^2}{Cy^2+1})},$$

where C denotes any function of $u(x, y)$, i.e. it is unaltered because $u(x, y)$ is an invariant. Therefore

$$\frac{3Cy}{Cy^2+1}dy = \frac{dy'}{y'},$$

or after integration we get

$$(uy^2+1)^{3/2} = y'C, \quad C = \text{const.}$$

Following the most general form of ODE that admits the symmetry group with the infinitesimal generator $\mathbf{U}$ is

$$y' = \left(\left(\frac{1}{x^2} - \frac{1}{y^2}\right)y^2 + 1\right)^{3/2} F\left(\frac{1}{x^2} - \frac{1}{y^2}\right),$$

where F is an arbitrary function of $u(x, y)$.

Problem 5.3.14 *Let a one-parameter group of point transformations (also called a scaling group of transformations) be generated by the operator*

$$\mathbf{U} = x\frac{\partial}{\partial x} + 2y\frac{\partial}{\partial y}.$$

The first order ordinary differential equation has the form

$$xy' = y(\ln y - 2\ln x). \tag{5.3.14}$$

Prove that this equation admits this scaling symmetry group (use the idea of the frame of equation or the solution manifold related to equation).

Solution Using the formula for the first prolongation of $\mathbf{U}$ we obtain

$$\text{pr}^{(1)}\mathbf{U} = \frac{\partial}{\partial x} + 2y\frac{\partial}{\partial y} + \eta^{(1)}\frac{\partial}{\partial y'},$$

where

$$\eta^{(1)}(x,y,y') = \frac{\partial \eta}{\partial x} - y'\frac{\partial \xi}{\partial y} = 2y' - y' = y',$$

we obtain

$$\mathrm{pr}^{(1)}\mathbf{U} = \frac{\partial}{\partial x} + 2y\frac{\partial}{\partial y} + y'\frac{\partial}{\partial y'},$$

Now, let us show that $\mathrm{pr}^{(1)}\mathbf{U}u(x,y,y') = 0$ on the frame $xy' = y(\ln y - 2\ln x)$ (i.e. on the solution manifold of this equation). This will prove that our ODE admits the scaling symmetry group. Indeed,

$$\begin{aligned}\mathrm{pr}^{(1)}\mathbf{U}u(x,y,y')|_{xy'=y(\ln y-2\ln x)} &= xy' + \frac{2yx}{x} - 2y\ln y - \frac{2y^2}{y} + 4y\ln x + xy' \\ &= 2xy' - 2y\ln y + 4y\ln x \\ &= 2(xy' - y(\ln y - 2\ln x)) = 0\end{aligned}$$

because on the frame we have $xy' = y(\ln y - 2\ln x)$.

Problem 5.3.15 *Let a one-parameter group of point transformations be given by*

$$\mathbf{U} = \frac{1}{x\cos(x-y)}\frac{\partial}{\partial y}.$$

The first order ordinary differential equation has the form

$$y' = 1 + \frac{\tan(x-y)}{x}. \tag{5.3.15}$$

Prove that this equation admits the symmetry group generated by $\mathbf{U}$ *(use the idea of the frame of equation or the solution manifold related to equation).*

Solution Using the formula for the first prolongation of $\mathbf{U}$ we obtain

$$\mathrm{pr}^{(1)}\mathbf{U} = \frac{1}{x\cos(x-y)}\frac{\partial}{\partial y} + \eta^{(1)}\frac{\partial}{\partial y'},$$

where

$$\begin{aligned}\eta^{(1)}(x,y,y') &= \frac{\partial \eta}{\partial x} - y'\frac{\partial \xi}{\partial y} \\ &= \frac{-x\sin(x-y)(1-y') + \cos(x-y)}{x^2\cos^2(x-y)}.\end{aligned}$$

Respectively

$$\mathrm{pr}^{(1)}\mathbf{U} = \frac{1}{x\cos(x-y)}\frac{\partial}{\partial y} + \frac{x\sin(x-y)(y'-1) + \cos(x-y)}{x^2\cos^2(x-y)}\frac{\partial}{\partial y'}.$$

Now, let us show that $\mathrm{pr}^{(1)}\mathbf{U}v(x,y,y') = 0$ on the frame $y' = 1 + \frac{\tan(x-y)}{x}$. This will prove, that our ODE admits this symmetry group. Indeed,

$$\begin{aligned}
\mathrm{pr}^{(1)}\mathbf{U}v(x,y,y')|_{y'=1+\frac{\tan(x-y)}{x}} &= \left(\frac{1}{x\cos(x-y)}\frac{\partial}{\partial y}\right)(xy' - x - \tan(x-y)) \\
&+ \left(\frac{x\sin(x-y)(y'-1)+\cos(x-y)}{x^2\cos^2(x-y)}\right)\frac{\partial}{\partial y'}(xy' - x - \tan(x-y)) \\
&= -\frac{1}{x\cos(x-y)}\frac{1}{\cos^2(x-y)} + \frac{x\sin(x-y)(y'-1)+\cos(x-y)}{x\cos^2(x-y)} \\
&= \frac{\sin(x-y)}{x\cos^2(x-y)}\left(-\frac{1}{\sin(x-y)\cos(x-y)} + \frac{\cos(x-y)}{\sin(x-y)} + x(y'-1)\right) \\
&= \frac{\sin(x-y)}{x\cos^2(x-y)}(-\tan(x-y) + xy' - x) = 0
\end{aligned}$$

because on the frame $y' = 1 + \frac{\tan(x-y)}{x}$.

Problem 5.3.16 *The group of transformations is given by the infinitesimal generator*

$$\mathbf{U} = (y^2 + x^2)\frac{\partial}{\partial x} - 2xy\frac{\partial}{\partial y}, \tag{5.3.16}$$

or in polar coordinates

$$\mathbf{U} = -2r^3\cos\theta\frac{\partial}{\partial r} - r\sin\theta\frac{\partial}{\partial \theta}. \tag{5.3.17}$$

Find the second differential invariant using the Lie theorem.

Solution *We already know the orbits and the first invariant due to the fact that we have calculated them before*

$$u(x,y) = \frac{y}{x^2+y^2},$$

$$v(x,y,y') = \frac{y}{x} - \arctan y'.$$

We can find the second invariant using the formula

$$i_2(x,y,y',y'') = \frac{dv}{du},$$

where in polar coordinates

$$u(\theta,r) = \frac{\sin\theta}{r}, \quad v(\theta,r,r') = \tan\theta - \theta - \frac{r}{r'}.$$

So, for the second differential invariant we get

$$i_2(\theta, r, r', r'') = \frac{(\sec^2\theta - 1)d\theta - \frac{1}{r'}dr + \frac{1}{r'^2}dr'}{\frac{\cos\theta}{r}d\theta - \frac{\sin\theta}{r^2}dr}.$$

We can simplify that expression using the formulas for $u(\theta, r)$ and $v(\theta, r, r')$ and obtain the following expression

$$i_2(\theta, r, r', r'') = \frac{r'^2(1 - 2\cos^2\theta) + rr''\cos^2\theta}{(\frac{r'}{r})^2\cos^2\theta - r'\sin\theta}.$$

Problem 5.3.17 *The one-parameter group of point transformations is given by the infinitesimal generator*

$$\mathbf{U} = 2x\frac{\partial}{\partial x} - 3y\frac{\partial}{\partial y}.$$

Find the second differential invariant using the Lie theorem.

Solution At first we need to calculate the first prolongation using the formula for $\eta^{(1)}$

$$\eta^{(1)}(x, y, y') = \frac{d\eta}{dx} - y'\frac{d\xi}{dx} = -3y' - 2y' = -5y'.$$

Thus for $\mathbf{U}^{(1)}$ we obtain

$$\mathbf{U}^{(1)} = 2x\frac{\partial}{\partial x} - 3y\frac{\partial}{\partial y} - 5y'\frac{\partial}{\partial y'}.$$

Now we need to find the orbits $u(x, y)$ and the first differential invariant $v(x, y, y')$. For these purposes we need to solve the characteristic system of equations

$$\frac{dx}{2x} = \frac{dy}{-3y} = \frac{dy'}{-5y'}.$$

The orbits can be found from the first equation

$$\frac{dx}{2x} = \frac{dy}{-3y} \Rightarrow \ln x^3 = \ln y^{-2} + C, \quad C = \text{const.},$$
$$x^3 = uy^{-2} \Rightarrow u(x, y) = x^3y^2.$$

Whereas $v(x, y, y')$ can be calculated from the second equation

$$\frac{dy}{-3y} = \frac{dy'}{-5y'} \Rightarrow \ln y^5 = \ln y'^3 + c$$
$$y^5 = cy'^3 \Rightarrow v(x, y, y') = \frac{y'^3}{y^5}.$$

We can find the second differential invariant using the formula

$$i_2(x, y, y', y'') = \frac{dv}{du}.$$

So, we obtain for this invariant the following expression

$$\begin{aligned} i_2(x, y, y', y'') &= \frac{d(\frac{y'^3}{y^5})}{x^3y^2} = \frac{3(y')^2y^{-5}dy' - 5y^{-6}(y')^3dy}{3x^2y^2dx + 2yx^3dy} \\ &= \frac{3(y')^2y^{-5}y'' - 5y^{-6}(y')^4}{3x^2y^2 + 2yx^3y'}. \end{aligned}$$

Using the formulas for $u(x, y)$ and $v(x, y, y')$ we can simplify this expression

$$i_2(x, y, y', y'') = \frac{3\frac{1}{y'}vy'' - 5\frac{1}{y}y'v}{3\frac{1}{x}u + 2\frac{1}{y}y'u}.$$

We notice that if we multiply an invariant with another invariant the obtained expression will be an invariant as well. Thus we can simplify expressions for invariants and rewrite $i_2(x, y, y', y'')$ in the following form

$$i_2(x, y, y', y'') = \frac{x(3y''y - 5y'^2)}{y'(3y + 2y'x)}.$$

It is only natural to ask whether it is the simplest form for the second invariant. We can find numerous representations for the second invariant. For example, if we build the second prolongation of the given infinitesimal generator according to the standard procedure we can add one more equation to our characteristic system and obtain the following expression

$$i_2(x, y, y', y'') = \frac{y''^5}{y'^7}. \tag{5.3.18}$$

This expression for $i_2(x, y, y', y'')$ and the previous one are equivalent, we can choose one of them as $i_2(x, y, y', y'')$.

Problem 5.3.18 *The group of transformations is given by the infinitesimal generators*

$$\mathbf{U}_1 = \frac{\partial}{\partial y}, \quad \mathbf{U}_2 = x\frac{\partial}{\partial x} + y\frac{\partial}{\partial y}.$$

Find the most general form of the second order ordinary differential equation that admits both given symmetries.

Solution At first we will calculate the second prolongation of the given infinitesimal generators. For the first generator we get

$$\mathbf{U}_1{}^{(2)} = \frac{\partial}{\partial y}, \tag{5.3.19}$$

and for the second

$$\begin{aligned}\mathbf{U}_2{}^{(2)} &= x\frac{\partial}{\partial x} + y\frac{\partial}{\partial y} + \eta^{(1)}\frac{\partial}{\partial y'} + \eta^{(2)}\frac{\partial}{\partial y''}\\ &= x\frac{\partial}{\partial x} + y\frac{\partial}{\partial y} - y''\frac{\partial}{\partial y''},\end{aligned}$$

where we used

$$\begin{aligned}\eta^{(1)} &= \frac{d\eta}{dx} - y'\frac{d\xi}{dx} = y' - y' = 0\\ \eta^{(2)} &= \frac{d\eta^{(1)}}{dx} - y''\frac{d\xi}{dx} = -y''.\end{aligned}$$

The most general form of an ordinary differential equation that admits the symmetry of $\mathbf{U}_1$ is

$$y'' = f(x, y'). \tag{5.3.20}$$

But due to the fact that our equation should also admit the second symmetry generated by $\mathbf{U}_2$ the following equation should be satisfied

$$\begin{aligned}\mathbf{U}_2{}^{(2)}(y'' - f(x,y'))|_{y''=f(x,y')} &= x\frac{\partial}{\partial x}[y'' - f(x,y')]\\ + y\frac{\partial}{\partial y}[y'' - f(x,y')] &- y'\frac{\partial}{\partial y'}[y'' - f(x,y')]\\ = -x\frac{\partial}{\partial x}f(x,y') - y''|_{y''=f(x,y')} &\Rightarrow -xf_x(x,y') = f(x,y').\end{aligned}$$

Thus, it should fulfill

$$(xf(x,y'))_x = 0$$

and we obtain that $f(x,y') = \frac{F(y')}{x}$ where g is an arbitrary function. Finally,

$$y'' = \frac{F(y')}{x}$$

is the most general second order ODE that admits both symmetries.

Problem 5.3.19 *Solve the PDE*

$$x\frac{\partial u}{\partial x} + y\frac{\partial u}{\partial y} + z^2(x-3y)\frac{\partial u}{\partial z} = 0,$$

with a boundary condition

$$u(x,y) = \frac{x^2}{y} \quad by \quad 3xz = 1.$$

Solution As usual we need to start the solution of the problem from the characteristic system

$$\frac{dx}{x} = \frac{dy}{y} = \frac{dz}{z^2(x-3y)}.$$

Solving the first equation we obtain

$$x = C_1 y, \quad C_1 = \text{const.} \tag{5.3.21}$$

The second equation takes the form

$$\begin{aligned}\frac{dy}{y} &= \frac{dz}{z^2(x-3y)} \\ \Rightarrow ydz - z^2(x-3y)dy &= 0 \\ \Rightarrow ydz - z^2(C_1-3)ydy &= 0.\end{aligned}$$

Integrating this equation we get the second invariant and the general solution for the initial PDE

$$\begin{aligned}C_2 &= \frac{1}{z} + (C_1-3)y = \frac{1}{z} + x - 3y \Rightarrow \\ u(x,y) &= f(C_1, C_2) = f\left(\frac{x}{y}, \frac{1}{z} + x - 3y\right).\end{aligned}$$

Now, we need to express x, y and z through C_1 and C_2 using boundary conditions. It means we have to solve the following system of usual algebraic equations

$$3xz = 1, \quad C_1 = \frac{x}{y}, \quad C_2 = \frac{1}{z} + x - 3y.$$

After the simple calculations, we obtain

$$x = \frac{C_1C_2}{4C_1-3}, \quad y = \frac{C_2}{4C_1-3}, \quad z = \frac{4C_1-3}{3C_1C_2}.$$

Substituting expressions for x and y into the boundary condition we get the formula for u

$$u(x,y) = \frac{x^2}{y} = \frac{\left(\frac{C_1 C_2}{4C_1 - 3}\right)^2}{\frac{C_2}{4C_1 - 3}} = \frac{{C_1}^2 C_2}{4C_1 - 3}.$$

Recalling the expressions for C_1 and C_2 and simplifying the last fraction, we obtain the final expression for the solution

$$u(x,y) = \frac{x^2 + x^3 z - 3x^2 yz}{4xyz - 3y^2 z}.$$

5.4 Exercises

1. Find the first prolongation of the infinitesimal generator of the linear fractional transformation group

$$\mathbf{U} = x^2 \frac{\partial}{\partial x} + y^2 \frac{\partial}{\partial y}.$$

2. Find the second prolongation of the following infinitesimal generator

$$\mathbf{U} = (x-y)\frac{\partial}{\partial x} + (x+y)\frac{\partial}{\partial y}.$$

3. Find the differential invariant of the second order for the group with the infinitesimal generator

$$\mathbf{U} = x\frac{\partial}{\partial x} - 2\frac{\partial}{\partial y}.$$

   Prove that this differential invariant can be obtained by a differentiation of the first differential invariant.

Chapter 6

Short overview of the Lie algebra properties

As we have mentioned before, this particular book is mostly written for beginners who want to get a general understanding of the theory and learn how they can apply it to their everyday tasks. It is not only impossible to describe all the properties of Lie algebras in this book but also not necessary. Therefore, for the convenience of readers, in this chapter we provide the most important definitions and features of Lie algebras which are used in the next chapters, yet omit a lot of details that are interesting but not important for our purposes. A more fundamental description of the theory, if need be, can be found in [59], [60], [40], [39], [61] or [75].

In the previous chapters we have already seen that the closed forms of point transformations of a group $G(\mathbb{R}^2)$ are complicated expressions that are usually nonlinear. On the other hand we have seen that we can describe the group action locally by a linear differential operator $\mathbf{U}$. It is much easier to study the properties of the linear operators than the properties of nonlinear point transformations. Sophus Lie first had an idea to replace the investigations of properties of nonlinear point transformations with the study of properties of linear operators which locally represent the same transformations. This idea was and is very fruitful not only in applications to the theory of PDEs. Sophus Lie [51] has proved that if we find r infinitesimal generators $\mathbf{U}_i, i = 1, 2, \dots, r$ then their linear span is closed under commutation and they build an algebra L_r. This algebra generates r-parameter symmetry group of the studied equation.

First of all in order to apply the Lie group analysis to a differential equation we need to determine the Lie algebra and correspondingly the Lie group admitted by this equation. We need to know the internal structure of the algebra to use these symmetries in the most appropriate way. We get different possibilities to reduce this equation to a simpler one depending on the algebraic structure of the admitted Lie symmetry algebra. That is why we obtain different classes of invariant solutions for the original equation. In some cases we are even able to find the fundamental solution of this equation or to solve it explicitly. The idea to use the internal algebraic structure of the admitted Lie algebra was first introduced by Sophus Lie [53] and was developed further, for instance, in [60], [59] and [61].

In Section 6.1 we introduce the general idea of a linear space and give some typical examples of such spaces. Then in Section 6.2 we define a bilinear operation with certain properties on a linear space. This operation in fact defines the Lie algebraic structure on the linear space. In Section 6.3 we provide the classifications of two-, three-, and four–dimensional real solvable Lie algebras often arising in applications.

6.1 Linear space of differential operators

In the previous chapters we mostly worked with a one-parameter group $G(\mathbb{R}^2)$ of point transformations on the plane. The action of this group was locally represented by an infinitesimal generator of the group, i.e. by a linear differential operator $\mathbf{U}$. Some of differential equations can admit many one-parameter groups of point transformations, so we obtain a set of linear differential operators $\mathbf{U}_i$, $i = 1, 2, \ldots, r$. To understand the structure of this set we need to introduce the ideas of a linear space and an algebra.

Let us look at a set L of all linear differential operators of the form

$$\mathbf{U}_i = \xi_i \frac{\partial}{\partial x} + \eta_i \frac{\partial}{\partial y}, \quad i = 1, 2, \ldots. \tag{6.1.1}$$

All properties of a linear space are fulfilled. In fact

1. $\mathbf{U}_i + \mathbf{U}_j = (\xi_i + \xi_j)\frac{\partial}{\partial x} + (\eta_i + \eta_j)\frac{\partial}{\partial y} \in L$,

2. $\alpha \in \mathbb{R} \Rightarrow \alpha \mathbf{U}_i \in L$,

3. the neutral element $0 \in L$: $\mathbf{U}_i + 0 = 0 + \mathbf{U}_i = \mathbf{U}_i$.

The addition of the elements of L is commutative and associative

$$\mathbf{U}_i + \mathbf{U}_j = \mathbf{U}_j + \mathbf{U}_i, \qquad \mathbf{U}_i + (\mathbf{U}_j + \mathbf{U}_k) = (\mathbf{U}_i + \mathbf{U}_j) + \mathbf{U}_k,$$

and the multiplication with a real number $\alpha \in \mathbb{R}$ yields

$$\begin{aligned}(\alpha + \beta)\mathbf{U} &= \alpha \mathbf{U} + \beta \mathbf{U} \in L, \\ (\alpha\beta)\mathbf{U} &= \alpha\,(\beta \mathbf{U}) \in L, \\ \alpha\,(\mathbf{U}_i + \mathbf{U}_j) &= \alpha \mathbf{U}_i + \alpha \mathbf{U}_j \in L.\end{aligned}$$

The described set L is called a linear space or a vector space.

It is just one example of linear spaces. The well known linear space of $(n \times n)$-matrices or the space of continuous functions $f(x), g(x) \in C[a,b]$ corresponding to the usual addition operation are other examples of linear spaces.

On a vector space we can introduce one or more additional structures. The vector space endowed with an additional structure will be called correspondingly to this structure. There are many books on functional analysis where the theory of such spaces is studied, for the beginners we can suggest for instance [68]. In Example 6.1.1 we look at the most familiar representatives of linear spaces with an additional structure.

Example 6.1.1

1. If we introduce the measure of distance between two elements of a linear space we get a metric space. For instance, when the elements of the linear space $L = C[a,b]$ are continuous functions on an interval $[a,b]$, i.e. $f(x), g(x) \in C[a,b]$ we can introduce a distance between two functions in the form

$$\mu_{f,g} = \max_{x \in [a,b]} |f(x) - g(x)|,$$

   and obtain *a metric space.*

2. If we introduce a scalar product $(\cdot,\cdot)$ on the same linear space defined as

$$(f,g) = \int f(x)g(x)\,dx,$$

   we get a *pre-Hilbert space.*

3. If we take a linear space of n–dimensional vectors $c, d \in \mathbb{R}^n$ and introduce a scalar product after

$$a \in \mathbb{R}^n, b \in \mathbb{R}^n : \ (a,b) = \sum_{i=1}^{n} a_i b_i.$$

   we get *a Hilbert space.*

4. If we introduce a bilinear operation with special properties we get an algebraic structure and the space will be called *algebra*, for instance an algebra of linear differential operators. Let us discuss this particular case in detail in Section 6.2.

◇

6.2 Introduction to Lie algebras

Now let the underlying vector space be the space of linear differential operators. Suppose we have a finite set of linear differential operators $\mathbf{U}_i, \ i = 1, 2, \ldots, r$. All linear combinations of the operators from the set create a linear space L, i.e. we build a linear span $L = \langle \mathbf{U}_1, \mathbf{U}_2, \ldots, \mathbf{U}_r \rangle$. Now we enrich this space and introduce a bilinear operation on elements of the linear space L.

We introduce a bilinear operation $\circ$ later denoted by $[\cdot, \cdot]$

$$\mathbf{U}_i, \mathbf{U}_j \in L : \quad \mathbf{U}_i \circ \mathbf{U}_j \equiv [\mathbf{U}_i, \mathbf{U}_j], \quad i, j = 1, 2, \ldots, r.$$

This operation should have the following properties for any element of the linear space $\mathbf{U}_1, \mathbf{U}_2, \mathbf{U}_3 \in L$

1. $(\alpha_1 \mathbf{U}_1 + \alpha_2 \mathbf{U}_2) \circ \mathbf{U}_3 = \alpha_1 (\mathbf{U}_1 \circ \mathbf{U}_3) + \alpha_2 (\mathbf{U}_2 \circ \mathbf{U}_3)$,
   $\mathbf{U}_1 \circ (\beta_1 \mathbf{U}_2 + \beta_2 \mathbf{U}_3) = \beta_1 (\mathbf{U}_1 \circ \mathbf{U}_2) + \beta_2 (\mathbf{U}_1 \circ \mathbf{U}_3)$,
2. (skew-symmetry) $\mathbf{U}_1 \circ \mathbf{U}_2 = -\mathbf{U}_2 \circ \mathbf{U}_1$,
3. (Jacobi identity) $\mathbf{U}_1 \circ (\mathbf{U}_2 \circ \mathbf{U}_3) + \mathbf{U}_2 \circ (\mathbf{U}_3 \circ \mathbf{U}_1) + \mathbf{U}_3 \circ (\mathbf{U}_1 \circ \mathbf{U}_2) = 0$.

In general, if a bilinear operation with the properties as described above is introduced on a linear space (also called vector space), we obtain an algebra, in this particular case a Lie algebra. The dimension of the algebra is defined by the dimension of the underlying linear space L.

Let P be a field of real ($P = \mathbb{R}$) or complex ($P = \mathbb{C}$) numbers and L be a finite-dimensional linear space with $\dim L = k$. We introduce now a definition for a Lie algebra $\mathfrak{G}$.

Definition 6.2.1 (Lie Algebra) *A Lie algebra over P is a vector (linear) space $\mathfrak{G}$ with a bilinear operation*

$$[\cdot, \cdot] : \mathfrak{G} \times \mathfrak{G} \to \mathfrak{G}. \tag{6.2.2}$$

The operation $[\cdot, \cdot]$ is called the Lie bracket for $\mathfrak{G}$ and satisfies the axioms

1. *bilinearity*

$$[c\mathbf{V} + \tilde{c}\tilde{\mathbf{V}}, \mathbf{W}] = c[\mathbf{V}, \mathbf{W}] + \tilde{c}[\tilde{\mathbf{V}}, \mathbf{W}],$$
$$[\mathbf{V}, c\mathbf{W} + \tilde{c}\tilde{\mathbf{W}}] = c[\mathbf{V}, \mathbf{W}] + c[\mathbf{V}, \tilde{\mathbf{W}}],$$

2. *skew-symmetry*

$$[\mathbf{V}, \mathbf{W}] = -[\mathbf{W}, \mathbf{V}],$$

3. *Jacobi-identity*

$$[\mathbf{U}, [\mathbf{V}, \mathbf{W}]] + [\mathbf{V}, [\mathbf{W}, \mathbf{U}]] + [\mathbf{W}, [\mathbf{U}, \mathbf{V}]] = 0,$$

for all elements $\mathbf{V}, \tilde{\mathbf{V}}, \mathbf{W}, \tilde{\mathbf{W}} \in \mathfrak{G}$ *and for all* $c, \tilde{c} \in P$.

If $P = \mathbb{R}$, *then* $\mathfrak{G}$ *is called a real Lie algebra, if* $P = \mathbb{C}$ *it is called a complex Lie algebra.*

Let us now introduce the most important features and properties of Lie algebras, which we will use later in this book.

Definition 6.2.2 (Abelian algebra) *A Lie algebra* $\mathfrak{G}$ *is called Abelian (or commutative) algebra, if*

$$[\mathbf{V}, \mathbf{W}] = 0 \quad \text{for all} \quad \mathbf{V}, \mathbf{W} \in \mathfrak{G}. \tag{6.2.3}$$

Example 6.2.1 We can take the space of all real $n \times n$-matrices as a linear space $\mathfrak{G}$ and the usual matrix commutator as a bilinear operation (called Lie commutator or Lie bracket). Then this vector space, endowed with the new operation, is called a general real Lie algebra. The notation for the general real Lie algebra is $gl(n, \mathbb{R})$. Similarly, a general Lie algebra over a field P is denoted as $gl(n, P)$. Further we have $\dim(gl(n, \mathbb{R})) = n^2$ and $gl(n, \mathbb{R}) \approx \mathbb{R}^{n^2}$.
◇

Example 6.2.2 Let us consider the space $\mathfrak{G}$ as a space of a skew-symmetric $n \times n$-matrices ($A^T = -A$) over the field P, then $[A, B]$ is skew-symmetric as well and this set of matrices is a subspace (closed under the operation $[\cdot, \cdot]$) of the space of all $n \times n$-matrices over the field P. The Lie algebra of all skew-symmetric matrices is called orthogonal Lie algebra and is denoted by $o(n, P)$.
◇

In the examples we realized the abstract algebraic structure on the linear space of $n \times n$-matrices. The same algebraic structure can be represented also on a vector space of linear differential operators or other vectors spaces. This means, we can obtain different representations of the same algebraic structure. Our goal is to describe the applications of Lie group analysis to ODEs and PDEs. For our applications the space of linear differential operators is the most important one. We use mostly $P = \mathbb{R}$ as an underlying field; however, in some cases we work with $P = \mathbb{C}$.

Now let us introduce the definition of a Lie algebra whose elements belong to the space of linear differential operators.

Definition 6.2.3 (Lie algebra) *A Lie algebra $\mathfrak{G}$ is a vector space of linear differential operators*

$$\mathbf{U} = \sum_{i=1}^{n} a_i(x_1, x_2, \ldots, x_n)\frac{\partial}{\partial x_i} \tag{6.2.4}$$

with the following property:
If $\mathbf{U}_1 = \sum_{i=1}^{n} a_i \frac{\partial}{\partial x_i}$ and $\mathbf{U}_2 = \sum_{i=1}^{n} b_i \frac{\partial}{\partial x_i}$ are elements of $\mathfrak{G}$, then their commutator is equal to

$$[\mathbf{U}_1, \mathbf{U}_2] = \mathbf{U}_1\mathbf{U}_2 - \mathbf{U}_2\mathbf{U}_1 = \sum_{i=1}^{n} (\mathbf{U}_1(b_i) - \mathbf{U}_2(a_i))\frac{\partial}{\partial x_i}.$$

The commutator is also an element of $\mathfrak{G}$ and defines a corresponding Lie bracket.

The dimension of the Lie algebra $\mathfrak{G}$ ($\dim \mathfrak{G}$) is the dimension of the vector space L. If $\dim \mathfrak{G} = r$ we denote the r–dimensional Lie algebra as $\mathfrak{G}_r$.

The bilinearity, skew-symmetry and Jacobi identity can be easily proved.
If the linear space L is finite–dimensional, we can choose a basis in this space

$$\{\mathbf{U}_\alpha\}, \ \alpha = 1, \ldots, r, \ \dim L = r.$$

Then any $\mathbf{U} \in L$ can be represented as a linear combination of basis elements

$$\mathbf{U} = \sum_{\alpha=1}^{r} c^\alpha \mathbf{U}_\alpha, \tag{6.2.5}$$

where $c^\alpha \in P$ are constants.

Example 6.2.3 Let $\dim L = r = 2$, $\mathbf{U}_1 = \frac{\partial}{\partial x}$, $\mathbf{U}_2 = x\frac{\partial}{\partial y}$ and L is the linear span of $\mathbf{U}_1$ and $\mathbf{U}_2$: $L = \langle \mathbf{U}_1, \mathbf{U}_2 \rangle$. Then any element of this space can be represented as a linear combination of the basis elements

$$\mathbf{U} = c_1\mathbf{U}_1 + c_2\mathbf{U}_2.$$

Now we can calculate the commutator

$$\left[\frac{\partial}{\partial x}, x\frac{\partial}{\partial y}\right] = \frac{\partial}{\partial y}.$$

We see that L is a linear space, but not a Lie algebra, because the commutator of two basis operators is the operator $\frac{\partial}{\partial y}$ and it cannot be represented as a linear combination of operators $\mathbf{U}_1, \mathbf{U}_2$. ◇

Example 6.2.4 If we enlarge the linear space L from the previous example with the element $\mathbf{U}_3 = \frac{\partial}{\partial y}$ and therefore obtain the three-dimensional space L as a linear span of three operators $L = \langle \mathbf{U}_1, \mathbf{U}_2, \mathbf{U}_3 \rangle$ (correspondingly $\dim L = r = 3$), then we obtain a Lie algebra. We can calculate

$$[\mathbf{U}_1, \mathbf{U}_3] = \left[\frac{\partial}{\partial x}, \frac{\partial}{\partial y}\right] = 0$$
$$[\mathbf{U}_2, \mathbf{U}_3] = \left[x\frac{\partial}{\partial y}, \frac{\partial}{\partial y}\right] = 0$$
$$[\mathbf{U}_1, \mathbf{U}_2] = \mathbf{U}_3,$$

we see that all commutators are elements of the same space. ◇

If the linear space L_r is a Lie algebra $\mathfrak{G}_r$ with the Lie brackets $[\cdot, \cdot]$, then $[\mathbf{U}_\alpha, \mathbf{U}_\beta] \in \mathfrak{G}_r = L_r$ for all $\alpha, \beta = 1, \ldots, r$. We can then present each commutator as a linear combination of basis elements of the space. The constant coefficients in the linear combinations have special properties and are called *structure constants.*

Definition 6.2.4 (Structure constants) *The constants $c^\gamma_{\alpha\beta}$, $\alpha, \beta, \gamma = 1, \ldots, r$ are defined by*

$$[\mathbf{U}_\alpha, \mathbf{U}_\beta] = \sum_{\gamma=1}^{r} c^\gamma_{\alpha\beta}\mathbf{U}_\gamma, \quad \alpha, \beta = 1, \ldots, r. \tag{6.2.6}$$

and are called the structure constants of $\mathfrak{G}_r$.

Remark 6.2.1 *If the basis of algebra and the corresponding structure constants are known, we can recover the Lie algebra completely using (6.2.6) and the linearity of the Lie bracket.*

Example 6.2.5 Let $\mathbf{U}_1 = \frac{\partial}{\partial x}$, $\mathbf{U}_2 = \frac{\partial}{\partial y}$ be basis elements of the linear space L and therefore the linear span is a span $L = \langle \mathbf{U}_1, \mathbf{U}_2 \rangle$. Then any element $\mathbf{U}$ of L can be represented as a linear combination of these elements

$$\mathbf{U} \subset \langle \mathbf{U}_1, \mathbf{U}_2 \rangle, \quad \mathbf{U} = c_1\frac{\partial}{\partial x} + c_2\frac{\partial}{\partial y}.$$

The vector space L is an algebra $\mathfrak{G}$ if we introduce the Lie bracket as the commutator relations between two operators

$$\left[\frac{\partial}{\partial x}, \frac{\partial}{\partial y}\right] = 0 \quad \Rightarrow \quad c^1_{12} = c^2_{12} = 0,$$
$$\left[\frac{\partial}{\partial x}, \frac{\partial}{\partial x}\right] = 0 \quad \Rightarrow \quad c^1_{11} = c^2_{11} = 0,$$
$$\left[\frac{\partial}{\partial y}, \frac{\partial}{\partial y}\right] = 0 \quad \Rightarrow \quad c^1_{22} = c^2_{22} = 0.$$

Thus, the Lie algebra $\mathfrak{G}$ is a 2–dimensional real Abelian Lie algebra and all structure constants are just zero $c_{ij}^k = 0$, $i, j, k = 1, 2$. ◇

Theorem 6.2.1 *The structure constants of a r–dimensional Lie algebra $\mathfrak{G}_r$ have the following properties for all $\alpha, \beta, \gamma, \tau = 1, \ldots, r$:*

1. *skew-symmetry*
$$c_{\alpha\beta}^{\gamma} = -c_{\beta\alpha}^{\gamma},$$
2. *Jacobi-identity*
$$\sum_{\delta=1}^{r} \left(c_{\alpha\beta}^{\delta} c_{\gamma\delta}^{\tau} + c_{\beta\gamma}^{\delta} c_{\alpha\delta}^{\tau} + c_{\gamma\alpha}^{\delta} c_{\beta\delta}^{\tau} \right) = 0,$$
3. *if we choose a new basis*
$$\hat{V}_\alpha = \sum_{\beta=1}^{r} a_{\alpha\beta} V_\beta$$
of $\mathfrak{G}$, then
$$\hat{c}_{\alpha\beta}^{\gamma} = \sum_{\delta,\tau,\sigma=1}^{r} a_{\alpha\delta} a_{\beta\tau} b_{\sigma\gamma} c_{\delta\tau}^{\sigma},$$
where the matrix $(b_{ij}) = (a_{ij})^{-1}$.

Corollary 6.2.1 *The set of structure constants determines the same Lie algebra if and only if they are related by the properties 1 to 3 in Theorem 6.2.1.*

Since for many applications of Lie group analysis to differential equations the choice of a system of subalgebras plays an important role, let us introduce the idea of subalgebra and give some information on the related properties of different subalgebras.

Definition 6.2.5 (Subalgebra) *Let a linear space L be a Lie algebra $\mathfrak{G}$ with the Lie bracket $[\cdot,\cdot]$. Let $K \subset L$ be a subspace (if $K_1, K_2 \in K \;\Rightarrow\; \langle K_1, K_2 \rangle \subset K$, i.e., K is closed under the linear operation "+"). If*

$$K_1, K_2 \in K \quad \Rightarrow \quad [K_1, K_2] \subset K, \tag{6.2.7}$$

i.e. K is closed under commutation, then K is a subalgebra of $\mathfrak{G}$. As a shortcut we write $[K, K] \subset K$.

Definition 6.2.6 (Ideal) *The subalgebra K is called an ideal of $\mathfrak{G}$ if*

$$[K, L] \subset K. \tag{6.2.8}$$

Lemma 6.2.1 *Any Lie algebra contains a one–dimensional subalgebra.*

Proof: Let $\mathbf{V} \in L$, then $K = \langle \mathbf{V} \rangle = \{c\mathbf{V} | c \in \mathbb{R}\}$. We can calculate $[c_1\mathbf{V}, c_2\mathbf{V}] = 0$ and thus K is a subalgebra. □

Theorem 6.2.2 *Any Lie algebra $\mathfrak{G}_r$, $r > 2$, contains a two–dimensional subalgebra. Given an arbitrary element $\mathbf{V}_1 \in \mathfrak{G}_r$, there exists an element $\mathbf{V}_2 \in \mathfrak{G}_r$, such that*

$$[\mathbf{V}_1, \mathbf{V}_2] = \alpha_1 \mathbf{V}_1 + \alpha_2 \mathbf{V}_2,$$

but in general case $\alpha_1, \alpha_2 \in \mathbb{C}$.

Remark 6.2.2 *In the cases where we have vector spaces of linear differential operators and Lie algebras with such elements we will denote the space and the algebra with the same letter L. The notation $\mathfrak{G}$ is used mostly for a Lie algebra with elements of an arbitrary type, for instance matrices.*

Example 6.2.6 Let $\mathbf{U}_1 = \frac{\partial}{\partial y}$, $\mathbf{U}_2 = y\frac{\partial}{\partial y}$, $\mathbf{U}_3 = y^2\frac{\partial}{\partial y}$ and L is a linear span of these operators hence

$$L = \left\langle \frac{\partial}{\partial y}, y\frac{\partial}{\partial y}, y^2\frac{\partial}{\partial y} \right\rangle.$$

We prove that L is closed under commutation operation and is an algebra $\mathfrak{G}_3$. We calculate all commutators of operators $\mathbf{U}_1, \mathbf{U}_2, \mathbf{U}_3$ and obtain

$$\left[\frac{\partial}{\partial y}, y\frac{\partial}{\partial y}\right] = \frac{\partial}{\partial y},$$
$$\left[\frac{\partial}{\partial y}, y^2\frac{\partial}{\partial y}\right] = 2y\frac{\partial}{\partial y},$$
$$\left[y\frac{\partial}{\partial y}, y^2\frac{\partial}{\partial y}\right] = 2y^2\frac{\partial}{\partial y} - y^2\frac{\partial}{\partial y} = y^2\frac{\partial}{\partial y}.$$

We collect the results in the following commutator table:

	$\mathbf{U}_1$	$\mathbf{U}_2$	$\mathbf{U}_3$
$\mathbf{U}_1$	0	$\mathbf{U}_1$	$2\mathbf{U}_2$
$\mathbf{U}_2$	$-\mathbf{U}_1$	0	$\mathbf{U}_3$
$\mathbf{U}_3$	$-2\mathbf{U}_2$	$-\mathbf{U}_3$	0

In this table in the first column and the first row we list the basis of operators. Where the i-th row crosses the j-th column stands the value of the commutator $[\mathbf{U}_i, \mathbf{U}_j]$. Because of the skew-symmetry property of the commutator relation $[\mathbf{U}_i, \mathbf{U}_j] = -[\mathbf{U}_j, \mathbf{U}_i]$ we obtain for any Lie algebra a skew-symmetric commutator table.

We can easily see that in this example $K = \langle \mathbf{U}_1, \mathbf{U}_2 \rangle$ and $M = \langle \mathbf{U}_2, \mathbf{U}_3 \rangle$ are two–dimensional subalgebras, but there are no ideals in the three–dimensional Lie algebra L_3. ◇

Example 6.2.7 We consider four operators $\mathbf{U}_1 = \frac{\partial}{\partial x}, \mathbf{U}_2 = x\frac{\partial}{\partial x}, \mathbf{U}_3 = \frac{\partial}{\partial y}, \mathbf{U}_4 = y\frac{\partial}{\partial y}$ and a linear span of them $L = \langle \mathbf{U}_1, \mathbf{U}_2, \mathbf{U}_3, \mathbf{U}_4 \rangle$. The commutator table is

	$\mathbf{U}_1$	$\mathbf{U}_2$	$\mathbf{U}_3$	$\mathbf{U}_4$
$\mathbf{U}_1$	0	$\mathbf{U}_1$	0	0
$\mathbf{U}_2$	$-\mathbf{U}_1$	0	0	0
$\mathbf{U}_3$	0	0	0	$\mathbf{U}_3$
$\mathbf{U}_4$	0	0	$-\mathbf{U}_3$	0

We see that L is closed under commutation and build an algebra. It is easy to see that $K = \langle \mathbf{U}_1, \mathbf{U}_3 \rangle$ is an ideal of this algebra L_4. ◇

Theorem 6.2.3 *Let $\mathfrak{G}$ be a Lie algebra and K, M ideals of $\mathfrak{G}$, then $[K, M]$ is also an ideal of $\mathfrak{G}$.*

Proof: Let $\mathbf{V} \in [K, M]$. Then we have

$$\mathbf{V} = \sum_{i=1}^{n} [\mathbf{U}_i, \mathbf{W}_i],$$

where $\mathbf{U}_i \in K$ and $\mathbf{W}_i \in M$. If $\tilde{\mathbf{V}}$ is an element of $\mathfrak{G}$, then the Jacobi identity yields

$$\begin{aligned}
\left[\tilde{\mathbf{V}}, \sum_{i=1}^{n} [\mathbf{U}_i, \mathbf{W}_i]\right] &= \sum_{i=1}^{n} \left[\tilde{\mathbf{V}}, [\mathbf{U}_i, \mathbf{W}_i]\right] \\
&= -\sum_{i=1}^{n} \left[\mathbf{U}_i, [\mathbf{W}_i, \tilde{\mathbf{V}}]\right] - \sum_{i=1}^{n} \left[\mathbf{W}_i, [\tilde{\mathbf{V}}, \mathbf{U}_i]\right] \\
&= \sum_{i=1}^{n} \Big[\mathbf{U}_i, \underbrace{[\tilde{\mathbf{V}}, \mathbf{W}_i]}_{\in M}\Big] + \sum_{i=1}^{n} \Big[\underbrace{[\tilde{\mathbf{V}}, \mathbf{U}_i]}_{\in K}, \mathbf{W}_i\Big].
\end{aligned}$$

Because K and M are ideals, the right side of this equation is an element of $[K, M]$. □

Definition 6.2.7 (Derived algebra) *Let L be a Lie algebra. The linear span of all commutators*

$$\langle[\mathbf{U}_i, \mathbf{U}_j]\rangle, \quad \mathbf{U}_i, \mathbf{U}_j \in L \tag{6.2.9}$$

forms a vector space and an algebra. This algebra is denoted by

$$L^{(1)} = [L, L] \tag{6.2.10}$$

and is called the derived algebra (or first derived algebra) of the Lie algebra L. (Further we often denote the linear space and the corresponding algebra with the same letter L if it is a linear space of differential operators).

Corollary 6.2.2 *$L^{(1)}$ is an ideal of L.*

The recursion of this construction gives us higher derived algebras

$$L^{(s+1)} = \left(L^{(s)}\right)^{(1)} = [L^{(s)}, L^{(s)}], \quad s = 1, 2, \ldots \tag{6.2.11}$$

using the idea of derived algebra we can introduce the following definition of an Abelian algebra.

Definition 6.2.8 (Abelian algebra) *The Lie algebra L is Abelian if*

$$L^{(1)} = 0, \tag{6.2.12}$$

i.e., all elements commute.

Example 6.2.8 Let we have a linear span of four operators $\mathbf{U}_1 = \frac{\partial}{\partial x}$, $\mathbf{U}_2 = x\frac{\partial}{\partial x}$, $\mathbf{U}_3 = \frac{\partial}{\partial y}$ and $\mathbf{U}_4 = y\frac{\partial}{\partial y}$

$$L = \left\langle \frac{\partial}{\partial x}, x\frac{\partial}{\partial x}, \frac{\partial}{\partial y}, y\frac{\partial}{\partial y} \right\rangle.$$

We calculate the first derived algebra and obtain

$$L^{(1)} = \langle \mathbf{U}_1, \mathbf{U}_3 \rangle = \left\langle \frac{\partial}{\partial x}, \frac{\partial}{\partial y} \right\rangle$$

and the second derived algebra is

$$L^{(2)} = 0, \quad \text{because} \quad \left[\frac{\partial}{\partial x}, \frac{\partial}{\partial y}\right] = 0.$$

◇

Example 6.2.9 We examine the algebra which describes the rotations in $\mathbb{R}^3$. The infinitesimal generators of corresponding rotations around three axes in $\mathbb{R}^3$ are

$$\mathbf{U}_1 = -y\frac{\partial}{\partial x} + x\frac{\partial}{\partial y}, \quad \mathbf{U}_2 = y\frac{\partial}{\partial z} - z\frac{\partial}{\partial y}, \quad \mathbf{U}_3 = -z\frac{\partial}{\partial x} + x\frac{\partial}{\partial z}.$$

The commutators are equal to

$$\begin{aligned}
[\mathbf{U}_1, \mathbf{U}_2] &= x\frac{\partial}{\partial z} - z\frac{\partial}{\partial x} = \mathbf{U}_3, \\
[\mathbf{U}_2, \mathbf{U}_3] &= \mathbf{U}_1, \\
[\mathbf{U}_1, \mathbf{U}_3] &= -\mathbf{U}_2
\end{aligned}$$

and we obtain that $L^{(1)} = L$. ◇

As we have seen, the derived algebra $L_r^{(1)}$ can be defined easily by the table of commutators. We just need to take the linear span of all operators "inside" the commutator table.

The next idea may be the most important one for the applications to differential equations.

Definition 6.2.9 (Solvable Lie algebra) *A Lie algebra L_r, $r < \infty$, is said to be solvable if there is a sequence*

$$L_r \supset L_{r-1} \supset \ldots \supset L_1 \tag{6.2.13}$$

of subalgebras of respective dimension $r, r-1, \ldots, 1$, such that L_k is an ideal in L_{k+1}, $k = 1, \ldots, r-1$.

Sophus Lie called such algebras solvable because the differential equations that admit such algebra as a symmetry algebra can be solved with a clear step-by-step procedure. In the next chapters we will discuss how one can use solvability to find a solution of different ODEs and PDEs.

Theorem 6.2.4 *A Lie algebra L_r is solvable if and only if its derived algebra of a finite order s vanishes*

$$L_r^{(s)} = 0, \quad 0 < s < \infty. \tag{6.2.14}$$

Proof: Let $L_r^{(s)} = 0$. Then there exists a chain of ideals

$$L_r \supset L_r^{(1)} \supset L_r^{(2)} \supset \ldots \supset L_r^{(s)} = 0$$

and $L_r^{(k)} \subset L_r^{(k-1)}$. All subalgebras (and also subspaces) $L_r^{(k)}$ are ideals of $L_r^{(k-1)}$. On the other hand we can add an element of $L_r^{(k)}$ to $L_r^{(k-1)}$ and obtain an ideal one dimension higher than $\dim L_r^{(k-1)}$. If we continue this procedure we fill out all gaps in the chain of ideals. □

Corollary 6.2.3 *For solvable Lie algebras we have*

$$L_r^{(r)} = \left(L_r^{(1)}\right)^{(r-1)} \subset L_{r-1}^{(r-1)} \subset \ldots \subset L_1^{(1)} = 0. \tag{6.2.15}$$

In the next example we demonstrate the procedure finding the chain of ideals.

Example 6.2.10 Let us have a six-dimensional linear space spanned by infinitesimal generators

$$\mathbf{U}_1 = \frac{\partial}{\partial x}, \quad \mathbf{U}_2 = \frac{\partial}{\partial y}, \quad \mathbf{U}_3 = \frac{\partial}{\partial z}, \quad \mathbf{U}_4 = y\frac{\partial}{\partial z},$$
$$\mathbf{U}_5 = x\frac{\partial}{\partial x} + 3z\frac{\partial}{\partial z}, \quad \mathbf{U}_6 = y\frac{\partial}{\partial y} - 2z\frac{\partial}{\partial z}.$$

The commutator table is

	$\mathbf{U}_1$	$\mathbf{U}_2$	$\mathbf{U}_3$	$\mathbf{U}_4$	$\mathbf{U}_5$	$\mathbf{U}_6$
$\mathbf{U}_1$	0	0	0	0	$\mathbf{U}_1$	0
$\mathbf{U}_2$	0	0	0	$\mathbf{U}_3$	0	$\mathbf{U}_2$
$\mathbf{U}_3$	0	0	0	0	$3\mathbf{U}_3$	$-2\mathbf{U}_3$
$\mathbf{U}_4$	0	$-\mathbf{U}_3$	0	0	$3\mathbf{U}_4$	$-3\mathbf{U}_4$
$\mathbf{U}_5$	$-\mathbf{U}_1$	0	$-3\mathbf{U}_3$	$-3\mathbf{U}_4$	0	0
$\mathbf{U}_6$	0	$-\mathbf{U}_2$	$2\mathbf{U}_3$	$3\mathbf{U}_4$	0	0

From this table we can see directly that

$$\begin{aligned} &L_6^{(1)} = \langle \mathbf{U}_1, \mathbf{U}_2, \mathbf{U}_3, \mathbf{U}_4 \rangle && \dim L_6^{(1)} = 4, \\ &L_6^{(2)} = \langle \mathbf{U}_3 \rangle && \dim L_6^{(2)} = 1, \\ &L_6^{(3)} = 0 \end{aligned}$$

and get the following relations

$$L_5 \supset L_6^{(1)} \supset L_6^{(2)} \supset L_6^{(3)} = 0.$$

Now we can fill out the gaps in this chain of ideals and define

$$\begin{aligned} &L_5 = \langle \mathbf{U}_1, \mathbf{U}_2, \mathbf{U}_3, \mathbf{U}_4, \mathbf{U}_5 \rangle && \text{ideal in } L_6, \\ &L_4 = \langle \mathbf{U}_1, \mathbf{U}_2, \mathbf{U}_3, \mathbf{U}_4 \rangle = L_6^{(1)} && \text{ideal in } L_5, \\ &L_3 = \langle \mathbf{U}_1, \mathbf{U}_2, \mathbf{U}_3 \rangle && \text{ideal in } L_6^{(1)}, \\ &L_2 = \langle \mathbf{U}_2, \mathbf{U}_3 \rangle && \text{ideal in } L_3, \\ &L_1 = \langle \mathbf{U}_3 \rangle = L_6^{(2)} \end{aligned}$$

and obtain

$$L_6 \supset L_5 \supset L_4 \supset L_3 \supset L_2 \supset L_1.$$

It proves that we have a solvable Lie algebra L_6. ◇

Definition 6.2.10 (Central element, center) *The element* $\mathbf{V} \in L$ *is called a central element if*

$$[\mathbf{V}, \mathbf{U}] = 0, \qquad \forall \mathbf{U} \in L. \tag{6.2.16}$$

The union $\mathfrak{Z}$ *of all central elements is called the center of the Lie algebra* L.

Theorem 6.2.5 *The center* $\mathfrak{Z}$ *is an ideal in* L.

Proof: Let $z_1, z_2 \in \mathfrak{Z}$. Then $a_1 z_1 + a_2 z_2 \in \mathfrak{Z}$ and $[z_1, z_2] = 0$. Therefore, $\mathfrak{Z}$ is a subalgebra and an ideal. □

Definition 6.2.11 (Lower central series) *Lie's powers of the Lie algebra* L *are defined by*

$$L^{[2]} = [L, L], \quad L^{[3]} = [L^{[2]}, L], \quad L^{[k+1]} = [L^{[k]}, L]. \tag{6.2.17}$$

All algebras $L^{[k+1]}$ *are ideals in* L *(we remark that* $L^{[2]} = L^{(1)}$*). The series*

$$L \supset L^{[2]} \supset L^{[3]} \supset \ldots \supset L^{[n]} \supset \ldots \tag{6.2.18}$$

is called the lower central series of the Lie algebra L.

For a r*-dimensional Lie algebra* L*, its lower central series terminates either with the null-ideal* $L^{[n]} = 0$ *or with* $L^{[n]}$ *for which* $L^{[n-1]} \neq L^{[n]} = L^{[n+1]}$.

Example 6.2.11 We take the same generators $\mathbf{U}_1, \ldots, \mathbf{U}_6$, as in Example 6.2.10 and obtain Lie's powers of the Lie algebra L_6

$$L^{[2]} = L^{(1)} = \langle \mathbf{U}_1, \mathbf{U}_2, \mathbf{U}_3, \mathbf{U}_4 \rangle$$
$$\Rightarrow L^{[3]} = [L^{[2]}, L] = \langle \mathbf{U}_1, \mathbf{U}_2, \mathbf{U}_3, \mathbf{U}_4 \rangle .$$

We see that the series of Lie's powers terminates with $L^{[2]}$. ◇

Definition 6.2.12 (Nilpotent) *The Lie algebra* L *is called nilpotent if its lower central series ends with the null-ideal.*

Lemma 6.2.2 *Any nilpotent Lie algebra is solvable.*

Proof: With the Jacobi identity we get

$$[L^{[k]}, L^{[l]}] \subset L^{[k+l]}, \quad \forall k, l.$$

Hence

$$L^{(k)} \subset L^{[2^k]}, \quad \forall k = 1, 2, \ldots$$

and if L is nilpotent (that is $L^{[2^k]} = 0$ for some n) then $L^{(n)} = 0$. □

Definition 6.2.13 (Isomorphic Lie algebras) *Let L and M be two Lie algebras with* $\dim L = \dim M$. *A linear one-to-one map f of L onto M is called an isomorphism (automorphism if $L = M$) if it preserves commutators*

$$f([\mathbf{U}_1, \mathbf{U}_2]) = [f(\mathbf{U}_1), f(\mathbf{U}_2)], \quad \forall \mathbf{U}_1, \mathbf{U}_2 \in L. \tag{6.2.19}$$

If the Lie algebras L and M can be related by an isomorphism, they are called isomorphic Lie algebras.

Theorem 6.2.6 *Two finite–dimensional Lie algebras are isomorphic if and only if one can choose bases for the algebras, such that they have equal structure constants in these bases, i.e. the same table of commutators.*

We skip the proof of this and the next theorem as they can be easily found in one of the cited books.

Remark 6.2.3 *The concept of isomorphisms is independent of the realization of Lie algebras.*

Theorem 6.2.7 *(Ado, 1947 [2])*
Every finite–dimensional complex Lie algebra is isomorphic to some matrix algebra. Every finite-dimensional complex Lie algebra is isomorphic to a subalgebra of $gl(n)$ for some n.

For algebras of differential operators there exists a second way to be related to each other.

Definition 6.2.14 (Similar algebras) *The Lie algebras L_r and $\tilde{L}_r$ are similar if one is obtained from the other by a change of variables, i.e. if the elements of the algebras are*

$$\mathbf{U} = \sum_{i=1}^{n} a_i(x) \frac{\partial}{\partial x_i} \in L, \quad \tilde{\mathbf{U}} = \sum_{i=1}^{n} \tilde{a}_i(\tilde{x}) \frac{\partial}{\partial \tilde{x}_i} \in \tilde{L} \tag{6.2.20}$$

then being similar means that there exist a transformation $\tilde{x}_i = \tilde{x}_i(x)$, with

$$\tilde{a}_i(\tilde{x}) = \sum_{k=1}^{n} a_k(x) \frac{\partial \tilde{x}_i}{\partial x_k}. \tag{6.2.21}$$

Lemma 6.2.3 *Let L_r and $\tilde{L}_r$ be the Lie algebras of linear differential operators with the same number of variables. It is necessary that L_r and $\tilde{L}_r$ are isomorphic if they are similar.*

Proof: Let

$$[\mathbf{U}_\alpha, \mathbf{U}_\beta] = \sum_{\gamma=1}^{r} c_{\alpha\beta}^{\gamma} \mathbf{U}_\gamma$$

and

$$\mathbf{U}_\alpha = \sum_{i=1}^{n} a_\alpha^i \frac{\partial}{\partial x_i}, \quad \tilde{\mathbf{U}}_\alpha = \sum_{i=1}^{n} \tilde{a}_\alpha^i \frac{\partial}{\partial \tilde{x}_i}, \quad \alpha = 1, \ldots, r,$$

be bases, then

$$[\tilde{\mathbf{U}}_\alpha, \tilde{\mathbf{U}}_\beta] = \widetilde{[\mathbf{U}_\alpha, \mathbf{U}_\beta]} = \sum_{\gamma=1}^{r} \widetilde{c_{\alpha\beta}^{\gamma} \mathbf{U}_\gamma} = \sum_{\gamma=1}^{r} c_{\alpha\beta}^{r} \tilde{\mathbf{U}}_\gamma.$$

Because the algebras have the same structure constants $c_{\alpha\beta}^{\gamma}$ they are isomorphic. □

Remark 6.2.4 *We can provide such mnemonic table*

similar	$\Rightarrow$	*isomorphic*
isomorphic	$\not\Rightarrow$	*similar*

Definition 6.2.15 (Equivalent elements of algebra) *Let K be an ideal of L, $[K, L] \subset L$. We say that $\mathbf{U}_1$ and $\mathbf{U}_2$ are equivalent, $\mathbf{U}_1 \sim \mathbf{U}_2$, if $\mathbf{U}_1 - \mathbf{U}_2 \in K$.*

Definition 6.2.16 (Coset) *The collection of all elements equivalent to $\mathbf{U}$ with $\mathbf{U} \in L$ is called coset and is denoted by $\mathbf{U} + K$. It means that for any element $\mathbf{U}_i$ of the coset $\mathbf{U} + K$ we have*

$$\mathbf{U} + K \quad \Leftrightarrow \quad \mathbf{U}_i \in \mathbf{U} + K, \quad \text{if} \quad \mathbf{U}_i = \mathbf{U} + K_i, \ K_i \in K. \tag{6.2.22}$$

The family of pairwise disjoint cosets is endowed with an algebraic structure

$$\alpha_1 \mathbf{U}_1 + K + \alpha_2 \mathbf{U}_2 + K = (\alpha_1 \mathbf{U}_1 + \alpha_2 \mathbf{U}_2) + K,$$
$$[\mathbf{U}_1 + K, \mathbf{U}_2 + K] = [\mathbf{U}_1, \mathbf{U}_2] + \underbrace{[\mathbf{U}_1, K] + [K, \mathbf{U}_2] + [K, K]}_{\in K}.$$

Definition 6.2.17 (Quotient algebra) *The algebra of cosets is called the quotient algebra of the Lie algebra L by its ideal K*

$$L/K.$$

6.3 An optimal system of subalgebras

Later we will use the symmetry properties of differential equations to solve them or to show that there exist simpler reductions of them. To get reductions we usually use some subalgebras of the symmetry algebra. To be able to obtain the complete list of non-equivalent reductions we need some classification of subalgebras. The problem is how to choose the optimal system of subalgebras (and correspondingly subgroups) which give us non-conjugate invariant solutions. This problem was solved in 1982 by Ovsiannikov (see [60]). We follow the algorithm of determining an optimal system of subalgebras that is proposed there. We use the notation h_i^g for the subalgebras of the Lie algebra L and H_i^g for the corresponding subgroups of the group G.

Here we provide optimal systems of subalgebras for some solvable 3- and 4-dimensional real Lie algebras, which arise from the study of different PDEs. The classification of all solvable 2-, 3-, 4-dimensional real Lie algebras and corresponding optimal systems of subalgebras was provided in a series of papers by Patera and Winternitz (see [61] for further details). We give just a short overview of the results for special types of 3- and 4-dimensional Lie algebras which we will use later in our examples. Furthermore we remark that F. Oliveri provided a package **SymboLie** (a **Mathematica** package) [58] which calculates an optimal system of subalgebras of a given dimension for a finite-dimensional real Lie algebra.

We will use these optimal systems for reductions of original PDEs to different ODEs in Chapter 9. Solutions of ODEs are invariant solutions of the original PDEs. We notice that the provided optimal system of subalgebras are not unique, one can provide another optimal system of subalgebras for each case which will be equivalent to the first one. For our purposes it is sufficient to use any one of the optimal systems of subalgebras to obtain the complete set of non-equivalent invariant solutions. The form of reduced equations and corresponding solutions will certainly depend on the optimal system chosen. The reduced equations and their solutions which we obtain using different optimal systems can be transformed to each other using the admitted symmetry group. All optimal systems of subalgebras provided in this section are from [61].

In Chapter 9 we present several financial models and the admitted Lie symmetry algebra and a table of optimal systems of subalgebras for each of them. We can see that for many of the studied models the admitted algebras are isomorph and have correspondingly equivalent optimal systems of subalgebras. Yet the generators of algebras often have different forms, so we provide the tables of optimal systems of subalgebras for each model separately in order to make the structure more transparent and understandable.

All these systems are actually isomorph to the system given here.

Three-dimensional Lie algebra
Let a Lie algebra L_3 possess a non-zero commutator relation $[\mathbf{U}_1, \mathbf{U}_2] = -\mathbf{U}_2$. We suppose that the Lie algebra L_3 has a two–dimensional subalgebra $L_2 = \langle \mathbf{U}_1, \mathbf{U}_2 \rangle$ spanned by the generators $\mathbf{U}_1, \mathbf{U}_2$. The algebra L_3 is a decomposable Lie algebra and can be represented as a semi-direct sum $L_3 = L_2 \bigoplus \mathbf{U}_3$. Then the optimal systems of one- and two–dimensional subalgebras are listed in Table 6.1.

Theorem 6.3.1 *[61]. The optimal system of subalgebras of L_3 contains the subalgebras shown in Table 6.1.*

Table 6.1: The optimal system of one- and two–dimensional subalgebras of L_3, where parameters are $\kappa \in \mathbb{R}$, $0 \leq \phi \leq \pi$, $\epsilon = \pm 1$.

Dimension of the subalgebra	Subalgebras
1	$h_1^g = \{\mathbf{U}_2\}$, $h_2^g = \{\mathbf{U}_1 \cos(\varphi) + \mathbf{U}_3 \sin(\varphi)\}$, $h_3^g = \{\mathbf{U}_2 + \epsilon \mathbf{U}_3\}$
2	$h_4^g = (\mathbf{U}_2, \mathbf{U}_3)$, $h_5^g = (\mathbf{U}_1, \mathbf{U}_3)$, $h_6^g = (\mathbf{U}_1 + \kappa \mathbf{U}_3, \mathbf{U}_2)$

Table 6.2: The optimal system of one-, two- and three–dimensional subalgebras of L_4. Here $\kappa \in \mathbb{R}$, $0 \leq \phi \leq \pi$, $\epsilon = \pm 1$ are parameters.

Dimension of the subalgebra	Subalgebras
1	$h_1 = \{\mathbf{U}_3\}$, $h_2 = \{\mathbf{U}_1 \cos\phi + \mathbf{U}_4 \sin(\phi)\}$, $h_3 = \{\mathbf{U}_2 + \kappa(\mathbf{U}_1 \cos(\phi) + \mathbf{U}_4 \sin(\phi))\}$, $h_4 = \{\mathbf{U}_3 + \epsilon(\mathbf{U}_1 \cos(\phi) + \mathbf{U}_4 \sin(\phi))\}$
2	$h_5 = \{\mathbf{U}_2 + \kappa(\mathbf{U}_1 \cos(\phi) + \mathbf{U}_4 \sin(\phi)), \mathbf{U}_3\}$, $h_6 = \{\mathbf{U}_2 + \kappa(\mathbf{U}_1 \cos(\phi) + \mathbf{U}_4 \sin(\phi)),$ $\mathbf{U}_1 \sin(\phi) - \mathbf{U}_4 \cos(\phi)\}$, $h_7 = \{\mathbf{U}_1, \mathbf{U}_4\}$, $h_8 = \{\mathbf{U}_3 + \epsilon(\mathbf{U}_1 \cos(\phi) + \mathbf{U}_4 \sin(\phi)), \mathbf{U}_1 \sin(\phi) - \mathbf{U}_4 \cos(\phi)\}$, $h_9 = \{\mathbf{U}_3, \mathbf{U}_1 \sin(\phi) - \mathbf{U}_4 \cos(\phi))\}$
3	$h_{10} = (\mathbf{U}_2, \mathbf{U}_1, \mathbf{U}_4)$, $h_{11}^2 = (\mathbf{U}_3, \mathbf{U}_1, \mathbf{U}_4)$, $h_{12} = (\mathbf{U}_2 + \kappa(\mathbf{U}_1 \cos(\phi) + \mathbf{U}_4 \sin(\phi)), \mathbf{U}_1 \sin(\phi)$ $-\mathbf{U}_4 \cos(\phi), \mathbf{U}_3)$

Four-dimensional Lie algebra
We take a Lie algebra L_4. We suppose that the algebra has the following properties. It has a non-zero commutator relation, $[\mathbf{U}_1, \mathbf{U}_3] = -\mathbf{U}_3$. The Lie algebra L_4 has a two–dimensional subalgebra $L_2^4 = \langle \mathbf{U}_1, \mathbf{U}_3 \rangle$ spanned by generators $\mathbf{U}_1, \mathbf{U}_3$. The algebra L_4 is a decomposable Lie algebra and can be represented as a semi-direct sum $L_3 = L_2 \bigoplus \mathbf{U}_2 \bigoplus \mathbf{U}_4$. In the paper [61] the Lie algebra L_4 is denoted by L_4^2. Then for this algebra one can prove the following theorem.

Theorem 6.3.2 *[61]. The optimal system of one–dimensional subalgebras of L_4 contains four subalgebras, the optimal system of two–dimensional subalgebras contains five subalgebras and the optimal system of three–dimensional subalgebras includes three subalgebras. All subalgebras from these systems are listed in Table 6.2.*

6.4 Problems with solutions

The five problems in this section will allow our reader to become more familiar with the ideas of a commutator, an algebra and subalgebras of different types. As examples, we use several well known differential equations and their symmetry algebras.

Problem 6.4.1 *Find the commutator of the infinitesimal generator of the Lorentz group*

$$\mathbf{U}_1 = y\frac{\partial}{\partial x} + x\frac{\partial}{\partial y}$$

and of the linear fractional transformation group

$$\mathbf{U}_2 = x^2\frac{\partial}{\partial x} + y^2\frac{\partial}{\partial y}.$$

Solution The commutator of the operators $\mathbf{U}_1$ and $\mathbf{U}_2$ is given by

$$\begin{aligned}
[\mathbf{U}_1, \mathbf{U}_2] &= \mathbf{U}_1\mathbf{U}_2 - \mathbf{U}_2\mathbf{U}_1 \\
&= \left(y\frac{\partial}{\partial x} + x\frac{\partial}{\partial y}\right)\left(x^2\frac{\partial}{\partial x} + y^2\frac{\partial}{\partial y}\right) \\
&\quad - \left(x^2\frac{\partial}{\partial x} + y^2\frac{\partial}{\partial y}\right)\left(y\frac{\partial}{\partial x} + x\frac{\partial}{\partial y}\right) \\
&= y \cdot 2x\frac{\partial}{\partial x} + x \cdot 2y\frac{\partial}{\partial y} - \left(x^2\frac{\partial}{\partial y} + y^2\frac{\partial}{\partial x}\right) \\
&= y(2x - y)\frac{\partial}{\partial x} + x(2y - x)\frac{\partial}{\partial y}.
\end{aligned}$$

Problem 6.4.2 *Equation $y''' = 0$ admits the 7-parametric group L_7 of point transformations. The corresponding Lie algebra L_7 has the following infinitesimal generators*

$$\mathbf{U}_1 = \frac{\partial}{\partial x}, \quad \mathbf{U}_2 = \frac{\partial}{\partial y}, \quad \mathbf{U}_3 = x\frac{\partial}{\partial x}, \quad \mathbf{U}_4 = x\frac{\partial}{\partial y},$$

$$\mathbf{U}_5 = x^2\frac{\partial}{\partial y}, \quad \mathbf{U}_6 = y\frac{\partial}{\partial y}, \quad \mathbf{U}_7 = x^2\frac{\partial}{\partial x} + 2xy\frac{\partial}{\partial y}.$$

Study the algebraic structure of the corresponding algebra (subalgebras, ideals, center, derived algebras, low central series).

Solution The commutator table of this algebra is

	$\mathbf{U}_1$	$\mathbf{U}_2$	$\mathbf{U}_3$	$\mathbf{U}_4$	$\mathbf{U}_5$	$\mathbf{U}_6$	$\mathbf{U}_7$
$\mathbf{U}_1$	0	0	$\mathbf{U}_1$	$\mathbf{U}_2$	$2\mathbf{U}_4$	0	$2(\mathbf{U}_3+\mathbf{U}_6)$
$\mathbf{U}_2$	0	0	0	0	0	$\mathbf{U}_2$	$2\mathbf{U}_4$
$\mathbf{U}_3$	$-\mathbf{U}_1$	0	0	$\mathbf{U}_4$	$2\mathbf{U}_5$	0	$\mathbf{U}_7$
$\mathbf{U}_4$	$-\mathbf{U}_2$	0	$-\mathbf{U}_4$	0	0	$\mathbf{U}_4$	$\mathbf{U}_5$
$\mathbf{U}_5$	$-2\mathbf{U}_4$	0	$-2\mathbf{U}_5$	0	0	$\mathbf{U}_5$	0
$\mathbf{U}_6$	0	$-\mathbf{U}_2$	0	$-\mathbf{U}_4$	$-\mathbf{U}_5$	0	0
$\mathbf{U}_7$	$-2(\mathbf{U}_3+\mathbf{U}_6)$	$-2\mathbf{U}_4$	$-\mathbf{U}_7$	$-\mathbf{U}_5$	0	0	0

Example of ideals are $\langle\mathbf{U}_2, \mathbf{U}_4, \mathbf{U}_5\rangle$, and $\langle\mathbf{U}_2, \mathbf{U}_4, \mathbf{U}_5, \mathbf{U}_6\rangle$. There is no center in the algebra. The first derived algebra is

$$L_7^{(1)} = \langle\mathbf{U}_1, \mathbf{U}_2, \mathbf{U}_3 + \mathbf{U}_6, \mathbf{U}_4, \mathbf{U}_5, \mathbf{U}_7\rangle = L_7^{(2)},$$

it coincides with the second derived algebra. The low central series of this algebra is

$$L_7^{(1)} = L_7^{[2]}.$$

Examples of subalgebras are

$$\langle\mathbf{U}_1, \mathbf{U}_2\rangle, \ \langle\mathbf{U}_1, \mathbf{U}_2, \mathbf{U}_3\rangle, \ \langle\mathbf{U}_1, \mathbf{U}_2, \mathbf{U}_3, \mathbf{U}_4\rangle,$$
$$\langle\mathbf{U}_1, \mathbf{U}_2, \mathbf{U}_3, \mathbf{U}_4, \mathbf{U}_5\rangle, \ \langle\mathbf{U}_1, \mathbf{U}_2, \mathbf{U}_3, \mathbf{U}_4, \mathbf{U}_5, \mathbf{U}_6\rangle, \ldots.$$

Problem 6.4.3 *Prove that the Lie algebra with infinitesimal generators $\mathbf{U}_1$, $\mathbf{U}_2$, $\mathbf{U}_3$ such that $[\mathbf{U}_1, \mathbf{U}_2] = -\mathbf{U}_3$, $[\mathbf{U}_1, \mathbf{U}_3] = -\mathbf{U}_2$, $[\mathbf{U}_2, \mathbf{U}_3] = -\mathbf{U}_1$ has a two-dimensional subalgebra.*

Solution Any element of the algebra L_3 can be represented as a linear sum of the basis elements $\mathbf{U}_1, \mathbf{U}_2, \mathbf{U}_3$. Let us take the differential operators (also called vector fields) $\mathbf{V}_1, \mathbf{V}_2$ which span the two-dimensional subalgebra in the form

$$\mathbf{V}_1 = a_1\mathbf{U}_1 + a_2\mathbf{U}_2 + a_3\mathbf{U}_3, \ \ \mathbf{V}_2 = b_1\mathbf{U}_1 + b_2\mathbf{U}_2 + b_3\mathbf{U}_3.$$

If they span a subalgebra then the commutator of them should be represented as a linear sum of these $\mathbf{V}_1, \mathbf{V}_2$. We should verify that there exists such constants $a_1, a_2, a_3, b_1, b_2, b_3, \lambda_1, \lambda_2$ that

$$[\mathbf{V}_1, \mathbf{V}_2] = \lambda_1\mathbf{V}_1 + \lambda_2\mathbf{V}_2.$$

For simplicity we take $b_1 = 1, b_2 = b_3 = 0$ (if we find a solution then we proved that the subalgebra exists, if not then we should regard a more complicated case). Then the commutator has the form

$$\begin{aligned}
[\mathbf{V}_1, \mathbf{V}_2] &= [a_1\mathbf{U}_1 + a_2\mathbf{U}_2 + a_3\mathbf{U}_3, \mathbf{U}_1] \\
&= a_1[\mathbf{U}_1, \mathbf{U}_1] + a_2[\mathbf{U}_1, \mathbf{U}_2] + a_3[\mathbf{U}_1, \mathbf{U}_3] \\
&= -a_2\mathbf{U}_3 - a_3\mathbf{U}_2.
\end{aligned}$$

On the other hand,

$$\begin{aligned}
[\mathbf{V}_1, \mathbf{V}_2] &= \lambda_1\mathbf{V}_1 + \lambda_2\mathbf{V}_2 \\
&= \lambda_1 a_1\mathbf{U}_1 + \lambda_1 a_2\mathbf{U}_2 + \lambda_1 a_3\mathbf{U}_3 + \lambda_2\mathbf{U}_1.
\end{aligned}$$

The coefficients by $\mathbf{U}_1, \mathbf{U}_2, \mathbf{U}_3$ can be gathered in the following system

$$\begin{aligned}
\lambda_1 a_1 + \lambda_2 &= 0, \\
\lambda_1 a_2 &= -a_3, \\
\lambda_1 a_3 &= -a_2.
\end{aligned}$$

From the last two equations we have $\lambda_1^2 a_3 = a_3$, i.e. $\lambda_1 = \pm 1$ and $a_2 = \pm a_3$. Taking into account the first equation, we obtain $\lambda_2 = \pm a_1$. Consequently, the vector field $\mathbf{V}_1$ takes the form $\mathbf{V}_1 = a_1\mathbf{U}_1 + a_2(\mathbf{U}_2 - \mathbf{U}_3)$. We found a two -dimensional subalgebra spanned by $\mathbf{V}_1, \mathbf{V}_2$.

Problem 6.4.4 *The equation of a sub-sonic gas motion*

$$u_x u_{xx} + u_{yy} = 0$$

admits the 6–dimensional symmetry group. The corresponding infinitesimal generators of the Lie algebra L_6 are

$$\begin{aligned}
\mathbf{U}_1 &= \tfrac{\partial}{\partial x}, \ \ \mathbf{U}_2 = \tfrac{\partial}{\partial y}, \ \mathbf{U}_3 = \tfrac{\partial}{\partial u}, \ \mathbf{U}_4 = y\tfrac{\partial}{\partial u} \\
\mathbf{U}_5 &= x\tfrac{\partial}{\partial x} + 3u\tfrac{\partial}{\partial u}, \quad \mathbf{U}_6 = y\tfrac{\partial}{\partial y} - 2u\tfrac{\partial}{\partial u}.
\end{aligned}$$

Is L_6 a solvable Lie algebra? If L_6 is a solvable algebra then present the corresponding chain of ideals.

Solution The commutator table is

	$\mathbf{U}_1$	$\mathbf{U}_2$	$\mathbf{U}_3$	$\mathbf{U}_4$	$\mathbf{U}_5$	$\mathbf{U}_6$
$\mathbf{U}_1$	0	0	0	0	$\mathbf{U}_1$	0
$\mathbf{U}_2$	0	0	0	$\mathbf{U}_3$	0	$\mathbf{U}_2$
$\mathbf{U}_3$	0	0	0	0	$3\mathbf{U}_3$	$-2\mathbf{U}_3$
$\mathbf{U}_4$	0	$-\mathbf{U}_3$	0	0	$3\mathbf{U}_4$	$-3\mathbf{U}_4$
$\mathbf{U}_5$	$-\mathbf{U}_1$	0	$-3\mathbf{U}_3$	$-3\mathbf{U}_4$	0	0
$\mathbf{U}_6$	0	$-\mathbf{U}_2$	$2\mathbf{U}_3$	$3\mathbf{U}_4$	0	0

The algebra L_6 spanned by generators $\mathbf{U}_1, \ldots, \mathbf{U}_6$ is solvable since the derived algebra of a finite order vanishes

$$L_6^{(1)} = \langle \mathbf{U}_1, \mathbf{U}_2, \mathbf{U}_3, \mathbf{U}_4 \rangle, \ L_6^{(2)} = \langle \mathbf{U}_3 \rangle, \ L_6^{(3)} = 0.$$

The corresponding chain of ideals is

$$0 \subset \langle \mathbf{U}_3 \rangle \subset \langle \mathbf{U}_1, \mathbf{U}_3 \rangle \subset \langle \mathbf{U}_1, \mathbf{U}_2, \mathbf{U}_3 \rangle \subset \langle \mathbf{U}_1, \mathbf{U}_2, \mathbf{U}_3, \mathbf{U}_4 \rangle \\ \subset \langle \mathbf{U}_1, \mathbf{U}_2, \mathbf{U}_3, \mathbf{U}_4, \mathbf{U}_5 \rangle \subset L_6.$$

Problem 6.4.5

The nonlinear filtration equation

$$u_t + e^{u_x} u_{xx} = 0 \tag{6.4.23}$$

admits a 5–dimensional Lie algebra L_5 spanned by the infinitesimal generators

$$\mathbf{U}_1 = \frac{\partial}{\partial x}, \quad \mathbf{U}_2 = \frac{\partial}{\partial t}, \quad \mathbf{U}_3 = \frac{\partial}{\partial u}$$
$$\mathbf{U}_4 = 2t\frac{\partial}{\partial t} + x\frac{\partial}{\partial x} + u\frac{\partial}{\partial u},$$
$$\mathbf{U}_5 = -t\frac{\partial}{\partial t} + x\frac{\partial}{\partial u}.$$

Study the algebraic structure of the corresponding algebra (commutator table, subalgebras, ideals, center, derived algebras, low central series). Is L_5 a solvable algebra? If L_5 is a solvable algebra then present the corresponding chain of ideals.

Solution The commutator table is

	$\mathbf{U}_1$	$\mathbf{U}_2$	$\mathbf{U}_3$	$\mathbf{U}_4$	$\mathbf{U}_5$
$\mathbf{U}_1$	0	0	0	$\mathbf{U}_1$	$\mathbf{U}_3$
$\mathbf{U}_2$	0	0	0	$2\mathbf{U}_2$	$-\mathbf{U}_2$
$\mathbf{U}_3$	0	0	0	$\mathbf{U}_3$	0
$\mathbf{U}_4$	$-\mathbf{U}_1$	$-2\mathbf{U}_2$	$-\mathbf{U}_3$	0	0
$\mathbf{U}_5$	$-\mathbf{U}_3$	$\mathbf{U}_2$	0	0	0

There are infinitely many subalgebras in the algebra. For example we obtain a subalgebra if we take a linear span of

$$\begin{aligned} &\text{any subset of operators } \langle \mathbf{U}_1, \mathbf{U}_2, \mathbf{U}_3, \mathbf{U}_4 \rangle \\ \cup\ &\text{any subset of operators } \langle \mathbf{U}_2, \mathbf{U}_3, \mathbf{U}_4, \mathbf{U}_5 \rangle \\ \cup\ &\text{any set of operators containing } \langle \mathbf{U}_1, \mathbf{U}_3 \rangle. \end{aligned}$$

As an example of a one-dimensional subalgebra we can take a linear span of $\mathbf{U}_3$.
Indeed, one can notice that

- $[\mathbf{U}_i, \mathbf{U}_j] = 0$ and $[\mathbf{U}_4, \mathbf{U}_i] = \mathbf{U}_i$ for $i, j = 1, \ldots, 3$. That is why $\mathbf{U}_4$ with any set of $\mathbf{U}_1, \mathbf{U}_2, \mathbf{U}_3$ forms the subalgebra;

- $[\mathbf{U}_5, \mathbf{U}_1] = -\mathbf{U}_3$. That is why if subalgebra contains $\mathbf{U}_1$ and $\mathbf{U}_5$ it must contain $\mathbf{U}_3$ as well;

- $[\mathbf{U}_5, \mathbf{U}_i] = \mathbf{U}_i$ for $i \neq 1$. That is why if arbitrary subalgebra G does not contain $\mathbf{U}_1$ then $G \cup \mathbf{U}_5$ is a subalgebra as well;

- $[\mathbf{U}_2, \mathbf{U}_i] = 0$ or $\mathbf{U}_2$, and $[\mathbf{U}_i, \mathbf{U}_j] = 0, \mathbf{U}_1$ or $\mathbf{U}_2$. That is why any set that contains $\mathbf{U}_1$ and $\mathbf{U}_3$ is a subalgebra.

In a similar way we obtain ideals if we take

$$\langle \mathbf{U}_2 \rangle \cup \langle \mathbf{U}_3 \rangle \cup \langle \mathbf{U}_1, \mathbf{U}_3 \rangle \cup \text{any set containing } \langle \mathbf{U}_2, \mathbf{U}_3 \rangle.$$

This is clear from the following observations:

- $\mathbf{U}_2$ and $\mathbf{U}_3$ are one-dimensional ideals. Since the linear span of the two ideals is also an ideal, we can add $\mathbf{U}_2$ and $\mathbf{U}_3$ to any ideal set.

- $[\mathbf{U}_i, \mathbf{U}_j] = \mathbf{U}_2$ or $\mathbf{U}_3$ if $i, j \neq 1$. This means that if the set contains both $\mathbf{U}_2$ and $\mathbf{U}_3$, it is an ideal.

- If the ideal contains $\mathbf{U}_1$, it must contain $\mathbf{U}_3$ as well, since $[\mathbf{U}_1, \mathbf{U}_5] = \mathbf{U}_3$.

There is no row with only zeroes, that is why the center is empty. Derived algebras

$$L_5^{(1)} = [L_5, L_5] = \langle \mathbf{U}_1, \mathbf{U}_2, \mathbf{U}_3 \rangle$$

Furthermore, $L_5^{(2)} = 0$, because $\langle \mathbf{U}_1, \mathbf{U}_2, \mathbf{U}_3 \rangle$ is Abelian algebra, that is why only first derived subalgebra $L_5^{(1)}$ is non-trivial. Having this, we can simply construct the chain of ideals and show that L_5 is solvable

$$\begin{aligned} \langle \mathbf{U}_1 \rangle &\subset \langle \mathbf{U}_1, \mathbf{U}_2 \rangle \subset \langle \mathbf{U}_1, \mathbf{U}_2, \mathbf{U}_3 \rangle \subset \\ &\subset \langle \mathbf{U}_1, \mathbf{U}_2, \mathbf{U}_3, \mathbf{U}_4 \rangle \subset \langle \mathbf{U}_1, \mathbf{U}_2, \mathbf{U}_3, \mathbf{U}_4, \mathbf{U}_5 \rangle. \end{aligned}$$

The lower central series is

$$\begin{aligned} L_5^{[1]} &= L_5^{(1)} = \langle \mathbf{U}_1, \mathbf{U}_2, \mathbf{U}_3 \rangle \\ L_5^{[2]} &= [\langle \mathbf{U}_1, \mathbf{U}_2, \mathbf{U}_3 \rangle, L_5] = \langle \mathbf{U}_1, \mathbf{U}_2, \mathbf{U}_3 \rangle, \end{aligned}$$

and $L_5^{[i]} = \langle \mathbf{U}_1, \mathbf{U}_2, \mathbf{U}_3 \rangle$ for $i > 2$.

Chapter 7

High order ODEs

The main goal of using Lie group analysis in applications to PDEs or ODEs is to find a way to simplify the given equation: either to solve it or to reduce its order to a lower one. In this chapter we explain some steps of this procedure using the examples of higher order ODEs.

Here we can demonstrate all the main features of Lie group analysis in application to differential equations. On one hand the high order ODEs are more complicated than first order ODEs, so they provide deeper insights of all typical procedures needed to find the admitted symmetry group. On the other hand the calculations are not overloaded with technical details as in the case of PDEs so all the steps can be performed without using one of the computer packages. The relative simplicity of calculations is especially important for beginners since the usage of computer packages often obscures typical mistakes based on misunderstanding.

Later in Chapter 9 we will discuss some financial models described by nonlinear PDEs. Using Lie group analysis we find reductions of these PDEs to ODEs and sometimes we can even find some special exact analytical solutions. All studied PDEs possess algebras of the solvable type, that is why in this chapter we pay most attention to this case and for other types of algebras we give just a sketch that shows how symmetries can be used to solve the given equation.

In this chapter we deal with the n-th order differential equations

$$y^{(n)} = f(x, y, y', \ldots, y^{(n-1)}). \tag{7.0.1}$$

We will show that the way of using admitted symmetry group $G_r(\mathbb{R}^2)$ depends on their structure (or on the structure of the corresponding algebra Lie algebra $L_r = \langle \mathbf{U}_1, \ldots, \mathbf{U}_r \rangle$). Many models in financial mathematics have a form of a second order PDE. After reductions we naturally obtain second order ODEs. That is why in this chapter we pay main attention to the symmetry properties of such equations. It is also important to note that the study of high order ODEs is more time consuming and seems less promising in the case of nonlinear ODEs, that is why we provide just a sketch description of Lie group analysis in the case of higher order ODEs.

In this chapter we follow the ideas of Sophus Lie represented in his book [51] and later in [52], [53]. Very similar descriptions of used procedures can also be found in more recent books [75], [39], [40], [59] and [60].

7.1 Operator $\mathcal{A}$ for a high order ordinary differential equation

Before we start with the generalization of the main tools of Lie group analysis to the case of high order ODEs we recall some important details from the theory of first order ODE in this context.

Let us reconsider an infinitesimal generator $\mathbf{U}$ of a one-parameter group $G(\mathbb{R}^2)$ of point transformations admitted by an ODE of the first order and the linear operator $\mathcal{A}$ corresponding to this ODE. We saw that the equation $\mathbf{U}u(x,y) = 0$ defines the invariant functions of the group $G(\mathbb{R}^2)$. Whereas to get the first order differential invariant $i_1(x,y,y') = v(x,y,y')$ we use the first prolongation $\mathbf{U}^{(1)}$ of this operator, i.e. the equation $\mathbf{U}^{(1)}v(x,y,y') = 0$ defines all first order differential invariants of the given group.

In the case of a first order differential equation given in the form $y' = \frac{\mathcal{Y}}{\mathcal{X}}$ we define a linear differential operator $\mathcal{A}$ associated with this equation as

$$\mathcal{A} = \mathcal{X}\frac{\partial}{\partial x} + \mathcal{Y}\frac{\partial}{\partial y}.$$

If the first order ODE is given in the form $y' = f(x,y)$ we can take respectively $\mathcal{Y} = f(x,y)$, $\mathcal{X} = 1$ and obtain the operator $\mathcal{A}$ in the form

$$\mathcal{A} = \frac{\partial}{\partial x} + f(x,y)\frac{\partial}{\partial y}.$$

The form of the operator $\mathcal{A}$ is not unique but this operator is helpful if we study the symmetry properties of the equation.

In fact to prove that a first order ODE admits a given group of point transformations $G(\mathbb{R}^2)$ we can use the infinitesimal generator $\mathbf{U}$ and prove the relation $[\mathbf{U},\mathcal{A}] = \lambda(x,y)\mathcal{A}$ or using the first prolongation we can also see if the condition $\mathbf{U}^{(1)}v(x,y,y')\big|_{y'=\frac{\mathcal{Y}}{\mathcal{X}}} = 0$ is satisfied.

To work in this manner with ODEs of higher order we should give an answer to the question: Which linear first order differential operator $\mathcal{A}$ corresponds to a higher order ordinary differential equation? The answer to this question is well known in the theory of PDEs. Here we just recall for the convenience of the reader the main ideas: first we present a n-th order ODE as a system of first order ODEs, then we represent this system as a characteristic system of equations for a first order PDE and in this way obtain a possible form of the operator $\mathcal{A}$.

Let us look at the n-th order ODE given in the explicit form

$$y^{(n)} = f(x, y, y', \dots, y^{(n-1)}), \quad \text{or equivalent} \tag{7.1.2}$$
$$\Delta(x, y, y', \dots, y^{(n-1)}, y^{(n)}) = y^{(n)} - f(x, y, y', \dots, y^{(n-1)}) = 0.$$

If we introduce new variables $y^{(1)}, \dots, y^{(n-1)}$ this equation can be written equivalently as a system of ordinary differential equations first order on $y^{(1)}, \dots, y^{(n-1)}$

$$\frac{dy}{dx} = y^{(1)} \Rightarrow \frac{dy}{y^{(1)}} = dx,$$
$$\frac{dy^{(1)}}{dx} = y^{(2)} \Rightarrow \frac{dy^{(1)}}{y^{(2)}} = dx,$$
$$\vdots$$
$$\frac{dy^{(n-1)}}{dx} = f(x, y, y^{(1)}, \dots, y^{(n-1)}) \Rightarrow \frac{dy^{(n-1)}}{f(x, y, y^{(1)}, \dots, y^{(n-1)})} = dx,$$

or

$$\frac{dy}{y^{(1)}} = \frac{dy^{(1)}}{y^{(2)}} = \dots = \frac{dy^{(n-1)}}{f(x, y, \dots, y^{(n-1)})} = dx.$$

This system of equations is the characteristic system for the first order partial differential equation $\mathcal{A}u(x, y, y^{(1)}, \dots, y^{(n-1)}) = 0$ with the linear differential operator $\mathcal{A}$

$$\mathcal{A} = \left(\frac{\partial}{\partial x} + y^{(1)}\frac{\partial}{\partial y} + y^{(2)}\frac{\partial}{\partial y^{(1)}} + \dots + f(x, y, y^{(1)}, \dots, y^{(n-1)})\frac{\partial}{\partial y^{(n-1)}}\right). \tag{7.1.3}$$

To describe the form of this operator we have introduced new variables $y^{(1)}, \dots, y^{(n-1)}$ which denote the derivatives of the dependent variable y with respect to the independent variable x. Thus, the corresponding linear operator $\mathcal{A}$ of the n-th order differential equation (7.1.2) can be represented only in the jet bundle of the order $(n-1)$, i.e. in $J^{(n-1)}$. The form of the operator $\mathcal{A}$ is not unique but this operator cannot be represented in a lower order jet bundle.

In the jet bundle $J^{(n-1)}$ we have two linear operators, first of them is the $(n-1)$-th prolongation of the infinitesimal operator $\mathbf{U}$, i.e. the linear differential operator $\mathbf{U}^{(n-1)}$ and the second one is the operator $\mathcal{A}$, which describes the action of n-th order ODE. The operator $\mathcal{A}$ defines the integral surface of the differential equation in $J^{(n-1)}$ and $\mathbf{U}^{(n-1)}$ the motion along this surface if it describes locally an admitted symmetry group of the Equation (7.1.2). It is easy to prove that the generalization of condition known from the case of the first order ODEs

$$\mathbf{U}^{(n)}\Delta(x,y,y^{(1)},\ldots,y^{(n)})\big|_{\Delta=0}=0 \quad \Leftrightarrow \quad \left[\mathbf{U}^{(n-1)},\mathcal{A}\right]=\lambda(x,y)\mathcal{A} \quad (7.1.4)$$

holds for any order n of the differential equation. These equations are the determining equations for the case of n-th order ODEs. Using these equations we can either verify if the high order ODE is invariant under the group of point transformations with the infinitesimal generator $\mathbf{U}$ or use these equations to find the Lie group admitted by the given equation.

If we find r different one-parameter groups of point transformations admitted by the high order ODE then their infinitesimal generators build a Lie algebra L_r. This theorem was proven by Sophus Lie and can be used to verify our calculations: if the span of the found operators $\mathbf{U}_i$, $i=1,\ldots,r$ does not build an algebra then we lost one or several of them. We can add to the found list of infinitesimal generators their commutators if they differ from $\mathbf{U}_i$ $i=1,\ldots,r$. Then we can check if the the enlarged set provides the linear span which is closed under algebraic operation, i.e. in this case if all commutators belong to the linear span.

We use both versions of determining equations (7.1.4). First we use these equations in the form $\left[\mathbf{U}^{(n-1)},\mathcal{A}\right]=\lambda(x,y)\mathcal{A}$ to verify invariance of the n-th order ODE under the given group of point transformations. We explain using a simple example of a second order ODE how to prove if the given one-parameter group of point transformations is admitted by an ODE. Then in Example 7.1.2 we use the symmetry group to find a solution of a high order ODE. Finally, in Example 7.1.3 we use the determining equations in the form $\mathbf{U}^{(n)}\Delta(x,y,y^{(1)},\ldots,y^{(n)})\big|_{\Delta=0}=0$ to calculate the symmetry group of a second order ODE.

Example 7.1.1 We examine the second order differential equation

$$y''=y^2y'.$$

The linear differential operator $\mathcal{A}$ is given by

$$\mathcal{A}=\frac{\partial}{\partial x}+y'\frac{\partial}{\partial y}+y^2y'\frac{\partial}{\partial y'}.$$

From the general structure of $\mathcal{A}$ (7.1.3) we see that if $f(x,y,y') = f(x,y')$, i.e. f is a function of just two variables (x,y') then the corresponding equation admits a one-parameter group of point transformations with the infinitesimal generator $\mathbf{U} = \frac{\partial}{\partial y}$. If $f(x,y,y') = f(y,y')$, i.e. f is a function of just two variables (y,y') then the admitted symmetry group possesses the infinitesimal generator $\mathbf{U} = \frac{\partial}{\partial x}$.

Therefore, in our example we expect that the infinitesimal generator

$$\mathbf{U}_1 = \frac{\partial}{\partial x} \text{ and } \mathbf{U}_1^{(n)} = \mathbf{U}_1, \ \forall n$$

is admitted by the equation. Now we calculate the commutator

$$\left[\mathbf{U}_1^{(1)}, \mathcal{A}\right] = \left[\frac{\partial}{\partial x}, \frac{\partial}{\partial x} + y'\frac{\partial}{\partial y} + y^2y'\frac{\partial}{\partial y'}\right] = 0,$$

and thus obtain that $\lambda = 0$ in (7.1.4). It means $\mathbf{U}$ is an infinitesimal generator of a symmetry group or in other words the corresponding symmetry group is admitted by this equation.

With this procedure we can also find other symmetries. We suppose that the equation admits another one-parameter group of point transformations, for instance with an infinitesimal operator $\mathbf{U}_2$. We use the ansatz for the generator $\mathbf{U}_2$ in the following form

$$\mathbf{U}_2 = ax\frac{\partial}{\partial x} + by\frac{\partial}{\partial y}.$$

Then the first prolongation of $\mathbf{U}_2$ takes the form

$$\mathbf{U}_2^{(1)} = ax\frac{\partial}{\partial x} + by\frac{\partial}{\partial y} + (b-a)y'\frac{\partial}{\partial y'}$$

where correspondingly

$$\eta^{(1)} = \frac{d\eta}{dx} - y'\frac{d\xi}{dx} = by' - y'a = (b-a)y'.$$

Now we calculate the commutator of linear operators $\mathbf{U}_2^{(1)}$ and $\mathcal{A}$ and obtain

$$\begin{aligned}\left[\mathbf{U}_2^{(1)}, \mathcal{A}\right] &= \left[ax\frac{\partial}{\partial x} + by\frac{\partial}{\partial y} + (b-a)y'\frac{\partial}{\partial y'}, \frac{\partial}{\partial x} + y'\frac{\partial}{\partial y} + y^2y'\frac{\partial}{\partial y'}\right] \\ &= \left(ax\frac{\partial}{\partial x} + by\frac{\partial}{\partial y} + (b-a)y'\frac{\partial}{\partial y'}\right)\left(\frac{\partial}{\partial x} + y'\frac{\partial}{\partial y} + y^2y'\frac{\partial}{\partial y'}\right) \\ &\quad - \left(\frac{\partial}{\partial x} + y'\frac{\partial}{\partial y} + y^2y'\frac{\partial}{\partial y'}\right)\left(ax\frac{\partial}{\partial x} + by\frac{\partial}{\partial y} + (b-a)y'\frac{\partial}{\partial y'}\right)\end{aligned}$$

$$\begin{aligned}
&= 2by^2y'\frac{\partial}{\partial y'} + (b-a)y'\frac{\partial}{\partial y} + (b-a)y^2y'\frac{\partial}{\partial y'} \\
&\quad - a\frac{\partial}{\partial x} - by'\frac{\partial}{\partial y} - (b-a)y^2y'\frac{\partial}{\partial y'} \\
&= -a\frac{\partial}{\partial x} - ay'\frac{\partial}{\partial y'} + 2by^2y'\frac{\partial}{\partial y'}.
\end{aligned}$$

If the infinitesimal generator $\mathbf{U}_2$ represents an admitted symmetry group then the last expression under conditions (7.1.4) must be equal to

$$\lambda(x,y)\left(\frac{\partial}{\partial x} + y'\frac{\partial}{\partial y} + y^2y'\frac{\partial}{\partial y'}\right).$$

Therefore $\mathbf{U}_2$ is an infinitesimal generator of an admitted symmetry group if and only if the condition

$$2b = -a$$

is fulfilled. So we obtain that the second generator $\mathbf{U}_2$ is equal to

$$\mathbf{U}_2 = -2x\frac{\partial}{\partial x} + y\frac{\partial}{\partial y}.$$

We used a special form of the generator for the ansatz and were able to find a new admitted symmetry. ◇

We assume that we found at least one infinitesimal generator $\mathbf{U}$ of a one-parameter group of point transformations in the form

$$\mathbf{U} = \xi(x,y)\frac{\partial}{\partial x} + \eta(x,y)\frac{\partial}{\partial y}$$

admitted by the n-th order ODE. Now we want to solve or simplify an ordinary differential equation

$$y^{(n)} = f(x, y, y^{(1)}, \ldots, y^{(n-1)}) \tag{7.1.5}$$

using the infinitesimal generator $\mathbf{U}$ admitted by this equation. How can we use this generator to solve or simplify the original equation? One of the possible ways is the following.

We introduce the canonical variables $t(x,y)$, $s(x,y)$ for $\mathbf{U}$ solving the system of equations

$$\begin{aligned}
\mathbf{U}s(x,y) &= 1, \\
\mathbf{U}t(x,y) &= 0.
\end{aligned}$$

In the canonical variables the infinitesimal generator $\mathbf{U}$ takes its normal form $\mathbf{U} = \frac{\partial}{\partial s}$. In these new variables the Equation (7.1.5) takes the form

$$s^{(n)} = \tilde{f}(t, s, s^{(1)}, \dots, s^{(n-1)}).$$

Since the equation admits the symmetry group with the infinitesimal generator $\mathbf{U}$ this equation is invariant under the translation group described by the infinitesimal generator $\mathbf{U} = \frac{\partial}{\partial s}$. It means as we have seen before that the function $\tilde{f}$ is a function of a reduced number of variables

$$\tilde{f} = \tilde{f}(t, s^{(1)}, \dots, s^{(n-1)}).$$

Correspondingly, we can use $z = s^{(1)}$ as a new dependent variable and reduce the original equation to a $(n-1)$-th order ODE

$$z^{(n-1)} = \tilde{f}(t, z, z^{(1)}, \dots, z^{(n-2)}). \tag{7.1.6}$$

If we solve this equation and obtain the solution in the form $z = z(t, c_1, \dots, c_{n-1})$, where $c_1, \dots, c_{n-1}$ are integration constants then in the old variables (s, t) the solution of n-th order ODE looks like

$$s = \int z(t, c_1, \dots, c_{n-1})\, dt + c_n. \tag{7.1.7}$$

We have seen that originally we have had to integrate a n-th order ODE. Using the symmetry of the equation we split the problem into two simpler ones: at first one can solve a $(n-1)$-th order ODE and then integrate a first order ODE. Let us illustrate this idea once again using the same second order differential equation as in the previous example.

Example 7.1.2 We have to solve a second order ODE

$$y'' = y^2 y',$$

which admits translation symmetry with an infinitesimal generator $\mathbf{U} = \frac{\partial}{\partial x}$.

First let us define the canonical variables for the given $\mathbf{U}$. We solve the system of equations to define canonical variables $t(x, y)$ and $s(x, y)$

$$\begin{aligned} \mathbf{U}s(x,y) &= 1, \quad \Rightarrow \quad s_x = 1, \quad s = x, \\ \mathbf{U}t(x,y) &= 0, \quad \Rightarrow \quad t_x = 0, \quad t = y. \end{aligned}$$

It means we should change independent and dependent variables and take $x = x(y)$ as a new dependent variable to simplify the equation. Correspondingly we should replace the first derivative with

$$\frac{dy}{dx} = \frac{1}{\frac{dx}{dy}}$$

and the second derivative with

$$\frac{d^2y}{dx^2} = \frac{d}{dx}\left(\frac{1}{\frac{dx}{dy}}\right) = -\frac{1}{\left(\frac{dx}{dy}\right)^2} \cdot \frac{d^2x}{dy^2} \cdot \frac{dy}{dx} = -\frac{1}{\left(\frac{dx}{dy}\right)^3} \cdot \frac{d^2x}{dy^2}.$$

It means we transform the original equation to

$$-\frac{1}{(x')^3}x'' = y^2\frac{1}{x'}.$$

Using the substitution $z = x'(y)$ we reduce the order of the equation. Now it is an ODE with separable variables

$$-\frac{z'}{z^2} = y^2.$$

After the integration we obtain

$$-\frac{1}{z} = \frac{y^3}{3} + c_1 \quad \text{or} \quad z(y) = -\frac{3}{y^3 + c_1}.$$

Eventually we can solve this first order ODE for x and obtain after integration

$$x(y) = -\int \frac{3}{y^3 + c_1}\, dy = \alpha_1 \ln(y - \beta_1) + \alpha_2 \ln(y - \beta_2) + \alpha_3 \ln(y - \beta_3) + c_2,$$

or respectively

$$c_2 e^x = (y + c_1)^{\alpha_1}(y + \beta_1(c_1))^{\alpha_2}(y + \beta_2(c_1))^{\alpha_3}.$$

Here the values of the constants $\alpha_1, \alpha_2, \alpha_3$ depend on the values of the roots $\beta_1, \beta_2, \beta_3$ of a third order polynomial $y^3 + c_1 = 0$ and correspondingly on c_1. We skip the expressions for $\alpha_1, \alpha_2, \alpha_3$.

We use the translation symmetry of the original equation to reduce the original equation to a first order ODE , i.e. to the equation $z' = -y^2z^2$. We are able to solve this first order equation because it is an equation with separable variables. This circumstance hints to us that the original second order ODE admits a second symmetry. In fact it can be proved that the original equation admits the second symmetry with the infinitesimal generator $\mathbf{U}_2 = x\frac{\partial}{\partial x} - \frac{y}{2}\frac{\partial}{\partial y}$. ◇

In the next example we use the determining equations and calculate the admitted Lie group of point transformations for a high order ODE. To make the calculations transparent we take a simple second order ODE and explain each step.

Example 7.1.3 We have to find the Lie point symmetries of the second order differential equation

$$y'' = \exp(-y'). \tag{7.1.8}$$

Let us denote by $\Delta(x, y, y', y'') = y'' - e^{-y'}$. The partial derivatives of the function $\Delta(x, y, y', y'')$ defined in the second order jet bundle $J^{(2)}$ have the following form

$$\frac{\partial}{\partial x}\Delta = 0, \quad \frac{\partial}{\partial y}\Delta = 0, \quad \frac{\partial}{\partial y'}\Delta = e^{-y'}, \quad \frac{\partial}{\partial y''}\Delta = 1.$$

We are looking for the infinitesimal generator $\mathbf{U}$ where the second prolongation has the form

$$\mathbf{U}^{(2)} = \xi(x,y)\frac{\partial}{\partial x} + \eta(x,y)\frac{\partial}{\partial y} + \eta^{(1)}\frac{\partial}{\partial y'} + \eta^{(2)}(x,y)\frac{\partial}{\partial y''}.$$

Here

$$\begin{aligned}
\eta^{(1)} &= \frac{d\eta}{dx} - y'\frac{d\xi}{dx} = \eta_x + y'(\eta_y - \xi_x) - y'^2\xi_y, \\
\eta^{(2)} &= \frac{d\eta^{(1)}}{dx} - y''\frac{d\xi}{dx} \\
&= \eta_{xx} + y'(2\eta_{xy} - \xi_{xx}) + y'^2(\eta_{yy} - 2\xi_{xy}) - y'^3\xi_{yy} + y''(\eta_y - 2\xi_x - 3y'\xi_y).
\end{aligned}$$

The determining equations $\mathbf{U}^{(2)}\Delta|_{\Delta=0} = 0$ which describe the action of $\mathbf{U}^{(2)}$ on Δ restricted on the solution manifold $\Delta = 0$, should hold

$$\begin{aligned}
&(\eta_x + y'(\eta_y - \xi_x) - y'^2\xi_y)e^{-y'} + \eta_{xx} + y'(2\eta_{xy} - \xi_{xx}) + y'^2(\eta_{yy} - 2\xi_{xy}) \\
&\left. -y'^3\xi_{yy} + y''(\eta_y - 2\xi_x - 3y'\xi_y))\right|_{\Delta=0} = 0.
\end{aligned}$$

We replace y'' by $e^{-y'}$ then multiply the equation above by $e^{y'}$ and obtain

$$\begin{aligned}
&\eta_x + \eta_y - 2\xi_x + y'(\eta_y - \xi_x - 3\xi_y) - y'^2\xi_y + e^{y'}(\eta_{xx} + y'(2\eta_{xy} - \xi_{xx}) \\
&+y'^2(\eta_{yy} - 2\xi_{xy}) - y'^3\xi_{yy}) = 0.
\end{aligned}$$

Since the variables y, y' are independent now, we can separate that equation according to the degrees of y' into monomials. Then the coefficient of the each monomial should be equal to zero independently. The monomials and correspondent coefficients are given in Table 7.1. The set of equations on coefficients in the right column defines the functions $\xi(x, y)$ and $\eta(x, y)$, it defines also the infinitesimal generator $\mathbf{U}$.

Table 7.1: The determining equations for Equation (7.1.8).

	monomial	equation on the coefficient
(1)	$y'^3 e^{y'}$	$-\xi_{yy} = 0$
(2)	$y'^2 e^{y'}$	$\eta_{yy} - 2\xi_{xy} = 0$
(3)	$y' e^{y'}$	$2\eta_{xy} - \xi_{xx} = 0$
(4)	$e^{y'}$	$\eta_{xx} = 0$
(5)	y'^2	$\xi_y = 0$
(6)	y'	$\eta_y - \xi_x - 3\xi_y = 0$
(7)	y'^0	$\eta_x + \eta_y - 2\xi_x = 0$

From the Equation (5) in Table 7.1 we get that $\xi(x, y)$ does not depend on the argument y, i.e. $\xi(x, y) = \xi(x)$. Equation (4) in Table 7.1 leads to the linearity of the function $\eta(x, y)$ with respect to the argument x, i.e. $\eta(x, y) = a_1(y)x + a_2(y)$. Taking into consideration that $\xi(x)$ does not depend on y, we get from Equation (2) in Table 7.1 that $\eta(x, y)$ is also linear in the variable y. Thus, we have

$$\eta = (a_{11}x + a_{12})y + (a_{21}x + a_{22}).$$

From the Equation (3) in Table 7.1 then follows

$$\xi_{xx} = 2a_{11}, \quad \Rightarrow \quad \xi = a_{11}x^2 + b_1 x + b_2.$$

Till now we have not used equations (6) and (7) in Table 7.1. We can insert the formulas for ξ and η in these equations

$$a_{11}x + a_{12} - a_{11}x - b_1 - 3a_{11}x - 3a_{12} = 0, \quad \Rightarrow \quad a_{11} = 0, \ b_1 = a_{12}.$$
$$a_{11}y + a_{21} + a_{11}x + a_{12} - 2a_{11}x - 2b_1 = 0, \quad \Rightarrow$$
$$a_{11} = 0, \ a_{21} = 2b_1 - a_{12} = a_{12}.$$

Therefore,

$$\xi(x) = a_{12}x + b_2, \ \eta(x, y) = a_{12}y + a_{12}x + a_{22},$$

where $a_{12}, a_{22}, b_2 \in \mathbb{R}$ are arbitrary constants. We see that the equation $y'' = e^{-y'}$ admits three different one-parameter symmetry groups with the following infinitesimal generators

$$\mathbf{U}_1 = x\frac{\partial}{\partial x} + (x + y)\frac{\partial}{\partial y}, \ \mathbf{U}_2 = \frac{\partial}{\partial x}, \ \mathbf{U}_3 = \frac{\partial}{\partial y}.$$

The non-zero commutation relations between the generators are

$$[\mathbf{U}_1, \mathbf{U}_2] = -\mathbf{U}_1 - \mathbf{U}_2, \quad [\mathbf{U}_1, \mathbf{U}_3] = -\mathbf{U}_3.$$

It is easy to see that these generators span a three-dimensional Lie algebra $L_3 = \langle \mathbf{U}_1, \mathbf{U}_2, \mathbf{U}_3 \rangle$. The second order ODE admits the three-parameter Lie group of point transformations generated by this algebra. ◇

In the previous example we have seen that the second order ODE is invariant under three-parameter group of point transformations. In the first two examples we have seen that we can use one symmetry to reduce the order of equation by one. Is it possible to solve an ODE if we know a sufficient number of symmetries?

Usually the answer to this question is non-trivial and depends on the algebraic structure of the related symmetry algebra of the group admitted by the equation. In the next sections we give some hints on how to use the invariance of high order ODEs.

7.2 Second order ODEs with two symmetries

Sophus Lie provided a simple classification of two-dimensional real Lie groups of transformations on $\mathbb{R}^2$ based on the properties of their Lie algebras. These two–dimensional groups $G_2(\mathbb{R}^2)$ are especially relevant when we are solving second order ODEs with more than one–dimensional admitted Lie algebras. The main goal here is to reduce the second order ODE to simpler ones or to solve it using appropriate substitutions. Sophus Lie was also interested in transformations which reduce the given generators of the Lie algebra to a canonical form. Correspondingly a second order ODE will be reduced after the same variable transformations to a simpler one which can then be solved. In this way one gets either explicit or implicit solutions. We use such transformations many times in this book to provide the exact solutions in examples and problems discussed in this chapter and later in Chapter 9. The form of the substitutions depends on the algebraic and analytic properties of the generators $\mathbf{U}_1, \mathbf{U}_2$ of the admitted symmetry algebra. Let us denote the generators as

$$\mathbf{U}_1 = \xi_1(x, y)\frac{\partial}{\partial x} + \eta_1(x, y)\frac{\partial}{\partial y}, \quad \mathbf{U}_2 = \xi_2(x, y)\frac{\partial}{\partial x} + \eta_2(x, y)\frac{\partial}{\partial y}.$$

To describe the analytical properties of the two-dimensional algebras Sophus Lie introduced the function $\delta = \xi_1\eta_2 - \eta_1\xi_2$. The classification of two-dimensional symmetry algebras is given in the Table 7.2.

Table 7.2: In the first column of the table we provide the notations for group types using Sophus Lie classical notation. In the second and the third columns, the algebraic and analytic properties of the corresponding Lie algebras are displayed. In the fourth column the system of equations is presented which should be solved to find new variables $t(x, y), s(x, y)$. In the last column the original ODE $y'' = f(x, y)$ is presented in the new, reduced form in our new variables. Here we denoted $\delta(x, y) = \xi_1(x, y)\eta_2(x, y) - \eta_1(x, y)\xi_2(x, y)$.

Type	Commutator	δ	Substitution $(x, y) \to (t, s)$	Equation
G_2, Ia	$[\mathbf{U}_1, \mathbf{U}_2] = 0$	$\delta \neq 0$	$\mathbf{U}_1 s = 1, \mathbf{U}_1 t = 0$ $\mathbf{U}_2 s = 0, \mathbf{U}_2 t = 1$	$s'' = \omega(s')$
G_2, Ib	$[\mathbf{U}_1, \mathbf{U}_2] = 0$	$\delta = 0$	$\mathbf{U}_1 s = 1, \mathbf{U}_1 t = 0$ $\mathbf{U}_2 s = t, \mathbf{U}_2 t = 0$	$s'' = \omega(t)$
G_2, IIa	$[\mathbf{U}_1, \mathbf{U}_2] = \mathbf{U}_1$	$\delta \neq 0$	$\mathbf{U}_1 s = 1, \mathbf{U}_1 t = 0$ $\mathbf{U}_2 s = s, \mathbf{U}_2 t = t$	$s'' = t^{-1}\omega(s')$
G_2, IIb	$[\mathbf{U}_1, \mathbf{U}_2] = \mathbf{U}_1$	$\delta = 0$	$\mathbf{U}_1 s = 1, \mathbf{U}_1 t = 0$ $\mathbf{U}_2 s = s, \mathbf{U}_2 t = 0$	$s'' = s'\omega(t)$

To reduce the generators of a symmetry algebra to the canonical form he suggested introducing two new variables, a dependent variable $s(x, y)$ and an independent variable $t(x, y)$. The procedure of variable substitution in generators of Lie algebra is the same as described in Section 3.5. In Section 3.5 we proved that each generator can be reduced to the translation generator $\mathbf{U} = \frac{\partial}{\partial s}$. So far we deal with a two-dimensional Lie algebra, we cannot reduce both of the generators $\mathbf{U}_1, \mathbf{U}_2$ using the same transformation to the translation generators. We assume now that we choose new variables $s(x, y), t(x, y)$ so that the first generator takes the canonical form $\mathbf{U}_1 = \frac{\partial}{\partial s}$. Then the second one, $\mathbf{U}_2$, takes different forms depending on the algebraic and analytical structure of the admitted Lie algebra. To find the corresponding functions $s(x, y)$ and $t(x, y)$, we have to solve a system of (two by two) equations on these functions. These systems are presented in Table 7.2 in the fourth column.

After the transformation of the studied ODE to one of the canonical forms, listed in the last column of the Table 7.2, the equation can be integrated easily in two steps. To get the solutions in the original variables x, y one needs to just replace $t(x, y), s(x, y)$ using the known form of variable transformations $t = t(x, y)$ and $s = s(x, y)$.

7.3 A case of n-th order ODE which admits a high-dimensional solvable Lie algebra

First we consider the case when n-th order ODE admits r-parameter group of point transformations. Roughly speaking we have enough symmetries. It means that the dimension of the admitted symmetry group $G_r(\mathbb{R}^2)$ is at least n, i.e. $r \geq n$. Furthermore, we suppose that the corresponding algebra L_r possesses a solvable subalgebra of the order n. This subalgebra L_n will be the focus of our investigations.

As L_n is solvable, we can find a chain of ideals of the type

$$L_n \supset L_{n-1} \supset \ldots \supset L_1 \supset 0.$$

Because L_i, $i < n$, are not only subalgebras, but also ideals for all L_{i+1}, $i < n$, we can find one differential operator $\mathbf{U}_\alpha$ with the property

$$[\mathbf{U}_j, \mathbf{U}_\alpha] = \sum_{\substack{1 \leq \beta \leq i+1 \\ \beta \neq \alpha}} c^{\beta}_{j\alpha} \mathbf{U}_\beta, \quad j = 1, \ldots, i+1, \tag{7.3.9}$$

i.e. the right hand side does not contain the operator $\mathbf{U}_\alpha$. We can remunerate the basis elements of the algebra in the following way

$$\begin{aligned} L_n &= \langle \mathbf{U}_1, \mathbf{U}_2, \ldots, \mathbf{U}_n \rangle, \\ L_{n-1} &= \langle \mathbf{U}_2, \ldots, \mathbf{U}_n \rangle, \\ &\vdots \\ L_1 &= \langle \mathbf{U}_n \rangle. \end{aligned}$$

In general, if we find n symmetries admitted by a n-th order ODE, and the corresponding Lie algebra is solvable then we have the possibility to integrate the equation. This property of the algebra gave Sophus Lie the idea to call such algebras *solvable*.

This procedure might be carried out in different ways, but it can be represented in the following way: If we find n first integrals $\varphi_i(x, y, \ldots, y^{(n-1)})$, $i = 1, \ldots n$, i.e. the equation $\mathcal{A}\varphi_i(x, y, \ldots, y^{(n-1)}) = 0$ (the expression for the operator $\mathcal{A}$ is given in (7.1.3)) is satisfied for every i, then we can assume that these integrals are equal to constants

$$\varphi_i(x, y, \ldots, y^{(n-1)}) = c_i, \quad i = 1, \ldots, n. \tag{7.3.10}$$

Using this set of equations we can express all derivatives included in the original n-th order ODE and get a relation of the type

$$\omega(x, y, c_1, \ldots, c_n) = 0, \tag{7.3.11}$$

which represents the solution of the equation.

We look at this procedure the other way round. We try to find the first integrals and then use them as new variables. First we examine the admitted Lie algebra $L_n \supset L_{n-1}$ and take $\mathbf{U}_1 \in L_n$, $\mathbf{U}_1 \notin L_{n-1}$

$$[\mathbf{U}_i, \mathbf{U}_1] = \sum_{\beta=2}^{n} c_{i1}^{\beta} \mathbf{U}_{\beta}, \quad \mathbf{U}_{\beta} \in L_{n-1}. \tag{7.3.12}$$

Then we use one first integral, let us denote it by φ_1 as a new variable and represent the generator $\mathbf{U}_1$ as $\mathbf{U}_1 = \frac{\partial}{\partial \varphi_1}$, i.e. we take φ_1 as a canonical variable for this infinitesimal generator. This first integral φ_1 will be invariant under the action of all other infinitesimal generators $\mathbf{U}_{\beta}$, $\quad \mathbf{U}_{\beta} \in L_{n-1}$. Thus, we get the system of equations

$$\begin{aligned} \mathbf{U}_1 \varphi_1 &= 1, \\ \mathcal{A} \varphi_1 &= 0, \\ \mathbf{U}_2 \varphi_1 &= 0, \\ &\vdots \\ \mathbf{U}_n \varphi_1 &= 0. \end{aligned} \tag{7.3.13}$$

The first equation describes the condition that φ_1 is the new canonical variable, the second one is the condition on φ_1 to be a first integral and the remaining equations show that φ_1 is invariant under the action of the infinitesimal generators $\mathbf{U}_2, \ldots, \mathbf{U}_n$.

But up to this point we do not have any closed form expressions for first integrals φ_i, $1 \leq 1 \leq n$. Therefore, we should solve the system (7.3.13) first and find a non-constant solution to these n equations for the $(n+1)$ variables $x, y, y', \ldots, y^{(n-1)}$. The integrability condition for this system is the following one:

The set of first order differential operators $\{\mathcal{A}, \mathbf{U}_2, \ldots, \mathbf{U}_n\}$ has the property that

$$[\mathcal{A}, \mathbf{U}_i^{(n-1)}] = \sum_{\beta=2}^{n} \mu_i^{\beta} \mathbf{U}_{\beta}^{(n-1)} + \mu \mathcal{A}, \quad i = 2, \ldots, n. \tag{7.3.14}$$

Since the Lie group is admitted by this n-th order equation we know that

$$[\mathbf{U}_i^{(n-1)}, \mathcal{A}] = \lambda_i \mathcal{A}, \quad i = 1, \ldots, n, \tag{7.3.15}$$

$$[\mathbf{U}_i, \mathbf{U}_j] = \sum_{\beta=2}^{n} c_{ij}^{\beta} \mathbf{U}_{\beta}$$

(we will not get $\mathbf{U}_1$ in the last sum), and if we suggest now that

$$\mathbf{U}_1^{(n-1)} \neq \sum_{\beta=2}^{n} \varrho^b \mathbf{U}_b^{(n-1)} + \nu \mathcal{A},$$

the conditions (7.3.15) give us the full set of integrability conditions. Hence, we know that there exists a non-constant function φ_1 solving this system of equations. If we find φ_1, we should prove the condition $\mathbf{U}_1\varphi = 1$ and that this condition does not contradict any former conditions.

Let us take a closer look at the integrability conditions now. We should have

$$[\mathbf{U}_1^{(n-1)}, \mathcal{A}]\varphi_1 = 0,$$

but $\mathbf{U}_1$ is a symmetry admitted by $\mathcal{A}$, thus, this condition is fulfilled. Further we should have

$$[\mathbf{U}_1, \mathbf{U}_i]\varphi_1 = 0.$$

As $\mathbf{U}_1\varphi = 1$ and $[\mathbf{U}_1, \mathbf{U}_i] = \sum c_{1i}^{\beta}\mathbf{U}_\beta$ respectively $[\mathbf{U}_i, \mathbf{U}_j] = \sum c_{ij}^{\beta}\mathbf{U}_\beta$, we can ensure that this condition is fulfilled too, because L_{n-1} is an ideal in L_n.

In the case where the operators $\mathcal{A}$ and $\mathbf{U}_i^{(n-1)}$, $i = 1, \ldots, n$, are linear independent, it follows that the determinant of the system does not vanish. We obtain the first integral φ_1 and consequently the solution of the system (7.3.13) as the integral

$$\varphi_1(x, y, \ldots, y^{(n-1)}) = \int \frac{\det \begin{pmatrix} dx & dy & dy' & \ldots & dy^{(n-1)} \\ \xi_2 & \eta_2 & \eta_2^{(1)} & \ldots & \eta_2^{(n-1)} \\ \vdots & \vdots & \vdots & & \vdots \\ \xi_n & \eta_n & \eta_n^{(1)} & \ldots & \eta_n^{(n-1)} \\ 1 & y' & y'' & \ldots & f(x, y, y', \ldots, y^{(n-1)}) \end{pmatrix}}{\det \begin{pmatrix} \xi_1 & \eta_1 & \eta_1^{(1)} & \ldots & \eta_1^{(n-1)} \\ \xi_2 & \eta_2 & \eta_2^{(1)} & \ldots & \eta_2^{(n-1)} \\ \vdots & \vdots & \vdots & & \vdots \\ \xi_n & \eta_n & \eta_n^{(1)} & \ldots & \eta_n^{(n-1)} \\ 1 & y' & y'' & \ldots & f(x, y, y', \ldots, y^{(n-1)}) \end{pmatrix}} = c_1. \tag{7.3.16}$$

Now we introduce a new variable φ_1 instead of $y^{(n-1)}$ and get a representation of the algebra of infinitesimal generators $\mathbf{U}_i$ and the operator $\mathcal{A}$ in the form

$$\mathbf{U}_1 = \frac{\partial}{\partial \varphi_1},$$
$$\mathbf{U}_2^{(n-2)} = \xi_2 \frac{\partial}{\partial x} + \eta_2 \frac{\partial}{\partial y} + \ldots + \eta_2^{(n-2)} \frac{\partial}{\partial y^{(n-2)}},$$
$$\vdots$$
$$\mathcal{A} = \frac{\partial}{\partial x} + y' \frac{\partial}{\partial y} + \ldots + y^{(n-1)}(\varphi_1, x, y, \ldots, y^{(n-2)}) \frac{\partial}{\partial y^{(n-2)}},$$

with the variables $\varphi_1, x, y, \ldots, y^{(n-2)}$. The operators $\mathbf{U}_2^{(n-2)}, \mathbf{U}_3^{(n-2)}, \ldots, \mathbf{U}_n^{(n-2)}$ and $\mathcal{A}$ do not include any differentiation with respect to φ_1 variable.

The above system of equations corresponds to the differential equation of the order $(n-1)$

$$y^{(n-1)} = f(x, y, y', \ldots, y^{(n-2)}).$$

This equation admits $(n-1)$ symmetries $\mathbf{U}_2, \ldots, \mathbf{U}_n$ that depend on the constant φ_1 now.

At this point, we can repeat the procedure. We remark that $L_{n-1} \supset L_{n-2}$ and now $\mathbf{U}_2$ is the operator with the property

$$[\mathbf{U}_2, \mathbf{U}_i] = \sum_{\beta=3}^{n} c_{2i}^{\beta} \mathbf{U}_\beta,$$
$$[\mathbf{U}_i, \mathbf{U}_j] = \sum_{\beta=3}^{n} c_{ij}^{\beta} \mathbf{U}_\beta.$$

We get the system of equations to solve

$$\tilde{\mathbf{U}}_2 \varphi_2 = 1,$$
$$\mathcal{A} \varphi_2 = 0,$$
$$\tilde{\mathbf{U}}_3 \varphi_2 = 0,$$
$$\vdots$$
$$\tilde{\mathbf{U}}_n \varphi_2 = 0$$

where we denoted by $\tilde{\mathbf{U}}_i$ the operators $\mathbf{U}_i$ in new variables $\varphi_1, x, y, \ldots, y^{(n-2)}$. We suppose that also in this case the operators $\tilde{\mathbf{U}}_i$, $2 \leq i \leq n$ are linear independent and $\tilde{\mathbf{U}}_2^{(n-2)} \neq \sum \tilde{\varrho}^{\beta} \tilde{\mathbf{U}}_\beta^{(n-2)} + \tilde{\nu} \mathcal{A}$. We also know that all integrability conditions will be fulfilled, because L_{n-2} is an ideal in L_{n-1}. We continue the procedure along the chain of ideals and as a result we obtain a solution of the original equation. We demonstrate this procedure in the next example. This example was developed by Sophus Lie and is also described in [75].

Example 7.3.1 Let us consider the third order differential equation

$$y''' = \frac{3}{2}\frac{(y'')^2}{y'}.$$

It admits the symmetries

$$\mathbf{U}_1 = \frac{\partial}{\partial y}, \quad \mathbf{U}_2 = x\frac{\partial}{\partial x}, \quad \mathbf{U}_3 = \frac{\partial}{\partial x}$$

and the operator $\mathcal{A}$ has the form

$$\mathcal{A} = \frac{\partial}{\partial x} + y'\frac{\partial}{\partial y} + y''\frac{\partial}{\partial y'} + \frac{3}{2}\frac{(y'')^2}{y'}\frac{\partial}{\partial y''}.$$

The second prolongation of $\mathbf{U}_2$ is

$$\mathbf{U}_2^{(2)} = x\frac{\partial}{\partial x} - y'\frac{\partial}{\partial y'} - 2y''\frac{\partial}{\partial y''}.$$

Because the commutator table has the form

	$\mathbf{U}_1$	$\mathbf{U}_2$	$\mathbf{U}_3$
$\mathbf{U}_1$	0	0	0
$\mathbf{U}_2$	0	0	$-\mathbf{U}_3$
$\mathbf{U}_3$	0	$\mathbf{U}_3$	0

we see that the corresponding algebra is solvable. Hence, we can use the procedure described above to solve the equation. We start with the generator $\mathbf{U}_1$. We calculate

$$\begin{aligned}
\Xi &= \det \begin{pmatrix} \xi_1 & \eta_1 & \eta_1^{(1)} & \dots & \eta_1^{(n-1)} \\ \xi_2 & \eta_2 & \eta_2^{(1)} & \dots & \eta_2^{(n-1)} \\ \vdots & \vdots & \vdots & & \vdots \\ \xi_n & \eta_n & \eta_n^{(1)} & \dots & \eta_n^{(n-1)} \\ 1 & y' & y'' & \dots & f(x, y, y', \dots, y^{(n-1)}) \end{pmatrix} \\
&= \begin{vmatrix} 0 & 1 & 0 & 0 \\ x & 0 & -y' & -2y'' \\ 1 & 0 & 0 & 0 \\ 1 & y' & y'' & \frac{3}{2}\frac{(y'')^2}{y'} \end{vmatrix} \\
&= \begin{vmatrix} 1 & 0 & 0 \\ 0 & -y' & -2y'' \\ y' & y'' & \frac{3}{2}\frac{(y'')^2}{y'} \end{vmatrix} = -\frac{3}{2}(y'')^2 + 2(y'')^2 = \frac{(y'')^2}{2} \neq 0,
\end{aligned}$$

and thus

$$\varphi_1 = \int \frac{1}{\Xi} \begin{vmatrix} dx & dy & dy' & dy'' \\ x & 0 & -y' & -2y'' \\ 1 & 0 & 0 & 0 \\ 1 & y' & y'' & \frac{3}{2}\frac{(y'')^2}{y'} \end{vmatrix} = \int \frac{1}{\Xi} \begin{vmatrix} dy & dy' & dy'' \\ 0 & -y' & -2y'' \\ y' & y'' & \frac{3}{2}\frac{(y'')^2}{y'} \end{vmatrix}$$

$$= \int \frac{\frac{1}{2}(y'')^2\,dy - 2y'y''\,dy' + (y')^2\,dy''}{\frac{1}{2}(y'')^2} = \int dy + 2\left(\frac{y'}{y''}\right)^2 dy'' - 4\frac{y'}{y''}\,dy'$$

$$= y - 2\frac{(y')^2}{y''} = \varphi_1.$$

This is equivalent to

$$y'' = \frac{2(y')^2}{y - \varphi_1}.$$

Now we can use the operator $\mathbf{U}_3$ and calculate

$$\tilde{\Xi} = \begin{vmatrix} x & 0 & -y' \\ 1 & 0 & 0 \\ 1 & y' & y'' \end{vmatrix} = -(y')^2$$

and then

$$\varphi_2 = \int \frac{1}{\tilde{\Xi}} \begin{vmatrix} dx & dy & dy' \\ 1 & 0 & 0 \\ 1 & y' & y'' \end{vmatrix} = \int \frac{y'\,dy' - y''\,dy}{-(y')^2} = \int -\frac{1}{y'}\,dy' + \frac{2}{y - \varphi_1}\,dy$$

$$= -\ln y' + 2\ln(y - \varphi_1).$$

Equivalently we have

$$\tilde{\varphi}_2 = \frac{(y - \varphi_1)^2}{y'} \quad \Leftrightarrow \quad y' = \frac{(y - \varphi_1)^2}{\tilde{\varphi}_2}$$

and we can solve the last equation to obtain the solution of the original third order ODE

$$y = \varphi_1 - \frac{\tilde{\varphi}_2}{x + \tilde{\varphi}_3}, \quad \varphi_1, \tilde{\varphi}_2, \tilde{\varphi}_3 \in \mathbb{R}.$$

◇

7.4 A case of n-th order ODEs which are invariant under a non-solvable Lie algebra with a special property

In this section we consider a non-solvable algebra L_r. We look first at the first derived algebra $L_r^{(1)}$. Let this algebra be spanned by the infinitesimal generators $\mathbf{U}_{\alpha_i}$, $i = 1, \ldots, k$, and we can choose an operator $\mathbf{U}_{\alpha_i}$ that admits the maximal number of operators $\mathbf{U}_j$ with the property

$$[\mathbf{U}_j, \mathbf{U}_{\alpha_i}] = \mathbf{U}_j, \quad j = 1, \ldots, l. \tag{7.4.17}$$

We denote this operator as $\mathbf{U}_{\alpha_1}$ and introduce new variables $s(x, y)$ and $t(x, y)$ in the following way

$$\begin{aligned} \mathbf{U}_{\alpha_1} s &= 1, \\ \mathbf{U}_{\alpha_1} t &= 0. \end{aligned}$$

Using these variables we reduce the original Equation (7.0.1) to a $(n-1)$-th order equation

$$z^{(n-1)} = \tilde{f}(t, z, \ldots, z^{(n-2)}), \quad z = s',$$

the substitution $z = s'$ was possible because of translation invariance of Equation (the operator $\mathbf{U}_{\alpha_1}$ is reduced to the normal form in the new variables). The left operators $\mathbf{U}_{\alpha_i}$, $i = 2, \ldots, l$ from (7.4.17) build the $(l-1)$–dimensional algebra L_{l-1}, and they are the symmetry generators of the reduced equation. We can repeat this procedure and calculate the first derived algebra $L_l^{(1)}$ of this operators and apply the above described procedure one more and so on. The question now is: how many of our original generators can we use in this way? The answer to this question depends heavily on the structure of the algebra L_r. Usually not every admitted symmetry described by $\mathbf{U}_{\alpha_i}$,$i = 1, \ldots, l$ can be used in this way. Why does it happen?

Let us look at the procedure we used for the first operator $\mathbf{U}_{\alpha_1}$ more carefully. First we found the operators $\mathbf{U}_1, \ldots, \mathbf{U}_r$ that span an algebra L_r. Then we pick one of them $\mathbf{U}_{\alpha_1}$, let us denote them as $\mathbf{U}_1$ and change the variables to present this operator in the normal form $\mathbf{U}_1 = \frac{\partial}{\partial s}$. But now the other infinitesimal generators and the operator $\mathcal{A}$ of the original equation change as well. After variable substitution $(x, y) \to (t, s)$ they have the following form

$$\begin{aligned} \mathcal{A} &= \frac{\partial}{\partial t} + s' \frac{\partial}{\partial s} + s'' \frac{\partial}{\partial s'} + \ldots + \tilde{\omega}(t, s, \ldots) \frac{\partial}{\partial s^{(n-1)}}, \\ \mathbf{U}_1 &= \frac{\partial}{\partial s}, \\ \mathbf{U}_i^{(n-1)} &= \xi_i \frac{\partial}{\partial t} + \eta_i \frac{\partial}{\partial s} + \eta_i^{(1)} \frac{\partial}{\partial s'} + \ldots + \eta_i^{(n-1)} \frac{\partial}{\partial s^{(n-1)}}, \quad i = 2, \ldots, r. \end{aligned} \tag{7.4.18}$$

The original ODE in the variables t, s takes the form $s^{(n)} = \tilde{f}(t, s', \ldots, s^{(n-1)})$ because it admits the symmetry generated by $\mathbf{U}_1$. If we take in this ODE the variable s' instead of s as a new dependent variable, then we reduce the order of ODE by one. It means also that we look for $(n-1)$ first integrals $\varphi_j(t, s', \ldots, s^{(n-1)})$ (i.e. $\mathcal{A}\varphi_j = 0$, $j = 1, \ldots, n-1$) that do not depend on s. So we obtain the new form of the operator $\mathcal{A}$ which corresponds to the reduced ODE

$$\tilde{\mathcal{A}} = \frac{\partial}{\partial t} + s'' \frac{\partial}{\partial s'} + \ldots + \tilde{\omega}(t, s', \ldots) \frac{\partial}{\partial s^{(n-1)}}. \tag{7.4.19}$$

If we take a look at the generators $\mathbf{U}_i$, we also have to exclude the s-dependence and get

$$\tilde{\mathbf{U}}_i = \xi_i \frac{\partial}{\partial t} + \eta_i^{(1)} \frac{\partial}{\partial s'} + \ldots + \eta_i^{(n-1)} \frac{\partial}{\partial s^{(n-1)}}, \quad i = 2, 3, \ldots, n. \tag{7.4.20}$$

However there is a problem with the procedure described above. Excluding the s-dependence we can destroy the symmetry relations and the new operators $\tilde{\mathbf{U}}_i$, $i = 2, 3, \ldots, n$ should be admitted symmetries of $\tilde{\mathcal{A}}$, i.e.

$$[\mathbf{U}_i, \mathcal{A}] = \lambda_i \mathcal{A} \quad \not\Rightarrow \quad [\tilde{\mathbf{U}}_i, \tilde{\mathcal{A}}] = \tilde{\lambda}_i \tilde{\mathcal{A}}.$$

To make it visible we rewrite $[\mathbf{U}_i, \mathcal{A}]$ as $\left[\tilde{\mathbf{U}}_i + \eta_i \frac{\partial}{\partial s}, \tilde{\mathcal{A}} + s' \frac{\partial}{\partial s}\right]$ and calculate

$$\begin{aligned}
\lambda_i \mathcal{A} &= \left[\tilde{\mathbf{U}}_i + \eta_i \frac{\partial}{\partial s}, \tilde{\mathcal{A}} + s' \frac{\partial}{\partial s}\right] = [\tilde{\mathbf{U}}_i + \eta_i \mathbf{U}_1, \tilde{\mathcal{A}} + s' \mathbf{U}_1] \\
&= [\tilde{\mathbf{U}}_i + \eta_i \mathbf{U}_1, \tilde{\mathcal{A}}] + [\underbrace{\tilde{\mathbf{U}}_i + \eta_i \mathbf{U}_1}_{=\mathbf{U}_i}, s' \mathbf{U}_1] \\
&= [\tilde{\mathbf{U}}_i, \tilde{\mathcal{A}}] + [\eta_i \mathbf{U}_1, \tilde{\mathcal{A}}] + [\mathbf{U}_i, s' \mathbf{U}_1] \\
&= [\tilde{\mathbf{U}}_i, \tilde{\mathcal{A}}] + \eta_i \underbrace{[\mathbf{U}_1, \tilde{\mathcal{A}}]}_{=0} - (\tilde{\mathcal{A}} \eta_i) \mathbf{U}_1 + s' [\mathbf{U}_i, \mathbf{U}_1] + \underbrace{(\mathbf{U}_i s')}_{=\eta_i^{(1)}} \mathbf{U}_1 \\
&= [\tilde{\mathbf{U}}_i, \tilde{\mathcal{A}}] - (\tilde{\mathcal{A}} \eta_i) \mathbf{U}_1 - s' [\mathbf{U}_1, \mathbf{U}_i] + \eta_i^{(1)} \mathbf{U}_1 \\
&= \lambda_i (\tilde{\mathcal{A}} + s' \mathbf{U}_1).
\end{aligned}$$

If we suppose

$$[\tilde{\mathbf{U}}_i, \tilde{\mathcal{A}}] = \mu_i \tilde{\mathcal{A}}$$

now the relation

$$[\mathbf{U}_1, \mathbf{U}_i] = \tau_i \mathbf{U}_1 + \sigma_i \tilde{\mathcal{A}}$$

has to apply accordingly. We obtain the relation

$$\lambda_i (\tilde{\mathcal{A}} + s' \mathbf{U}_1) = \mu_i \tilde{\mathcal{A}} - (\tilde{\mathcal{A}} \eta_i) \mathbf{U}_1 - s' (\tau_i \mathbf{U}_1 + \sigma_i \tilde{\mathcal{A}}) + \eta_i^{(1)} \mathbf{U}_1,$$

or, correspondingly, the system of equations

$$\mu_i - s'\sigma_i = \lambda_i,$$
$$-\tilde{A}\eta_i - s'\tau_i + \eta_i^{(1)} = \lambda_i s'$$

should be satisfied. So $\sigma_i = 0$ and further $[\mathbf{U}_1, \mathbf{U}_i] = \tau_i \mathbf{U}_1$, where τ_i is a constant.

Therefore, we know that all symmetries which fulfilled this condition before are still symmetries after the change of variables, all other symmetries will be destroyed.

7.5 A case of a n-th order ODE with a general type of the admitted symmetry algebra

In this section we take a n-th order ODE with an algebra without any special properties, i.e. of the general type. We suppose that the algebra does not include any operator $\mathbf{U}_\alpha$ with the property

$$[\mathbf{U}_i, \mathbf{U}_\alpha] = \text{const.} \cdot \mathbf{U}_i, \quad i \neq \alpha, \quad i, \alpha = 1, 2, \ldots, r.$$

This means, we can use just one symmetry from the r–dimensional symmetry group $G_r(\mathbb{R}^2)$ to reduce the order of ODE by one like it was described in the previous section. We can use an arbitrary infinitesimal generator $\mathbf{U}_i$, but only one of them, because after the described variable substitutions all other symmetries will be destroyed.

However, there exists another method to simplify the Equation (7.1.2).

We examine the situation where we are able to calculate the r–dimensional symmetry algebra $L_r = \langle \mathbf{U}_1, \ldots, \mathbf{U}_r \rangle$. To solve or to simplify the n-th order ODE we can try to find all invariants and differential invariants of the corresponding group $G_r(\mathbb{R}^2)$, i.e. we try to solve the characteristic systems for the infinitesimal generators $\mathbf{U}_m$, $1 \leq m \leq r$ and their prolongations up to the order $k \leq n$

$$\begin{array}{lcl} \mathbf{U}_1 u(x,y) = 0, & & \mathbf{U}_1^{(1)} v(x,y,y') = 0, & & \mathbf{U}_1^{(k)} i_k(x,y,\ldots,y^{(k)}) = 0, \\ \mathbf{U}_2 u(x,y) = 0, & \Rightarrow & \mathbf{U}_2^{(1)} v(x,y,y') = 0, & \Rightarrow & \mathbf{U}_2^{(k)} i_k(x,y,\ldots,y^{(k)}) = 0, \\ \vdots & & \vdots & & \vdots \\ \mathbf{U}_r u(x,y) = 0, & & \mathbf{U}_r^{(1)} v(x,y,y') = 0, & & \mathbf{U}_r^{(k)} i_k(x,y,\ldots,y^{(k)}) = 0. \end{array}$$

Not all of these systems might have solutions. But as we take higher order jet bundles step-by-step, we add more and more variables and obtain a system of equations which may become a solvable one, i.e. we get a non-trivial solution to one of this systems.

Let us consider the case, where we find that the lowest order of the differential invariants is j, i.e. the system

$$\begin{aligned}
\mathbf{U}_m^{(j)} i_j(x, y, \ldots, y^{(j)}) &= 0, \quad m = 1, \ldots, r, \\
\mathbf{U}_m^{(j+1)} i_{j+1}(x, y, \ldots, y^{(j+1)}) &= 0, \quad m = 1, \ldots, r, \\
&\vdots \\
\mathbf{U}_m^{(n)} i_n(x, y, \ldots, y^{(n)}) &= 0, \quad m = 1, \ldots, r, \ j \leq n
\end{aligned}$$

has a solution. Then we have all differential invariants i_α from the order j up to the order n and can represent them as

$$i_{j+2} = \frac{di_{j+1}}{di_j}, \quad i_{j+3} = \frac{di_{j+2}}{di_j}, \quad \ldots$$

Thereafter, we start with the highest derivative $y^{(n)}$ in the given ODE and replace them and lower derivatives by derivatives of i_j, i_{j+1} to obtain the new equation

$$\omega(i_j, i_{j+1}, \ldots, i_{n-j}) = 0 \tag{7.5.21}$$

from Equation (7.1.2). It means, we get a differential equation of the order $(n - j)$. If we solve this, we get

$$i_{j+1}(x, y, \ldots, y^{(j+1)}) = \varphi(i_j(x, y, \ldots, y^{(j)})) \tag{7.5.22}$$

a differential equation of the order j with the symmetry group G_r and orbits of dimension j, $r > j$. In this way we split the original n-th order Equation (7.1.2) and obtain two simpler equations (7.5.21) of the order $(n - j)$ and (7.5.22) of order j instead of one of the order n. We use this procedure in the next example.

Example 7.5.1 We study the differential equation

$$y''' = yy'' - (y')^2.$$

We find that this equation admits two symmetries with generators

$$\mathbf{U}_1 = \frac{\partial}{\partial x}, \quad \mathbf{U}_2 = x\frac{\partial}{\partial x} - y\frac{\partial}{\partial y}.$$

The equation does not have any coefficient dependent on the variable x, because of that the first generator $\mathbf{U}_1$ will be admitted by the equation.

Then the differential invariants i_j, $j = 1, 2, \ldots$ cannot be functions of x also. We calculate the third prolongation of $\mathbf{U}_2$ and obtain

$$\mathbf{U}_2^{(3)} = x\frac{\partial}{\partial x} - y\frac{\partial}{\partial y} - 2y'\frac{\partial}{\partial y'} - 3y''\frac{\partial}{\partial y''} - 4y'''\frac{\partial}{\partial y'''},$$

therefore to find differential invariant of the third order we have to solve

$$\mathbf{U}_2^{(3)} i_3(y, y', y'', y''') = 0.$$

We can calculate the third order differential invariant if we first calculate invariants of the first order i_1 (we denoted i_1 by v) and the second order i_2

$$v(x, y, y') = \frac{y'}{y^2}, \quad i_2(x, y, y', y'') = \frac{y''}{y^3}$$

and obtain i_3 by invariant differentiation as described before

$$i_3 = \frac{di_2}{dv} = \frac{\frac{y'''}{y^3} - 3\frac{y'y''}{y^4}}{\frac{y''}{y^2} - 2\frac{(y')^2}{y^3}} = \frac{yy''' - 3y'y''}{y(yy'' - 2(y')^2)}.$$

Now we rearrange this expression

$$y''' = \frac{1}{y}(i_2' y(yy'' - 2(y')^2) + 3y'y'') = yy'' - (y')^2,$$

where we use the original equation in the last relation. If we put $i_2 = \frac{y''}{y^3}$ and $v = \frac{y'}{y^2}$ in the last term, we get

$$i_2'(y^4\frac{y''}{y^3} - 2(y')^2) + 3\frac{y^3}{y}\frac{y''}{y^3}y' = y^4\frac{y''}{y^3} - (y')^2,$$
$$\text{or} \quad i_2'(y^4 i_2 - 2y^4v^2) + 3y^4 i_2 v = y^4 i_2 - y^4 v^2.$$

It means we get a first order ODE

$$i_2'(i_2 - 2v^2) + 3i_2 v = i_2 - v^2. \tag{7.5.23}$$

We see that i_2 is a solution of this first order ODE and $i_2 = f(v)$. We have calculated i_2 and using its form we get the second equation in the form

$$\frac{y''}{y^3} = f\left(\frac{y'}{y^2}\right). \tag{7.5.24}$$

It is a second order ODE. It means we reduced the original third order equation to two simpler equations respectively of the first order (7.5.23) and the second order Equation (7.5.24). ◇

7.6 Problems with solutions

Here we provide numerous problems with detailed solutions. The solutions may be rather voluminous but we feel they will help the reader become familiar with the method. In Problems 7.6.1 and 7.6.2 we use the prolongation procedure for the given infinitesimal generators and check if a high order ODE is invariant under the action of this Lie group. We use the provided symmetries to solve high order ODEs in Problems 7.6.3, 7.6.5, 7.6.7 and 7.6.8.

The main practical skill of this chapter is the method of solving determining equations and the reader has the chance to practice this in Problems 7.6.4, 7.6.6, 7.6.9 and 7.6.13. Because the solution to such problems are usually very voluminous, we show the reader how to make use of the software package **IntroToSymmetry** from **Mathematica**. In the solution to Problem 7.6.10, we provide a screenshot of the calculations of the symmetry algebra using this package. In Problems 7.6.11 and 7.6.12, we assume that the reader knows how to use this package to find the infinitesimal generators of the symmetry algebra. We then provide the corresponding solutions.

We hope that after this practice the last two chapters of this book will be easily accessible for the reader.

Problem 7.6.1 *Prove using a prolongation of the infinitesimal generator* **U** *that the second order differential equation*

$$y'' = x(y - xy')^2$$

admits the symmetry group with the following infinitesimal generator

$$\mathbf{U} = x\frac{\partial}{\partial y}.$$

Solution Using the formula

$$\eta^{(n)} = \frac{d\eta^{(n-1)}}{dx} - y^{(n)}\frac{d\xi}{dx} \tag{7.6.25}$$

we can find the first prolongation of the infinitesimal generator **U**

$$\eta^{(1)}(x,y,y') = \frac{d\eta}{dx} - y'\frac{d\xi}{dx} = 1, \qquad \mathbf{U}^{(1)} = x\frac{\partial}{\partial y} + \frac{\partial}{\partial y'}.$$

An operator $\mathcal{A}$ corresponding to the given second order ODE looks as follows

$$\mathcal{A} = \frac{\partial}{\partial x} + y'\frac{\partial}{\partial y} + x(y - xy')^2\frac{\partial}{\partial y'}.$$

Now we can find the commutator of $\mathbf{U}^{(1)}$ and $\mathcal{A}$

$$[\mathbf{U}^{(1)},\mathcal{A}] = \left(x\frac{\partial}{\partial y}+\frac{\partial}{\partial y'}\right)\left(\frac{\partial}{\partial x}+y'\frac{\partial}{\partial y}+x(y^2-2xyy'+x^2y'^2)\frac{\partial}{\partial y'}\right)$$
$$-\left(\frac{\partial}{\partial x}+y'\frac{\partial}{\partial y}+x(y^2-2xyy'+x^2y'^2)\frac{\partial}{\partial y'}\right)\left(x\frac{\partial}{\partial y}+\frac{\partial}{\partial y'}\right)$$
$$=\left(2x^2y\frac{\partial}{\partial y'}-2x^2y'\frac{\partial}{\partial y'}+\frac{\partial}{\partial y}\right)-\left(2x^2y\frac{\partial}{\partial y'}-2x^2y'\frac{\partial}{\partial y'}+\frac{\partial}{\partial y}\right)=0.$$

This explicitly proves that the given equation admits the symmetry generated by $\mathbf{U}$.

Problem 7.6.2 *Prove that the second order differential equation*

$$yy''=(y')^2+y^2x^2+xyy'-y^2\ln y$$

admits the symmetry group with the following infinitesimal generator

$$\mathbf{U}=xy\frac{\partial}{\partial y}$$

using the commutation relation between operators $\mathbf{U}^{(1)}$ and $\mathcal{A}$.

Solution Functions $\xi(x,y)$ and $\eta(x,y)$ are the coefficients of the infinitesimal generator $\mathbf{U}$ and have the form

$$\xi(x,y)=0,\ \eta(x,y)=xy,\ \Rightarrow\ \eta^{(1)}(x,y,y')=\eta_x+\eta_y y'-y'(\xi_x+\xi_y y')=y+xy'.$$

The first prolongation of the infinitesimal generator $\mathbf{U}$ has the following form

$$\mathbf{U}^{(1)}=xy\frac{\partial}{\partial y}+(y+xy')\frac{\partial}{\partial y'}.$$

An operator $\mathcal{A}$ has the following form

$$\mathcal{A}=\frac{\partial}{\partial x}+y'\frac{\partial}{\partial y}+f(x,y,y')\frac{\partial}{\partial y'},\ f(x,y,y')=\frac{y'^2}{y}+yx^2+xy'-y\ln y.$$

Then the commutator is equal to

$$
\begin{aligned}
[\mathbf{U}^{(1)}, \mathcal{A}] &= \mathbf{U}^{(1)}\mathcal{A} - \mathcal{A}\mathbf{U}^{(1)} \\
&= \left(xy\frac{\partial}{\partial y} + (y + xy')\frac{\partial}{\partial y'}\right)\left(\frac{\partial}{\partial x} + y'\frac{\partial}{\partial y} + \left(\frac{y'^2}{y} + yx^2 + xy' - y\ln y\right)\frac{\partial}{\partial y'}\right) \\
&- \left(\frac{\partial}{\partial x} + y'\frac{\partial}{\partial y} + \left(\frac{y'^2}{y} + yx^2 + xy' - y\ln y\right)\frac{\partial}{\partial y'}\right)\left(xy\frac{\partial}{\partial y} + (y + xy')\frac{\partial}{\partial y'}\right) \\
&= \left(\frac{-xyy'^2}{y^2} + yx^3 - xy\ln y - xy\right)\frac{\partial}{\partial y'} + (y + xy')\frac{\partial}{\partial y} \\
&+ (y + xy')\left(2\frac{y'}{y} + x\right)\frac{\partial}{\partial y'} - y\frac{\partial}{\partial y} - y'\frac{\partial}{\partial y'} - y'x\frac{\partial}{\partial y} - y'\frac{\partial}{\partial y'} \\
&- x\left(\frac{y'^2}{y} + yx^2 + xy' - y\ln y\right)\frac{\partial}{\partial y'} \\
&= \left(\frac{-xy'^2}{y} + yx^3 - xy\ln y - xy + 2y' + \frac{2xy'^2}{y} + yx + x^2y' - \frac{xy'^2}{y} - yx^3\right. \\
&\left. - x^2y' + xy\ln y\right)\frac{\partial}{\partial y'} = 0.
\end{aligned}
$$

It means that there exists λ such that $[\mathbf{U}^{(1)}, \mathcal{A}] = \lambda\mathcal{A}$ and $\lambda = 0$. It means also that the second order differential equation admits the symmetry group.

Problem 7.6.3 *The second order differential equation*

$$y^{''} = y + x^2$$

admits two-dimensional symmetry algebra spanned by

$$\mathbf{U}_1 = e^x\frac{\partial}{\partial y} \quad \text{and} \quad \mathbf{U}_2 = e^{-x}\frac{\partial}{\partial y}.$$

Solve this equation using the symmetries (the problem setting is from [75]).

Solution We prove that the operators belong to the symmetry algebra admitted by this equation. We take the second prolongation from $\mathbf{U}_2^{(2)}$ and obtain $\mathbf{U}_2^{(2)} = e^{-x}\frac{\partial}{\partial y} - e^{-x}\frac{\partial}{\partial y'} - e^{-x}\frac{\partial}{\partial y''}$. From the determining equation

$$\mathbf{U}_1^{(2)}(y^{''} - y - x^2)|_{y''=y+x^2} = e^{-x} - e^{-x} = 0,$$

we obtain that this operator belongs to the symmetry algebra of the given equation. In a similar way it is also possible to prove that the operator $\mathbf{U}_1 = e^x\frac{\partial}{\partial y}$ describes a symmetry of the original equation. The operators $\mathbf{U}_1$

and $\mathbf{U}_2$ commute and build a symmetry algebra of the original equation of an Abelian type. In this case we can choose several ways to obtain a solution, but we will show only one of them.

Let us introduce the canonical variables $s(x,y)$ and $t(x,y)$ for the second infinitesimal generator

$$\begin{aligned} \mathbf{U}_2 s(x,y) &= 1, \\ \mathbf{U}_2 t(x,y) &= 0. \end{aligned}$$

Both equations are rather easy to solve, so, we obtain $s = ye^x$ and $t = x$. Now we can find the expression for the second derivative of y

$$\begin{aligned} y &= se^{-t}, \\ y' &= -se^{-t} + s'e^{-t}, \\ y'' &= se^{-t} - 2s'e^{-t} + s''e^{-t}, \end{aligned}$$

and rewrite our equation as follows

$$s'' = 2s' + t^2e^t.$$

If we use a substitution $z = s'$ we reduce it to a first order ODE

$$z' = 2z + t^2e^t.$$

This is an ordinary differential equation of the first order where the variables can be separated if we set $z = re^{2t}$ (where r is a new variable)

$$\begin{aligned} r'(t) &= t^2e^{-t}, \\ r(t) &= -e^{-t}(t^2 + 2t + 2) + C_1, \quad C_1 \in \mathbb{R}. \end{aligned}$$

If we return to our variable z we will obtain

$$z(t) = -e^t(t^2 + 2t + 2) + C_1e^{2t},$$

or after the second integration in terms of t, s

$$s(t) = -e^t(t^2 + 2) + C_1e^{2t} + C_2, \quad C_1, C_2 \in \mathbb{R},$$

and in terms of the original variables x and y

$$y = -x^2 - 2 + C_1e^x + C_2e^{-x}, \tag{7.6.26}$$

where C_1 and C_2 are arbitrary constants.

Problem 7.6.4 *Find the point symmetries of the following equation*

$$y'' - yy' + 2k(y')^2 = 0, \quad k \in \mathbb{R}.$$

Solution The determining equation for the symmetry group generators that admits our equation is

$$\mathbf{U}^{(2)}\Delta|_{\Delta=0} = 0, \quad \text{where} \quad \Delta = y'' - yy' + 2k(y')^2,$$

or

$$\left(\xi\frac{\partial}{\partial x} + \eta\frac{\partial}{\partial y} + \eta^{(1)}\frac{\partial}{\partial y'} + \eta^{(2)}\frac{\partial}{\partial y''}\right)\left(y'' - yy' + 2k(y')^2\right)|_{y''=yy'-2k(y')^2} = 0.$$

It means that we have to solve

$$-y'\eta - y\eta^{(1)} + 4ky'\eta^{(1)} + \eta^{(2)}|_{y''=yy'-2k(y')^2} = 0.$$

Now, we recall the explicit formulas for $\eta^{(1)}$ and $\eta^{(2)}$

$$\begin{aligned}
\eta^{(1)} &= \eta_x + \eta_y y' - y'\xi_x - {y'}^2\xi_y, \\
\eta^{(2)} &= \eta_{xx} + (2\eta_{xy} - \xi_{xx})y' \\
&+ (\eta_{yy} - 2\xi_{xy}){y'}^2 - \xi_{yy}{y'}^3 + (\eta_y - 2\xi_x)y'' - 3\xi_y y'y''
\end{aligned}$$

and substitute them into the determining equation. We obtain

$$\begin{aligned}
&- y'\eta - y(\eta_x + \eta_y y' - y'\xi_x - {y'}^2\xi_y) + 4ky'(\eta_x + \eta_y y' - y'\xi_x - {y'}^2\xi_y) \\
&+ \eta_{xx} + (2\eta_{xy} - \xi_{xx})y' + (\eta_{yy} - 2\xi_{xy}){y'}^2 - \xi_{yy}{y'}^3 \\
&+ (\eta_y - 2\xi_x)(yy' - 2k(y')^2) - 3\xi_y y'(yy' - 2k(y')^2) = 0.
\end{aligned}$$

Since this is a polynomial function with respect to y', each coefficient by monomials for different powers of y' should be equal to zero. We obtain the following system of equations

$$\begin{aligned}
(y')^3 \ &| \ 2k\xi_y - \xi_{yy} = 0, \\
(y')^2 \ &| \ \eta_{yy} - 2\xi_{xy} + 2k\eta_y - 2y\xi_y = 0, \\
y' \ &| \ -\eta - y\xi_x + 4k\eta_x + 2\eta_{xy} - \xi_{xx} = 0, \\
0 \ &| \ \eta_{xx} - y\eta_x = 0.
\end{aligned}$$

From the first Equation (the coefficient for the third power of y') we immediately obtain

$$\xi = e^{2ky}a(x) + b(x),$$

and the last Equation (for the coefficient by the zero power of y') gives us

$$\eta = e^{xy}c(y) + d(y),$$

where $a(x)$, $b(x)$, $c(y)$, $d(y)$ are arbitrary functions of their arguments. Now we can calculate partial derivatives of ξ and η

$$\begin{aligned}
\eta_y &= e^{xy}c'(y) + d'(y) + xe^{xy}c(y),\\
\eta_x &= ye^{xy}c(y),\\
\eta_{yy} &= e^{xy}c''(y) + d''(y) + 2xe^{xy}c'(y) + x^2e^{xy}c(y),\\
\eta_{xy} &= ye^{xy}c'(y) + e^{xy}c(y) + xye^{xy}c(y),\\
\xi_x &= e^{2ky}a'(x) + b'(x),\\
\xi_{xx} &= e^{2ky}a''(x) + b''(x),\\
\xi_{xy} &= 2ke^{2ky}a'(x),\\
\xi_y &= 2ke^{2ky}a(x),
\end{aligned}$$

and substitute them into the left equations of the system (for the the coefficients before the first and second power of y') to obtain the result as follows

$$\begin{aligned}
&e^{xy}c''(y) + d''(y) + 2xe^{xy}c'(y) + x^2e^{xy}c(y) - 4ke^{2ky}a'(x)\\
&+ 2ke^{xy}c'(y) + 2kd'(y) + 2kxe^{xy}c(y) - 4yke^{2ky}a(x) = 0, \qquad (7.6.27)\\
&- e^{xy}c(y) - d(y) - ye^{2ky}a'(x) - yb'(x) + 4kye^{xy}c(y) + 2ye^{xy}c'(y)\\
&+ 2e^{xy}c(y) + 2xye^{xy}c(y) - e^{2ky}a''(x) - b''(x) = 0. \qquad (7.6.28)
\end{aligned}$$

Let us look at the first Equation (7.6.27). We rewrite this equation in the form

$$\begin{aligned}
x^2c(y)e^{xy} + 2xe^{xy}(c'(y) + kc(y)) + e^{xy}(c''(y) + 2kc'(y))&\\
- 4yke^{2ky}(a'(x) + a(x)) + d''(y) + 2kd'(y) = 0.& \qquad (7.6.29)
\end{aligned}$$

We divide both parts of the equation by ye^{2ky}. Then this equation takes the form

$$\begin{aligned}
x^2e^{xy}F_1(y) + xe^{xy}F_2(y) + e^{xy}F_3(y) + F_4(x) + F_5(y) = 0, \qquad (7.6.30)\\
F_1(y) = c(y)y^{-1}e^{-2ky}, \quad F_2(y) = 2(c'(y) + kc(y))y^{-1}e^{-2ky},\\
F_3(y) = (c''(y) + 2kc'(y))y^{-1}e^{-2ky}, \quad F_4(x) = -4k(a'(x) + a(x)),\\
F_5(y) = (d''(y) + 2kd'(y))y^{-1}e^{-2ky}. \qquad (7.6.31)
\end{aligned}$$

Let us differentiate Equation (7.6.30) two times with respect to x, and one time with respect to y. Using all these equations we exclude x-dependence of the coefficients in the terms containing e^{xy}. In other words we reduce the equation to the following form

$$e^{xy}\Phi_1(y) + \Phi_2(y) = 0.$$

Consequently we should demand that each of the coefficients $\Phi_1(y)$, and $\Phi_2(y)$ is equal to zero because the variables x, y are now independent. From these last two equations $\Phi_1(y) = 0$, $\Phi_2(y) = 0$ we obtain some new restrictions on the functions $a(x), c(y), d(y)$. Then we insert new $a(x), c(y), d(y)$ in the last Equation (7.6.28) and repeat the procedure.

We can obtain the first symmetry in this way

$$\mathbf{U}_1 = \frac{\partial}{\partial x} \tag{7.6.32}$$

and do not lose the other possible symmetries. The symmetry algebra is in this case one-dimensional.

Problem 7.6.5 *Solve the equation*

$$4w^2w''' - 18ww'w'' + 15(w')^3 = 0, \quad w = w(q).$$

using the evident symmetries of this Equation (the problem setting is from [75]).

Solution The first obvious symmetry that is admitted by the equation is $\mathbf{U}_1 = \frac{\partial}{\partial q}$, because our equation does not contain the variable q explicitly. The canonical variables for $\mathbf{U}_1$ is simply $\zeta(q,w) = q$ and $r(q,w) = w$, and for derivatives we have

$$\begin{aligned}
w' &= \frac{dw}{dq} = \frac{1}{\frac{dq}{dw}} = \frac{1}{\zeta'},\\
w'' &= \frac{dw'}{dq} = \frac{d\frac{1}{\zeta'}}{dw} = -\frac{1}{\zeta'^2}\frac{d\zeta'}{d\zeta} = -\frac{\zeta''}{\zeta'^3},\\
w''' &= \frac{dw''}{dq} = \frac{d\left(-\frac{\zeta''}{\zeta'^3}\right)}{d\zeta} = -\frac{\frac{1}{\zeta'^3}\left(\frac{3\zeta'' d\zeta'}{\zeta'} - d\zeta''\right)}{d\zeta}\\
&= \frac{1}{\zeta'^3}\left(\frac{3\zeta''\zeta''}{\zeta'\zeta'} - \frac{\zeta'''}{\zeta'}\right) = \frac{1}{\zeta'^5}\left(3\zeta''^2 - \zeta'''\zeta'\right).
\end{aligned}$$

Substituting these formulas to the original equation, we obtain

$$4\frac{1}{\zeta'^5}\left(3\zeta''^2 - \zeta'''\zeta'\right)r^2 - 18\frac{r}{\zeta'}\left(-\frac{\zeta''}{\zeta'^3}\right) + \frac{15}{\zeta'^3} = 0.$$

Having $\zeta' = \tilde{y}$ and $r = \tilde{x}$ we rewrite it as

$$4\tilde{y}''\tilde{y}\tilde{x}^2 - 12(\tilde{y}')^2\tilde{x}^2 - 18\tilde{y}'\tilde{y}\tilde{x} - 15\tilde{y}^2 = 0.$$

The equation admits two symmetries $\mathbf{U}_1 = \tilde{x}\frac{\partial}{\partial \tilde{x}}$ and $\mathbf{U}_2 = \tilde{y}\frac{\partial}{\partial \tilde{y}}$, and $[\mathbf{U}_1, \mathbf{U}_2] = 0$, $\delta \neq 0$, so it is the group G_2, Ia using Sophus Lie classification presented in Table 7.2. Thus, it can be transformed to the form $\mathbf{U}_1 = \frac{\partial}{\partial s}$, $\mathbf{U}_2 = \frac{\partial}{\partial t}$ after we solve the system of equations

$$\begin{aligned}(\mathbf{U}_1 s)\frac{\partial}{\partial s} + (\mathbf{U}_1 t)\frac{\partial}{\partial t} &= \frac{\partial}{\partial s},\\ (\mathbf{U}_2 s)\frac{\partial}{\partial s} + (\mathbf{U}_2 t)\frac{\partial}{\partial t} &= \frac{\partial}{\partial t}\end{aligned}$$

or equivalent

$$\begin{aligned}\tilde{x}\frac{\partial s}{\partial \tilde{x}} = 1, \quad \tilde{x}\frac{\partial t}{\partial \tilde{x}} = 0,\\ \tilde{y}\frac{\partial s}{\partial \tilde{y}} = 0, \quad \tilde{y}\frac{\partial t}{\partial \tilde{y}} = 1.\end{aligned}$$

From the second and third equations it follows that $t = t(\tilde{y})$, $s = s(\tilde{x})$. The first and last equations $\tilde{x}s' = 1$, $\tilde{y}t' = 1$ give us the substitution $s = \ln\tilde{x}$, $t = \ln\tilde{y}$. In the new variables

$$\begin{aligned}\frac{\partial s}{\partial t} &= \frac{\tilde{y}d\tilde{x}}{\tilde{x}d\tilde{y}} = \frac{\tilde{y}}{\tilde{x}\tilde{y}'}\\ \frac{\partial^2 s}{\partial t^2} &= \frac{d\left(\frac{\tilde{y}}{\tilde{x}\tilde{y}'}\right)}{dt} = \frac{\tilde{y}}{\tilde{x}\tilde{y}'} - \frac{\tilde{y}^2}{\tilde{x}^2\tilde{y}'^2} - \frac{\tilde{y}^2\tilde{y}''}{\tilde{x}\tilde{y}'^3}.\end{aligned}$$

In the last equation, let us insert $\tilde{y}''$ from the equation

$$\tilde{y}'' = 3\frac{\tilde{y}'^2}{\tilde{y}} + \frac{9\tilde{y}'}{2\tilde{x}} + \frac{15\tilde{y}}{4\tilde{x}^2}$$

and replace $\frac{\tilde{y}}{\tilde{x}\tilde{y}'}$ by s'. We obtain for the function $s(t)$

$$s'' = s' - s'^2 - 3s' - \frac{9}{2}s'^2 - \frac{15}{4}s'^3,$$

simplifying

$$s'' = -\frac{15}{4}s'^3 - \frac{11}{2}s'^2 - 2s'.$$

Substitution $s' = z$ reduces the order of equation to the first one. In this equation we separate variables and integrate

$$\begin{aligned} \int \frac{dz}{15z^3 + 22z^2 + 8z} &= -4 \int dt + \text{const.}, \\ \ln z - 6\ln(2+3z) + 10\ln(4+5z) &= -32t + \text{const.} \end{aligned}$$

The solution of the original equation is given in an implicit form by the system of equations

$$\begin{gathered} w = e^s, \quad s' = z, \quad \tfrac{z(4+5z)^{10}}{(2+3z)^6} = C_1 e^{-32t}, \\ q = e^t + C_2, \quad C_1, C_2 = \text{const.} \end{gathered}$$

Problem 7.6.6 *Find all point symmetries of the second order differential equation* $y'' = y^2 (y')^2$.

Solution The determining equation for any admitted symmetry is

$$\mathbf{U}^{(2)} \Delta|_{\Delta=0} = 0, \quad \text{where} \quad \Delta = y'' - y^2 (y')^2,$$

or

$$\left(\xi \frac{\partial}{\partial x} + \eta \frac{\partial}{\partial y} + \eta^{(1)} \frac{\partial}{\partial y'} + \eta^{(2)} \frac{\partial}{\partial y''} \right) \left(y'' - y^2 (y')^2 \right) |_{y''=y^2 y'^2} = 0.$$

It means that the condition should be satisfied

$$-2\eta y (y')^2 - 2y' y^2 \eta^{(1)} + \eta^{(2)}|_{y''=y^2 y'^2} = 0.$$

Now, we recall the explicit formulas for $\eta^{(1)}$ and $\eta^{(2)}$

$$\begin{aligned} \eta^{(1)} &= \eta_x + \eta_y y' - y' \xi_x - y'^2 \xi_y, \\ \eta^{(2)} &= \eta_{xx} + (2\eta_{xy} - \xi_{xx}) y' \\ &+ (\eta_{yy} - 2\xi_{xy}) y'^2 - \xi_{yy} y'^3 + (\eta_y - 2\xi_x) y'' - 3\xi_y y' y'' \end{aligned}$$

and substitute them to the determining equation

$$\begin{aligned} -2\eta y (y')^2 &- 2y' y^2 \left(\eta_x + \eta_y y' - y' \xi_x - y'^2 \xi_y \right) \\ &+ \eta_{xx} + (2\eta_{xy} - \xi_{xx}) y' + (\eta_{yy} - 2\xi_{xy}) y'^2 \\ &- \xi_{yy} (y')^3 + y^2 (y')^2 (\eta_y - 2\xi_x) - 3y^2 (y')^3 \xi_y = 0. \end{aligned}$$

Since the above function is a polynomial function with respect to y', we reduce the equation to the system of equations for the coefficients of this polynomial

$$\begin{aligned} (y')^3 &\mid -\xi_{yy} - y^2\xi_y = 0, \\ (y')^2 &\mid \eta_{yy} - 2\xi_{xy} - y^2\eta_y - 2y\eta = 0, \\ y' &\mid -2y^2\eta_x + 2\eta_{xy} - \xi_{xx} = 0, \\ (y')^0 &\mid \eta_{xx} = 0. \end{aligned}$$

Going up from the last equation of the system we obtain

- The coefficient by the zero power of y' is equal to zero if

$$\eta_{xx} = 0 \Rightarrow \eta = a(y)x + b(y) \Rightarrow \eta_{xy} = a'(y).$$

- The third equation of the system (the coefficient by the first power of y') is equal to zero if

$$\begin{aligned} -2y^2a(y) + 2a'(y) - \xi_{xx} = 0 &\Rightarrow \xi_{xx} = 2a'(y) - 2y^2a(y), \\ &\Rightarrow \xi_x = 2a'(y)x - 2y^2a(y)x + c(y), \\ &\Rightarrow \xi = x^2a'(y) - x^2y^2a(y) + c(y)x + d(y). \end{aligned}$$

- Correspondingly the first equation of the system demands that

$$\begin{aligned} \xi_y &= a''(y)x^2 - 2ya(y)x^2 - y^2a'(y)x^2 + c'(y)x + d'(y), \\ \xi_{yy} &= a'''(y)x^2 - 2a(y)x^2 - 4ya'(y)x - y^2a''(y)x^2 + c''(y)x + d''(y), \end{aligned}$$

  and

$$\begin{aligned} a'''(y)x^2 &- 2a(y)x^2 - 4ya'(y)x^2 - y^2a''(y)x^2 \\ &+ c''(y)x + d''(y) + y^2a''(y)x^2 - 2y^3a(y)x^2 \\ &- y^4a'(y)x^2 + c'(y)y^2x + d'(y)y^2 = 0. \end{aligned}$$

  This is again the polynomial equation with respect to x, that is why we have the following system of Equation (the coefficients on different powers of x must be equal to zero simultaneously):

$$\begin{aligned} x^2 &\mid a''' - 2a - 4ya' - y^2a'' + y^2a'' - 2y^3a - y^4a' = 0, \\ x &\mid c'' + c'y^2 = 0, \\ (x)^0 &\mid d'' + d'y^2 = 0, \end{aligned}$$

or in a simplified form

$$\begin{aligned} a''' - a'(4y + y^4) - 2a(1 + y^3) &= 0, \\ c'' + c'y^2 &= 0, \\ d'' + d'y^2 &= 0. \end{aligned}$$

- We insert the special form ξ, η in the second Equation (coefficient by the second power of y'). For partial derivatives of η and ξ we get now

$$\begin{aligned} \eta_y &= a'(y)x + b'(y), \\ \eta_{yy} &= a''(y)x + b''(y), \\ \xi_x &= 2a'(y)x - 2y^2a(y)x + c(y), \\ \xi_{xy} &= 2a''(y)x - 4ya(y)x - 2y^2a'(y)x + c'(y). \end{aligned}$$

Substituting it into the equation $\eta_{yy} - 2\xi_{xy} - y^2\eta_y - 2y\eta = 0$ we obtain

$$\begin{aligned} &(a''x + b'') - 2(2a''x - 4yax - 2y^2a'x + c') - y^2(a'x + b) - 2y(ax + b) \\ &= -2y(ax + b) - y^2(a'x + b') + a''x + b'' - 4a''x + 8yax + 4y^2a'x - 2c' \\ &= 0. \end{aligned}$$

Because this equality must be satisfied for each value of x, each coefficient with different powers of x must be equal to 0. It means we have to solve the system of equations

$$\begin{aligned} &x \mid -2ya - y^2a + a'' - 4a'' + 8ya + 4y^2a' = 0, \\ &0 \mid -2yb - y^2b' + b'' - 2c' = 0. \end{aligned}$$

We have the final system of equations

$$\begin{aligned} a''' - a'(4y + y^4) - 2a(1 + y^3) &= 0, \\ c'' + c'y^2 &= 0, \\ d'' + d'y^2 &= 0, \\ a'' - y^2a' - 2ya &= 0, \\ b'' - y^2b' - 2yb - 2c' &= 0. \end{aligned} \tag{7.6.33}$$

The second equation immediately gives as the function $c(y)$

$$\frac{c''}{c'} = -y^2, \quad c' = C_1e^{-\frac{y^3}{3}}.$$

Similarly, we obtain $d' = D_1 e^{-\frac{y^3}{3}}$. For further calculations, we denote

$$\Phi(y) = \int e^{-\frac{y^3}{3}} \mathrm{dy} = 3^{-2/3} \mathcal{B}\left(\frac{1}{3}, \frac{\mathrm{y}^3}{3}\right) + \text{const.},$$

where $\mathcal{B}\left(\frac{1}{3}, \frac{y^3}{3}\right)$ is the incomplete gamma function, so we obtain

$$\begin{aligned} c(y) &= C_1\Phi(y) + C_2, \quad C_1, C_2 \text{ are constants,} \\ d(y) &= D_1\Phi(y) + D_2, \quad D_1, D_2 \text{ are constants.} \end{aligned}$$

Now, if we look closely to the initial formulas for $\xi(x, y)$ and $\eta(x, y)$

$$\begin{aligned} \xi(x, y) &= x^2 a'(y) - x^2 y^2 a(y) + c(y)x + d(y), \\ \eta(x, y) &= x a(y) + b(y), \end{aligned}$$

and equations (7.6.33). It is clear that $a = b = c = 0$ are solutions of the system and the corresponding symmetries are

$$\begin{aligned} \mathbf{U}_1 &= \frac{\partial}{\partial x}, \\ \mathbf{U}_2 &= \Phi(y)\frac{\partial}{\partial x}. \end{aligned}$$

Now we consider the special solution $d(y) = 0$. Also, we can choose $c(y)$ as a constant function. We see that $a = b = d = 0$, $c =$ const. is a solution of the system (7.6.33) and immediately obtain the third symmetry of the original equation with the generator

$$\mathbf{U}_3 = x\frac{\partial}{\partial x}.$$

We take a closer look at the first and fourth equations of the system (7.6.33). We see that if we differentiate the fourth equation one time and add the same equation multiplied by y^2 to the result of differentiation then we obtain the first one. It means the equations are dependent and we can exclude the first one from further investigations.

Now, let us solve the fourth equation in (7.6.33). It is easy to see that $a(y) = A_1 e^{\frac{y^3}{3}}$ is a particular solution. The second independent solution we try to find in the form $a(y) = A_1(y)e^{-\frac{y^3}{3}}$. After the substitution of $a(y)$ in this form to the fourth equation in (7.6.33) and collecting the terms we obtain an equation on $A_1(y)$

$$A_1''(y) + y^2 A_1'(y) = 0.$$

This equation coincides with the second equation in the system of equations (7.6.33) and we solved them before. It means $A_1 = \Phi(y)$ and the function $a(y)$ takes the form

$$a(y) = e^{\frac{y^3}{3}} \left(A_1 + A_2\Phi(y)\right), \quad A_1, \; A_2 \text{ are constants.}$$

Because A_1, A_2 are arbitrary constants we added to the set of symmetries two new ones with the infinitesimal generators

$$\mathbf{U}_4 = x^2\frac{\partial}{\partial x} + xe^{\frac{y^3}{3}}\Phi(y)\frac{\partial}{\partial y}, \quad \mathbf{U}_5 = xe^{\frac{y^3}{3}}\frac{\partial}{\partial y}.$$

Because any second order ODE can have 1,2,3 or 8 symmetries (see Theorem 5.2.2, [53],[33]) we have to find the last three. Solutions of the fifth equation in (7.6.33) provide us the generators in the form

$$\begin{aligned}
\mathbf{U}_6 &= e^{\frac{y^3}{3}}\Phi(y)\frac{\partial}{\partial y}, \quad \mathbf{U}_7 = e^{\frac{y^3}{3}}\frac{\partial}{\partial y}, \\
\mathbf{U}_8 &= C_1 x\, \Phi(y)\frac{\partial}{\partial x} + F(C_1, y)\frac{\partial}{\partial y},
\end{aligned}$$

where $F(C_1, y)$ is the special solution of the non-homogeneous fifth equation in the system (7.6.33). This also means that our equation can be reduced to the form $s'' = 0$ with the proper variables change.

Problem 7.6.7 *Solve the equation $3yy'' = 5y'$ using the symmetries*

$$\mathbf{U}_1 = \frac{\partial}{\partial x}, \quad \mathbf{U}_2 = x\frac{\partial}{\partial x} + y\frac{\partial}{\partial y}.$$

Solution We find the commutator for the given generators $[\mathbf{U}_1, \mathbf{U}_2] = \mathbf{U}_1$. Now we can decide which strategy is applicable in this case. We introduce canonical variables $t(x, y), s(x, y)$ in the way shown in Table 7.2 for the case G_2, IIa

$$\mathbf{U}_1 s(x, y) = 1, \quad \mathbf{U}_1 t(x, y) = 0.$$

In our case the canonical variables can be just chosen as the set $s = x$, $t = y$, and $\mathbf{U}_2$ now takes the form $\mathbf{U}_2 = t\frac{\partial}{\partial t} + s\frac{\partial}{\partial s}$. The linear differential operator corresponding to the given equation in original coordinates is equal to

$$\mathcal{A} = \frac{\partial}{\partial x} + y'\frac{\partial}{\partial y} + \frac{5y'}{3y}\frac{\partial}{\partial y'},$$

we need to rewrite it in the canonical variables. We transform the derivatives first

$$y' = \frac{dy}{dx} = \frac{1}{\frac{dx}{dy}} = \frac{1}{s'}, \quad y'' = \frac{dy'}{dx} = \frac{d\frac{1}{s'}}{ds} = -\frac{1}{s'^2}\frac{ds'}{ds} = -\frac{s''}{s'^3}.$$

The equation in the new variables, therefore, is

$$-3t\frac{s''}{s'^3} = 5\frac{1}{s'}, \quad \text{or} \quad s'' = -\frac{5s'^2}{3t}.$$

If we introduce a new variable $z = s'$, then we get the first order differential equation $z' = -\frac{5z^2}{3t}$, with the solution $z = (5/6 \ \ln t + C_1)^{-1}$. It means $s(t) = \int z(t) + C_2$, where C_1, C_2 are constants. We have the solution in an implicit form.

Let us now see if the second symmetry is left after transformation from the variables x, y to the variables s, t. The operator for the differential equation in variables s, t has the following form

$$\tilde{\mathcal{A}} = \frac{\partial}{\partial t} - \frac{5s'^2}{3t}\frac{\partial}{\partial s'}.$$

In order to check whether the change of the variables destroyed the second symmetry we calculate the commutator of the operator $\tilde{\mathcal{A}}$ with the infinitesimal generator $\mathbf{U}_2^{(1)}$. If as a result of the commutation we obtain an operator proportional to $\tilde{\mathcal{A}}$ the symmetry still holds. For $\mathbf{U}_2^{(1)}$ it follows

$$[\mathbf{U}_2^{(1)}, \tilde{\mathcal{A}}] = \left[t\frac{\partial}{\partial t} + s\frac{\partial}{\partial s}, \frac{\partial}{\partial t} - \frac{5s'^2}{3t}\frac{\partial}{\partial s'}\right] = -\tilde{\mathcal{A}}$$

that means that this symmetry still holds.

Problem 7.6.8 *Solve the equation*

$$y'''(1 + (y')^2) = (3y' + a)(y'')^2$$

using the symmetries $\mathbf{U}_1 = \frac{\partial}{\partial x}, \mathbf{U}_2 = \frac{\partial}{\partial y}$, $\mathbf{U}_3 = x\frac{\partial}{\partial x} + y\frac{\partial}{\partial y}$, $\mathbf{U}_4 = y\frac{\partial}{\partial x} - x\frac{\partial}{\partial y}$.

Solution The commutator table for the Lie algebra L_4 spanned by the given operators $L_4 = \langle \mathbf{U}_1, \mathbf{U}_2, \mathbf{U}_3, \mathbf{U}_4 \rangle$ is

	$\mathbf{U}_1$	$\mathbf{U}_2$	$\mathbf{U}_3$	$\mathbf{U}_4$
$\mathbf{U}_1$	0	0	$\mathbf{U}_1$	$-\mathbf{U}_2$
$\mathbf{U}_2$	0	0	$\mathbf{U}_2$	0
$\mathbf{U}_3$	$-\mathbf{U}_1$	$-\mathbf{U}_2$	0	$-\mathbf{U}_1$
$\mathbf{U}_4$	$\mathbf{U}_2$	0	$\mathbf{U}_1$	0

The first derived algebra is two–dimensional $L_4^{(1)} = \langle \mathbf{U}_1, \mathbf{U}_2 \rangle$, and $L_4^{(2)} = 0$. After the theorem of solvable algebras, the algebra L_4 is solvable. We can use the method as described before to find a solution of the given equation. Let us take the subalgebra $\langle \mathbf{U}_1, \mathbf{U}_2, \mathbf{U}_3 \rangle$. We calculate the second prolongations of all operators and provide the linear differential operator $\mathcal{A}$ which corresponds to the original ODE

$$\mathbf{U}_1^{(2)} = \frac{\partial}{\partial x}, \quad \mathbf{U}_2^{(2)} = \frac{\partial}{\partial y}, \quad \mathbf{U}_3^{(2)} = x\frac{\partial}{\partial x} + y\frac{\partial}{\partial y} - y''\frac{\partial}{\partial y''},$$

$$\mathcal{A} = \frac{\partial}{\partial x} + y'\frac{\partial}{\partial y} + y''\frac{\partial}{\partial y'} + \frac{(3y'+a)y''^2}{1+y'^2}\frac{\partial}{\partial y''}.$$

We calculate the determinant in the denominator of the fraction (7.3.16) for our equation and prove that it is not equal to zero identically

$$\Xi = \begin{vmatrix} 1 & 0 & 0 & 0 \\ 0 & 1 & 0 & 0 \\ x & y & 0 & -y'' \\ 1 & y' & y'' & \frac{(3y'+a)y''^2}{1+y'^2}. \end{vmatrix} = y''^2 \neq 0.$$

Then one of the first integrals can be found according to the formula (7.3.16) which takes in this case the form

$$\varphi_1 = \int \frac{1}{\Xi} \begin{vmatrix} dx & dy & dy' & dy'' \\ 1 & 0 & 0 & 0 \\ 0 & 1 & 0 & 0 \\ 1 & y' & y'' & \frac{(3y'+a)y''^2}{1+y'^2}. \end{vmatrix} = \frac{3}{2}\ln(1+y'^2) + a\arctan y' - \ln y''.$$

We express the second derivative of y by

$$y'' = (1+y'^2)^{\frac{3}{2}} e^{a\arctan y'} e^{-\varphi_1}.$$

In the next step we take as a two–dimensional subalgebra of $\langle \mathbf{U}_1, \mathbf{U}_2, \mathbf{U}_3 \rangle$ the subalgebra $\langle \mathbf{U}_1, \mathbf{U}_2 \rangle$, then if we apply the same procedure we get

$$\tilde{\Xi} = \begin{vmatrix} 1 & 0 & 0 \\ 0 & 1 & 0 \\ 1 & y' & y''. \end{vmatrix} = y'' \neq 0,$$

and the second of the first integrals has the form

$$\varphi_2 = \int \frac{1}{\tilde{\Xi}} \begin{vmatrix} dx & dy & dy' \\ 1 & 0 & 0 \\ 1 & y' & y'' \end{vmatrix} = \int \frac{1}{y''}(-y''dy + y'dy') = -y + \int \frac{y'}{y''}dy'$$

$$= -y + \int \frac{e^{\varphi_1} y' dy'}{(1+y'^2)^{\frac{3}{2}} e^{a\arctan y'}}.$$

We calculated two first integrals for the second order ODE and in this way solved the equation.

Problem 7.6.9 *Find the Lie point symmetries of the second order differential equations (the first Painleve equation)*

$$y'' - 6y^2 - x = 0.$$

Solution Let us denote for simplicity $\Delta = y'' - 6y^2 - x$. The derivatives of the equation are the following

$$\frac{\partial}{\partial x}\Delta = -1, \quad \frac{\partial}{\partial y}\Delta = -12y, \quad \frac{\partial}{\partial y'}\Delta = 0, \quad \frac{\partial}{\partial y''}\Delta = 1.$$

We are looking for the infinitesimal generator of the following form

$$\mathbf{U}^{(2)} = \xi(x,y)\frac{\partial}{\partial x} + \eta(x,y)\frac{\partial}{\partial y} + \eta^{(1)}\frac{\partial}{\partial y'} + \eta^{(2)}(x,y)\frac{\partial}{\partial y''},$$

where

$$\begin{aligned}
\eta^{(1)} &= \frac{d\eta}{dx} - y'\frac{d\xi}{dx} = \eta_x + y'(\eta_y - \xi_x) - y'^2\xi_y,\\
\eta^{(2)} &= \frac{d\eta^{(1)}}{dx} - y''\frac{d\xi}{dx}\\
&= \eta_{xx} + y'(2\eta_{xy} - \xi_{xx}) + y'^2(\eta_{yy} - 2\xi_{xy}) - y'^3\xi_{yy} + y''(\eta_y - 2\xi_x - 3y'\xi_y).
\end{aligned}$$

On the solution manifold $\Delta = 0$ the equation $\mathbf{U}^{(2)}\Delta|_{\Delta=0} = 0$ should hold, i.e.

$$\begin{aligned}
&-\xi - 12y\eta + \eta_{xx} + y'(2\eta_{xy} - \xi_{xx})\\
&\quad + y'^2(\eta_{yy} - 2\xi_{xy}) - y'^3\xi_{yy} + y''(\eta_y - 2\xi_x - 3y'\xi_y))\Big|_{\Delta=0} = 0,
\end{aligned}$$

it means we replace in this equation y'' with $6y^2 + x$ to stay on the solution manifold.

Since the variables y, y' are independent now, we can separate that equation according to the degrees of y' into monomials in a system of equations listed in Table 7.3 in the left column.

We solve the system of equation from the top to the bottom.
Equation (1) in Table 7.3 leads to the linearity of the function $\xi(x,y)$ with respect to y, i.e. $\xi(x,y) = a_1(x)y + a_2(x)$. Therefore, we get from Equation (2) in Table 7.3 that $\eta(x,y)$ has the form

$$\eta(x,y) = a_1(x)y^2 + b_1(x)y + b_2(x).$$

Table 7.3: The determining equations for the first Painleve equation.

	monomial	equation on the coefficient
(1)	y'^3	$-\xi_{yy} = 0$
(2)	y'^2	$\eta_{yy} - 2\xi_{xy} = 0$
(3)	y'	$2\eta_{xy} - \xi_{xx} - 3\xi_y(6y^2 + x) = 0$
(4)	y'^0	$-\xi - 12y\eta + \eta_{xx} + (\eta_y - 2\xi_x)(6y^2 + x) = 0$

We can insert the formulas for ξ and η in Equations (3) and (4) of Table 7.3 and obtain a system of equations

$$\begin{aligned}
&2(2a_1''(x)y + b_1'(x)) - a_1''(x)y - a_2''(x) - 3a_1(x)(6y^2 + x) = 0, \qquad (7.6.34)\\
&a_1'''(x)y^2 + b_1''(x)y + b_2''(x) - a_1(x)y - a_2(x) - 12y(a_1'(x)y^2 + b_1(x)y + b_2(x))\\
&+ (2a_1'(x)y + b_1(x) - 2a_1'(x)y - 2a_2'(x))(6y^2 + x) = 0.
\end{aligned}$$

Again, the equations can be separated into monomials according to the highest degree of the highest variable, which is the variable y now.

From the first Equation (7.6.34) we obtain a new system of equations listed in Table 7.4.

Therefore, the coefficients ξ and η have the simpler form

$$\xi = a_2(x), \quad \eta = b_1(x)y + b_2(x).$$

We insert these expressions in the last Equation (7.6.34) and obtain an updated system of equations listed in Table 7.5.

Taking into account the Equation (3) from Table 7.4 and the Equation (4) from Table 7.5, we get $a_2''(x) = 0 \Rightarrow a_2(x) = a_{21}x + a_{22}$. Correspondingly, from Equation (4) in Table 7.5 $b_1(x) = -2a_2'(x) = -2a_{21}$. Inserting $b_1(x)$ into Equation (5) in Table 7.5, we get $b_2(x) = 0$.

Table 7.4: The determining equations after the substitution (7.6.34).

	monomial	equation on the coefficient
(1)	y^2	$-18a_1(x) = 0$
(2)	y^1	$3a_1''(x) = 0$
(3)	y^0	$2b_1'(x) - a_2''(x) = 0$

Table 7.5: The determining equations in the case $\xi = a_2(x),\ \eta = b_1(x)y + b_2(x)$.

	monomial	equation on the coefficient
(4)	y^2	$-6b_1(x) - 12a_2'(x) = 0$
(5)	y^1	$b_1''(x) - 12b_2(x) = 0$
(6)	y^0	$b_2''(x) - a_2(x) + b_1(x)x - 2a_2'(x)x = 0$

Therefore, the formulas for ξ and η take the form

$$\xi = a_{21}x + a_{22},\ \eta = -2a_{21}y.$$

Till now, we have not used the restriction (6) in Table 7.5; let us insert the formulas for ξ, η in it

$$-a_{21}x - a_{22} - 2a_{21}x - 2a_{21}x = 0.$$

So we obtain that $a_{21} = a_{22} = 0$ and the equation

$$y'' - 6y^2 - x = 0$$

does not admit any one-parameter Lie symmetry group.

Problem 7.6.10 *Find the Lie algebra admitted by the equation*

$$x^2y'' - xy' + x^2 = 0$$

and solve this equation. To find the symmetries of the given equation use the **Mathematica** *package* **IntroToSymmetry**.

Solution We provide here the simulated screenshot of the calculations with comments to make the using of this package accessible. From the script below we see that we have found two infinitesimal generators of the type $\mathbf{U}_1 = \frac{\partial}{\partial y}$ and $\mathbf{U}_1 = x^2\frac{\partial}{\partial y}$. Using the first infinitesimal generator we substitute $z(x) = y'(x)$ and reduce the given second order ODE to a first order ODE. So we obtain the first order ODE $x^2z' - xz + x^2 = 0$ which we solve using the second symmetry and obtain the solution $y(x) = \frac{x^2}{4} - \frac{x^2}{2}\log x + C_1\frac{x^2}{2} + C_2,\quad C_1, C_2 \in \mathbb{R}$.

Simulated Mathematica script with comments
First load the package.

```
Needs["SymmetryAnalysis`IntroToSymmetry`"]
```
Enter the independent variables.
```
independentvariables={"x"};
```
Enter the dependent variables.
```
dependentvariables={"y"};
```
Enter the equation as a string, **without** == 0 at the end.
```
inputequation= "x^2 * y''[x]-x * y'[x]+x^2";
rulesarray  = {"x^2 * y''[x]  ->  x * y'[x]-x^2"};
```
Enter the array of names (constants) that must be preserved when the equation is converted to generic (x1,y1).
```
frozennames={ };
```
Enter the maximum derivative order of the equation.
```
p = 2;
```
If the search is for Lie point groups, we enter the following parameters:
```
r= 0; xseon= 1; internalrules= 0;
```
To find the determining equations for our ODE we call the function:
```
FindDeterminingEquations[independentvariables,
            dependentvariables,frozennames,p,r,xseon,
              inputequation,rulesarray,internalrules]
```

FindDeterminingEquations has finished executing. You can look at the output in the table zdeterminingequations. Each entry in this table is a determining equation in string format expressed in terms of z-variables. Rules for converting between z-variables and conventional variables are contained in the table ztableofrules. To view the determining equations in terms of conventional variables use the command ToExpression[zdeterminingequations]/.ztableofrules. There are two other items the user may wish to look at; the equation converted to generic variables (x1,x2,...,y1,y2,...) is designated equationgenericvariables and the various derivatives of the equation that appear in the invariance condition can be viewed in the table invarconditiontable. Rules for converting between z-variables and generic variables are contained in the table ztableofrulesxy.

Here are the determining equations expressed in terms of z-variables.
```
zdeterminingequations
{2*z1*xse1 [z1, z2] - z1*Derivative[1, 0][eta1][z1, z2]
+ z1^2*Derivative[2, 0] [eta1][z1, z2] == 0,
-xse1 [z1, z2] - z1* Derivative[0, 1][eta1][z1, z2]
- z1* Derivative[1, 0][xse11][z1, z2]
+ 2*z1^2*Derivative[1, 1][eta1][z1, z2]
-  z1^2*Derivative[2, 0][xse1][z1, z2] == 0,
2*z1*xse1[z1, z2] +  z1^2*Derivative[0, 1] [eta1][z1, z2]
```

```
- 2*z1^2*Derivative[1, 0][xse1][z1, z2] == 0,
z1*Derivative[0, 1] [xse1][z1, z2]
+  z1^2*Derivative[0, 2][eta1][z1, z2]
- 2*z1^2*Derivative[1, 1][xse1][z1, z2] ==  0,
- 3*z1^2*Derivative[0, 1][xse1][z1, z2] == 0,
- (z1^2*Derivative[0, 2][xse1][z1, z2]) == 0 }
```

We can see the number of determining equations we have.

```
Length[zdeterminingequations]
6
```

We can look at our determining equations in a more readable form.

```
TraditionalForm[ColumnForm[
           ToExpression[zdeterminingequations],Left, Below]]
eta1^(2,0) (z1, z2) z1^2 +2 xse1(z1, z2) z1
- eta1^{(1,0) (z1, z2) z1 = 0
2 eta1^(1,1)(z1, z2) z1^2 - xse1^(2,0) (z1, z2) z1^2
- eta1^(0,1) (z1, z2) z1 + xse1^{(1,0)} (z1, z2) z1
- xse1(z1, z2) = 0
eta1^(0,1)(z1, z2) z1^2 - 2 xse1^(1,0)(z1, z2) z1^2
+ 2 xse1(z1, z2) z1 = 0
eta1^(0,2)(z1, z2) z1^2 - 2 xse1^(1,1) (z1, z2) z1^2
+ xse1^{(0,1)}(z1, z2) z1 = 0
- 3 z1^2 xse1^(0,1)(z1, z2) = 0
- z1^2 xse1^(0,2)(z1, z2)  = 0
```

Now solve the determining equations in terms of multivariable polynomials of a selected order.

```
SolveDeterminingEquations[independentvariables,
     dependentvariables,r,xseon,zdeterminingequations,order=4]
```

The number of unknown polynomial coefficients = 30.
The number of equations for the polynomial coefficients = 69.
SolveDeterminingEquations has finished executing. You can look at the output in the tables xsefunctions and etafunctions. Each entry in these tables is an infinitesimal function in string format expressed in terms of z-variables and the group parameters. The output can also be viewed with the group parameters stripped away by looking at the table infinitesimalgroups. In either case you may wish to convert the z-variables to conventional variables using the table ztableofrule. Keep in mind that this function only finds solutions of the determining equations that are of algebraic form. The determining equations may admit solu-

tions that involve transcendental functions and/or integrals. Note that arbitrary functions may appear in the infinitesimals and that these can be detected by running the package function SolveDeterminingEquations for several polynomial orders. If terms of ever increasing order appear, then an arbitrary function is indicated.

The infinitesimals for the independent variables.

```
xsefunctions
{xse1[z1_, z2_]=0}
```

The infinitesimals for the dependent variables.

```
etafunctions
{eta1[z1_, z2_]=b10 + b12*z1^2}
```

Express the xse and eta functions in terms of variables x and y.

```
ToExpression[xsefunctions]/.ztableofrules
ToExpression[etafunctions]/.ztableofrules
{0}
{b10 +  b12 x^2}
```

Now we can look at the infinitesimal groups,

```
infinitesimalgroups
{{{0}, {1}}, {{0}, {z1^2}}}
```

and at the infinitesimal groups in terms of the variables x and y.

```
infinitesimalgroups1=infinitesimalgroups/.{z1->x,z2->y}
{{{0}, {1}}, {{0}, {x^2}}}
```

We can also make a commutator table.

```
MakeCommutatorTable[independentvariables,
    dependentvariables,infinitesimalgroups1]
```

MakeCommutatorTable has finished executing. You can look at the output in the table commutatortable. To present the output in the most readable form you may want view it as a matrix using MatrixForm[commutatortable]. Occasionally the entries in the commutatortable will have terms that cancel. To get rid of these terms use the function Simplify before viewing the table.

```
commutatortable
{{{{0}, {0}}, {{0}, {0}}}, {{{0}, {0}}, {{0}, {0}}}}
```

Problem 7.6.11 *Find the point symmetries of the following equation*

$$y'' + 3yy' + (y')^3 = 0.$$

Solution Using **Mathematica** as we did in the previous example, we find that the equation admits a one-dimensional algebra, with the following generator

$$\mathbf{U} = \frac{\partial}{\partial x}.$$

We prove that the operator admitted by the equation

$$\mathbf{U}^{(2)} = \frac{\partial}{\partial x}, \quad \Rightarrow \quad \mathbf{U}^{(2)}\Delta|_{\Delta=0} = 0$$

thus solving the problem.

Problem 7.6.12 *The differential equation*

$$3y'' + xyy' - y^2 = 0 \tag{7.6.35}$$

arises in the study of the expulsion of a headed compressible fluid from a long, slender tube. Find a scaling group admitted by this differential equation, construct invariants u, v and reduce the order of Equation (the problem setting is from [27]).

Solution Using **Mathematica** we see that the scaling operator has the form

$$\mathbf{U} = -\frac{x}{2}\frac{\partial}{\partial x} + y\frac{\partial}{\partial y}.$$

The first prolongation then is equal to

$$\mathbf{U}^{(1)} = -\frac{x}{2}\frac{\partial}{\partial x} + y\frac{\partial}{\partial y} + \frac{3}{2}y'\frac{\partial}{\partial y'}.$$

Invariants $u(x, y)$ and $v(x, y, y')$ are solutions of the Lie equations

$$\frac{dx}{\frac{-x}{2}} = \frac{dy}{y} = \frac{dy'}{\frac{3}{2}y'},$$

which we denote by

$$u = F(x^2 y), \; v = G(x^3 y').$$

We take the simplest form of functions F, G, i.e. $u = x^2 y, \; v = x^3 y'$. Now we calculate the derivative of the invariant v

$$v'(u) = \frac{dv}{du} = \frac{3x^2 y' + x^3 y''}{2xy + x^2 y'}.$$

We can rewrite this equation in the following way

$$y'' = \frac{v'(2xy + x^2y') - 3x^2y'}{x^3} = \frac{v'(2u+v) - 3v}{x^4}.$$

From the original Equation (7.6.35) we represent the second derivative of y in the form

$$y'' = \frac{1}{3}(y^2 - xyy') = \frac{v^2 - uv}{3x^4}.$$

Using both expressions for y'', we obtain a first order ODE

$$3v'(2u+v) - v^2 - 9v + uv = 0, \ v = v(u),$$

and so reduce the second order ODE (7.6.35) to the first order ODE.

Problem 7.6.13 *Find the Lie group admitted by equation*

$$x^2y'' + xy' - 1 = 0 \tag{7.6.36}$$

and solve this Equation (the problem setting is from [27]).

Solution Using **Mathematica** we can find two infinitesimal generators of symmetries of the second order ODE (7.6.36)

$$\mathbf{U}_1 = x\frac{\partial}{\partial x}, \ \mathbf{U}_2 = \frac{\partial}{\partial y}. \tag{7.6.37}$$

Let us look for a solution of the given equation using the symmetries (7.6.37). Any two–dimensional algebra is solvable. We will use the properties of the admitted symmetry algebra. The first prolongations of the generators $\mathbf{U}_1$ and $\mathbf{U}_2$ have the form

$$\mathbf{U}_1^{(1)} = x\frac{\partial}{\partial x} - y'\frac{\partial}{\partial y'}, \quad \mathbf{U}_2^{(1)} = \frac{\partial}{\partial y}.$$

Equation (7.6.36) can be rewritten as

$$y'' = \frac{1 - xy'}{x^2} = \frac{1}{x^2} - \frac{y'}{x}.$$

A linear operator, which corresponds to the Equation (7.6.36) has the form $\mathcal{A} = \frac{\partial}{\partial x} + y'\frac{\partial}{\partial y} + (\frac{1}{x^2} - \frac{y'}{x})\frac{\partial}{\partial y'}$. Thereafter, we calculate the determinant in the denominator of (7.3.16) for our equation

$$\Xi = \begin{vmatrix} x & 0 & -y' \\ 0 & 1 & 0 \\ 1 & y' & \frac{1}{x^2} - \frac{y'}{x}. \end{vmatrix} = x\left(\frac{1}{x} - \frac{y'}{x}\right) - (-y') = \frac{1}{x} \neq 0.$$

Consequently, the operators $\mathbf{U}^{(1)}, \mathbf{U}^{(2)}$ and $\mathcal{A}$ are independent. One of the first integrals can be found according to the formula

$$\varphi_1 = \int \frac{1}{\Xi} \begin{vmatrix} dx & dy & dy' \\ 0 & 1 & 0 \\ 1 & y' & \frac{1}{x^2} - \frac{y'}{x}. \end{vmatrix} = \int x \left(\left(\frac{1}{x^2} - \frac{y'}{x} \right) dx - dy' \right) = \ln x - 2xy'.$$

We express the first derivative of y using this expression and get

$$y' = \frac{\ln x - \varphi_1}{2x}.$$

It is an equation with separated variables and its solution has the form

$$y = \frac{1}{4} \ln^2 x - \frac{\varphi_1}{2} \ln x + C_2, \quad \varphi_1, C_2 \in \mathbb{R}$$

so we obtain the explicit solution of the original second order ODE.

7.7 Exercises

The solutions to these exercises are a bit tedious and voluminous but we recommend that the reader tries to solve at least two or three of them. If you have access to **Mathematica** you can also use the package **IntroToSymmetry** and check your solutions with the computer calculations. The packages to calculate Lie symmetries are now included in many modern computer systems. If the reader understands the main features of such calculations he or she can choose the best package for his or her own needs.

1. Find the Lie point symmetries of the equations

   (a) $y_{xx} = ay^4, a \in \mathbb{R}$,

   (b) $y_{xx} = ay^{-3}, a \in \mathbb{R}$,

   and solve the equations using the symmetries.

2. Find the point symmetries of the third order differential equation $y''' = \exp(-y'')$.

3. Solve the equation

   $$yy'' = y'^2$$

   using the symmetries $U_1 = \frac{\partial}{\partial x}$, $U_2 = x\frac{\partial}{\partial x}$, $U_3 = y\frac{\partial}{\partial y}$.

4. Find the Lie point symmetries of the equation

$$y_{xx} = y^2 y_x.$$

Hint: $\eta^{(2)} = \eta_{xx} + (2\eta_{xy} - \xi_{xx})y_x + (\eta_{yy} - 2\xi_{xy})y_x^2 - \xi_{yy}y_x^3 + (\eta_y - 2\xi_x)y_{xx} - 3\xi_y y_x y_{xx}$.

Chapter 8

Lie group analysis of PDEs

In this chapter we provide a rather short overview of the techniques used for studying PDEs with the Lie group method. The ideas behind these methods are the same as in the case of ODEs. Because we have more variables we have to deal with more complicated structures and need new notations, which are mostly just an extension of the existing ones. Lie group analysis in the case of PDEs is definitely more time consuming and advanced than in the case of ODEs (many examples and detailed discussions are provided in [32]). However, in financial mathematics one mostly works with different types of second order parabolic PDEs. Very rarely does one have to solve elliptic equations of the second order that describe the fair price of financial products, for instance, Russian options. That is why we introduce a minimum number of the ideas, and only those that are absolutely necessary for these investigations. When the definitions and the theorems in multi-dimensions are clear and easy to understand, we provide the reader with the most general formulations. If the definitions or theorems are too voluminous in a general case of multi-dimensions, we just work with the simplified cases that are enough to demonstrate the applications of Lie group analysis to financial mathematics covered in this particular book.

In this chapter we provide a detailed Lie group analysis of the heat equation and Black-Scholes model. Both equations are of parabolic type and are in fact connected to each other. They can be transformed to each other with a transformation which is well known in financial literature. We explain that this transformation provides a similarity transformation of the admitted symmetry algebras. Since the admitted symmetry algebras are six–dimensional we can find a lot of interesting reductions and corresponding families of invariant solutions. We also provide the famous solution for the call option from the Black-Scholes model and deal with different reductions in the last theoretical sections of this chapter.

In the last two sections as usual we provide problems with solutions and exercises which can help our reader to understand the ideas developed in this chapter.

8.1 Notations and terminology in the case of many independent and dependent variables

In the previous chapters we have seen that the variables x, y, and $y', \ldots$ play very different roles in our observations. If we can define the transformation of the variables x and y arbitrarily, the change of all "higher" variables $y', y'', \ldots$ will be defined uniquely by the transformation of the base variables x, y. There exists also a great difference between x and y as well - we assume from the very beginning that y depends on x.

Let us introduce the notations and ideas which we need for studying partial differential equations. The systems of partial differential equations can be studied in a very similar way (see for instance [59]), but they demand a more sophisticated notation. Many models of financial mathematics are based on the Black-Scholes model and are in fact nonlinear PDEs. That is why we restrict ourselves to the case of PDEs. We denote the independent variables as $(x_1, \ldots, x_p) \in X \cong \mathbb{R}^p$ and the dependent variable as $u \in V \cong \mathbb{R}$. Like before we call the space $\mathcal{M} = X \times V$ (or an open subset $\mathcal{M} \subset X \times V$) a base space.

Example 8.1.1 In the case of an ordinary differential equation we have $x \in \mathbb{R}$ and $y \in \mathbb{R}$ and the base space is $\mathcal{M} \subset X \times V \cong \mathbb{R} \times \mathbb{R} = \mathbb{R}^2$, $x \in X \cong \mathbb{R}, y \in V \cong \mathbb{R}$. ◇

To denote different partial derivatives we need to introduce the idea of a *multi-index.*

Definition 8.1.1 (Order of a multi-index) *Let $J = (j_1, \ldots, j_k)$ be an unordered k-tuple of integers with entries j_k, $1 \leq j_k \leq p$. The order of such a multi-index, which we denote as $\#J = |J| = k$, is defined as*

$$\#J = \#j_1 + \#j_2 + \ldots + \#j_k. \tag{8.1.1}$$

Further we denote

$$\tilde{J} = (\tilde{j}_1, \ldots, \tilde{j}_p), \tag{8.1.2}$$

where $\tilde{j}_i = \#j_i$. Then we also introduce $\tilde{J}! = \tilde{j}_1! \cdot \ldots \cdot \tilde{j}_p!$.

Example 8.1.2 If we consider $J_1 = (1, 1, 2, 2, 3)$, then

$$\#J_1 = 2 + 2 + 1 = 5 = k, \quad \tilde{J}_1 = (2, 2, 1),$$

and if we have $J_2 = (1, 1, 1, 1, 1)$, then

$$\#J_2 = 5, \quad \tilde{J}_2 = (5).$$

But if $J_3 = (4, 4, 4)$, $\tilde{J}_3$ looks like $\tilde{J}_3 = (0, 0, 0, 3)$. ◇

Now we want to consider the space V_1 of all first derivatives of the variable $u(x) \in V$, $V_1 \subseteq \mathbb{R}^p$,

$$\left(\frac{\partial u}{\partial x_1}, \frac{\partial u}{\partial x_2}, \ldots, \frac{\partial u}{\partial x_p}\right) \in V_1 \cong \mathbb{R}^p.$$

Then we describe the space V_2 of all second order derivatives of the variables $u(x)$, i.e.

$$\left(\frac{\partial^2 u}{\partial x_1 \partial x_1}, \frac{\partial^2 u}{\partial x_1 \partial x_2}, \ldots, \frac{\partial^2 u}{\partial x_p \partial x_p}\right) \in V_2 \cong \mathbb{R}^{N_2}, \quad N_2 = \binom{p+2-1}{2},$$

and so on up to the space V_k, the space of all k-th order derivatives of the variable $u(x)$,

$$\left(\frac{\partial^k u}{\partial {x_1}^k}, \ldots, \frac{\partial^k u}{\partial {x_p}^k}\right) \in V_k \cong \mathbb{R}^{N_k}, \quad N_k = \binom{p+k-1}{k}. \tag{8.1.3}$$

Further, we define the space $V^{(n)}$

$$V^{(n)} = V \times V_1 \times \ldots \times V_n, \quad dim V^{(n)} = \binom{p+n}{n} = N^{(n)} \tag{8.1.4}$$

to be the Cartesian product space, whose coordinates represent all the elements $u^{(n)}$ of $V^{(n)}$. Each element of the space $V^{(n)}$ has $N^{(n)} = (1 + N_1 + N_2 + \ldots + N_n)$ components u_J. Here the multi-index $J = (j_1, \ldots, j_k)$ with $1 \leq j_k \leq p$, $\#J = k$ and $0 \leq k \leq n$ denotes the type of the corresponding partial derivative. In the case $k = 0$, the value $u_{J=0}(x)$ coincides with the function $u(x)$ itself, i.e. $u_{J=0}(x) = u(x)$.

Example 8.1.3 Let $p = 1$. Therefore, we have the base space $\mathcal{M} = X \times V \cong \mathbb{R}^2$, with coordinates $(x, u) \in \mathcal{M}$, and calculate

$$V_1 \ni u' = \frac{du}{dx}, \quad V_2 \ni u'' = \frac{d^2u}{dx^2}, \quad V_k \ni u^{(k)} = \frac{d^k u}{dx^k}.$$

All V_k are one-dimensional spaces and

$$V^{(n)} = V \times V_1 \times \ldots \times V_n$$

is $(n+1)$–dimensional; a point of $V^{(n)}$ has components $(u, u', u'', \ldots, u^{(n)})$.

◇

Example 8.1.4 Let $p = 2$, $\mathcal{M} = X \times V \cong \mathbb{R}^3$. The space X is two-dimensional and we denote the coordinates $X \ni (x_1, x_2) = (x, y)$, $V \ni u$. We obtain then

$$V_1 \ni (u_x, u_y), \quad \dim V_1 = 2,$$
$$V_2 \ni (u_{xx}, u_{xy}, u_{yy}), \quad \dim V_2 = 3,$$
$$\vdots$$
$$V_k \ni (u_{\underbrace{x\ldots x}_{k}}, \ldots, u_{\underbrace{y\ldots y}_{k}}), \quad \dim V_k = k+1,$$
$$V^{(k)} = V \times V_1 \times \ldots \times V_k, \quad \dim V^{(k)} = \prod_{i=1}^{k}(i+1).$$

◇

Definition 8.1.2 (Jet bundle) *The total space $X \times V^{(n)}$, denoted by $\mathcal{M}^{(n)}$, whose coordinates represent the independent variables, the dependent variables and the derivatives of the dependent variables up to order n is called the n-th order jet bundle of the base space $\mathcal{M}$*

$$\mathcal{M}^{(n)} = \mathcal{M} \times V_1 \times \ldots \times V_n. \tag{8.1.5}$$

This is also called the n-th prolongation of $\mathcal{M}$.

We denote a partial derivative of the order k as a smooth, real-valued function $f(x) = f(x_1, x_2, \ldots, x_p)$ by

$$\partial_J f(x) = \frac{\partial^k f(x)}{\partial x_{j_1} \partial x_{j_2} \ldots \partial x_{j_k}}, \quad J = (j_1, j_2, \ldots, j_k), \quad \#J = k, \tag{8.1.6}$$

where $1 \le j_i \le p$, $i = 1, 2, \ldots, p$.

Definition 8.1.3 (n-th prolongation of f) *Given a smooth function $u = f(x)$, with $x \in X$ and $u \in V$, so $f : X \to V$. There exists an induced function*

$$u^{(n)} = \mathrm{pr}^{(n)} f(x)$$

called the n-th prolongation of $f(x)$, $x \in X \subset \mathbb{R}^p$ with values in $V^{(n)}$. Here we define the components $u^{(n)}$ by

$$u_J = \partial_J f(x), \tag{8.1.7}$$

i.e. $\mathrm{pr}^{(n)} f$ is a map from X to the space $V^{(n)}$.

Example 8.1.5 Let $p = 2$ and $(x, y) \in X$, $u \in V$. Then $u = f(x, y) : X \to V$ and $u^{(2)} = \mathrm{pr}^{(2)} f(x, y)$ is given by

$$(u; u_x, u_y; u_{xx}, u_{xy}, u_{yy}) = \left(f; \frac{\partial f}{\partial x}, \frac{\partial f}{\partial y}; \frac{\partial^2 f}{\partial x^2}, \frac{\partial^2 f}{\partial x \partial y}, \frac{\partial^2 f}{\partial y^2}\right).$$

Thereby all values are evaluated at the point (x, y). We can also calculate the Taylor series for $f(x, y)$ at the point $(x_0, y_0) = (0, 0)$ using

$$f(x,y) = f(0,0) + (f_x, f_y)|_{(0,0)} \cdot \begin{pmatrix} x \\ y \end{pmatrix} + \frac{1}{2}(x,y) \cdot \begin{pmatrix} f_{xx} & f_{xy} \\ f_{xy} & f_{yy} \end{pmatrix}\Bigg|_{(0,0)} \cdot \begin{pmatrix} x \\ y \end{pmatrix} + \dots.$$

◇

Now we introduce some definitions and notations for n-th order PDE, i.e. a solution subvariety, a natural projection in n- order jet bundle. They are very similar to the definitions introduced before for the analysis of ODEs. Their general form can be found in well known books for more advanced readers, for instance in [59] or [60]. We provide these definitions in the form adapted to our purposes.

Definition 8.1.4 (Partial differential equation of the n-th order) *We denote the n-th order PDE on dependent variable $u(x)$ and p independent variables as*

$$\Delta(x, u^{(n)}) = 0, \tag{8.1.8}$$

involving $x = (x_1, \dots, x_p)$, $u(x)$ and derivatives of $u(x)$ with respect to x_i, $i = 1, 2, \dots, p$ up to the order n.

We consider Δ as a smooth map from the jet bundle $\mathcal{M}^{(n)} = X \times V^{(n)}$ to some Euclidean space

$$\Delta : \mathcal{M}^{(n)} = X \times V^{(n)} \to \mathbb{R}.$$

The equality $\Delta = 0$ determines a subvariety

$$\mathcal{S}_\Delta = \{(x, u^{(n)}) : \Delta(x, u^{(n)}) = 0\} \subset \mathcal{M}^{(n)} = X \times V^{(n)}$$

of the total jet bundle. The differential Equation (8.1.8) is identified with its subvariety $\mathcal{S}_\Delta$. The subvariety is called the solution subvariety of the given differential equation.

Definition 8.1.5 (Solution) *A smooth solution of the given PDE is a smooth function $u = f(x)$, such that*

$$\Delta(x, \mathrm{pr}^{(n)} f(x)) = 0, \tag{8.1.9}$$

whenever x lies in the domain of $f(x)$.

Therefore we know that a solution has to be a smooth function $u = f(x)$, such that the graph of the n-th prolongation of f belongs to the subvariety $\mathcal{S}_\Delta$.

Example 8.1.6 We study the Laplace equation

$$\Delta(x, u^{(n)}) = u_{xx} + u_{yy} = 0, \quad n = 2, \ p = 2.$$

Then $(x, y) \in X$, $u \in V$ and $(x, y; u; u_x, u_y; u_{xx}, u_{xy}, u_{yy}) \in M^{(2)}$. The Laplace equation defines a linear subvariety or, in other words, the hyperplane $u_{xx} + u_{yy} = 0$ in $\mathcal{M}^{(2)}$. A solution $u = f(x, y)$ has to satisfy

$$\frac{\partial^2 f}{\partial x^2} + \frac{\partial^2 f}{\partial y^2} = 0, \qquad \forall (x, y) \in X.$$

Let $f(x, y) = ax + by$, a, b are constants. We calculate the second prolongation of the function $f(x, y)$

$$\mathrm{pr}^{(2)} f(x, y) = (ax + by; \underbrace{a}_{f_x}, \underbrace{b}_{f_y}; \underbrace{0}_{f_{xx}}, \underbrace{0}_{f_{xy}}, \underbrace{0}_{f_{yy}}),$$

and obtain

$$f_{xx} + f_{yy} = 0 - 0 = 0.$$

Thus any linear function is a solution of the Laplace equation.

Let now $\phi(x, y) = ax^3 + bx^2y + cxy^2 + dy^3$, a, b, c, d are constants. We calculate the second prolongation of $\phi(x)$

$$\mathrm{pr}^{(2)} \phi(x, y) = (ax^3 + bx^2y + cxy^2 + dy^3; \tag{8.1.10}$$

$$\underbrace{3ax^2 + 2bxy + cy^2}_{\phi_x}, \underbrace{bx^2 + 2cxy + 3dy^2}_{\phi_y}; \tag{8.1.11}$$

$$\underbrace{6ax + 2by}_{\phi_{xx}}, \underbrace{2bx}_{\phi_{xy}}, \underbrace{2cx + 6dy}_{\phi_{yy}}), \tag{8.1.12}$$

and obtain

$$\phi_{xx} + \phi_{yy} = 6ax + 2by + 2cx + 6dy$$

equal to zero only if $c = -3a$ and $b = -3d$. Thus any polynomial $\phi(x, y) = ax^3 - 3dx^2y - 3axy^2 + dy^3$, a, d are constants is a solution of the Laplace equation. ◇

Now let $u = f(x)$ be a smooth function defined at the point $x_0 \in \mathcal{M}$ and its graph be defined in $\mathcal{M}^{(n)}$ with

$$u^{(n)}(x_0) = \mathrm{pr}^{(n)} f(x_0),$$

i.e., the components are defined by $u_J(x_0) = \partial_J f(x)|_{x=x_0}$, $\#J = k$, $1 \leq k \leq n$. As an example we can consider the n-th order Taylor polynomial at $x_0 \in X$

$$f(x) = \sum_J \frac{1}{\bar{J}!} u_J(x_0) \cdot (x - x_0)^J, \quad \#J = k, \quad 1 \leq k \leq n,$$

where we denote $(x - x_0)^J = (x_{j_1} - x_{0,j_1})(x_{j_2} - x_{0,j_2}) \cdots (x_{j_k} - x_{0,j_k})$ with $J = (j_1, j_2, \ldots, j_k)$ and $x_0 = (x_{0,1}, x_{0,2}, \ldots, x_{0,p})$.

It is similar to the case where X was a one-dimensional space. Let g be an element of a transformation group $g \in G$. We assume that the group transformation $g \circ f$ is well-defined in the neighborhood of x_0 and we have

$$(\tilde{x}_0, \tilde{u}_0) = g \circ (x_0, u_0), \quad u_0 = f(x_0).$$

We determine the action of the prolonged group transformation $\text{pr}^{(n)} g$ at the point $(x_0, {u_0}^{(n)})$ evaluating the derivatives of the transformed function $g \circ f$ at $\tilde{x}_0$

$$\text{pr}^{(n)} g \circ (x_0, {u_0}^{(n)}) = (\tilde{x}_0, {\tilde{u}_0}^{(n)}),$$

where ${\tilde{u}_0}^{(n)} = (\text{pr}^{(n)} g \circ f)(\tilde{x}_0)$.

Example 8.1.7 Let $p = 1$ and $\mathcal{M} = X \times V \cong \mathbb{R}^2$, $\mathcal{M}^{(1)} \cong \mathbb{R}^3$. If $u = f(x)$, then $\text{pr}^{(1)} f(x) = (f(x); f'(x))$. Further we have $(x_0, u_0, u_{0,x}) \in X \times V \times V^{(1)} = \mathcal{M} \times V^{(1)} = \mathcal{M}^{(1)}$ and we obtain

$$\text{pr}^{(1)} g \circ (x_0, u_0, u_{0,x}) = (\tilde{x}_0, \tilde{u}_0, \tilde{u}_{0,x}).$$

$\diamond$

Definition 8.1.6 (Natural projection, [59]) *We call the map*

$$\Pi_k^n : \mathcal{M}^{(n)} \to \mathcal{M}^{(k)} \tag{8.1.13}$$

with

$$\Pi_k^n(x, u^{(n)}) = (x, u^{(k)}), \tag{8.1.14}$$

where $u^{(k)}$ only consists of the components $u_{\alpha,J}$, $\#J \leq k$, $k < n$, of $u^{(n)}$ itself, a natural projection.

Remark 8.1.1 *According to the definition a prolongation of the 0 order is $\text{pr}^{(0)} G = G$. If we pay attention only to the derivatives up to order $k \leq n$, namely $(x, u^{(k)})$, the action of $\text{pr}^{(k)}$ coincides with the earlier described prolongation $\text{pr}^{(k)} G$ and it holds*

$$\Pi_k^n \circ \text{pr}^{(n)} g = \text{pr}^{(k)} g, \quad \forall g \in G.$$

Theorem 8.1.1 *(Invariance of differential equations, [59])*
Let $\mathcal{M}$ be an open subset of $X \times V$ and suppose $\Delta(x, u^{(n)}) = 0$ is an n-th order PDE defined over $\mathcal{M}$ with corresponding subvariety $\mathcal{S}_\Delta \subset \mathcal{M}^{(n)}$. Suppose G is a local continuous group of point transformations acting on $\mathcal{M}$, which prolongation leaves $\mathcal{S}_\Delta$ invariant, which means that whenever $(x, u^{(n)}) \in \mathcal{S}_\Delta$, we have $\mathrm{pr}^{(n)} g \circ (x, u^{(n)}) \in \mathcal{S}_\Delta$ for all $g \in G$, such that this expression is defined.
Then G is a symmetry group of the given PDE.

Proof: Suppose $u = f(x)$ is a local solution to $\Delta(x, u^{(n)}) = 0$. Then the graph of this function which is denoted as $\Gamma_f^{(n)} = \{x, \mathrm{pr}^{(n)} f(x)\}$ lies entirely within $\mathcal{S}_\Delta$.

If $g \in G$ and $g \circ f$ is well-defined, then $\Gamma_{g\circ f}^{(n)} = \mathrm{pr}^{(n)} g(\Gamma_f^{(n)})$. Because $\mathcal{S}_\Delta$ is invariant under $\mathrm{pr}^{(n)} g$, we have again that $\mathrm{pr}^{(n)}(g \circ f)$ lies entirely in $\mathcal{S}_\Delta$ and is a solution to the system $\Delta(x, u^{(n)}) = 0$ as well. □

Definition 8.1.7 (Prolongation of vector fields) *Let $\mathcal{M} \subset X \times V$ be an open subspace and suppose $\mathbf{U}$ is a vector field (an infinitesimal generator of a transformation group) on $\mathcal{M}$ with a corresponding (local) one-parameter group $G = \exp(\varepsilon \mathbf{U})$. The n-th prolongation of $\mathbf{U}$, denoted by $\mathrm{pr}^{(n)}\mathbf{U}$ or $\mathbf{U}^{(n)}$, will be a vector field on the jet bundle $\mathcal{M}^{(n)}$, and it is defined to be the infinitesimal generator of the corresponding prolonged group $\mathrm{pr}^{(n)} G = G^{(n)} = \mathrm{pr}^{(n)}(\exp(\varepsilon \mathbf{U}))$*

$$\mathrm{pr}^{(n)}\mathbf{U}\big|_{(x,u^{(n)})} = \frac{d}{d\varepsilon}\left(\mathrm{pr}^{(n)} \exp(\varepsilon \mathbf{U})(x, u^{(n)})\right)\Big|_{\varepsilon=0} \tag{8.1.15}$$

for any $(x, u^{(n)}) \in \mathcal{M}^{(n)}$.

Remark 8.1.2 *A prolonged vector field (an infinitesimal generator) has the following structure*

$$\mathbf{U}^{(n)} = \mathrm{pr}^{(n)}\mathbf{U} = \sum_{i=1}^{p} \xi_i \frac{\partial}{\partial x_i} + \sum_{\substack{J \\ 0 \le \#J \le n}} \Phi^J \frac{\partial}{\partial u_J}, \tag{8.1.16}$$

where $x = (x_1, \ldots, x_p)$ are independent variables, $u = u(x)$ denotes the dependent variable. The coefficients ξ_i, $i = 1, 2, \ldots, p$ depend on variables $x_i, i = 1, 2, \ldots, p$ and u, the function Φ^J depends on variables $x_i, i = 1, 2, \ldots, p$ and derivatives of u up to the order $\#J$.

Example 8.1.8 Let $p = 1$ and $x = x_1$, $u = y$. Then we have

$$\mathrm{pr}^{(n)}\mathbf{U} = \xi\frac{\partial}{\partial x} + \sum_{J=0}^{n} \Phi^J \frac{\partial}{\partial u_J} = \xi\frac{\partial}{\partial x} + \underbrace{\eta}_{\Phi^0}\frac{\partial}{\partial y} + \underbrace{\eta^{(1)}}_{\Phi^1}\frac{\partial}{\partial y'} + \ldots + \underbrace{\eta^{(n)}}_{\Phi^n}\frac{\partial}{\partial y^{(n)}}.$$

◇

Example 8.1.9 Let $p = 2$, and $x = (x_1, x_2) = (z, t)$, $u = u(x) = u(x_1, x_2) = u(z, t)$. Further we have a vector field $\mathbf{U}$ defined on $\mathcal{M}$ as

$$\mathbf{U} = \xi(z, t, u)\frac{\partial}{\partial x} + \tau(z, t, u)\frac{\partial}{\partial t} + \Phi(z, t, u)\frac{\partial}{\partial u}.$$

Then the second prolongation of $\mathbf{U}$ has the following structure

$$\mathrm{pr}^{(2)}\mathbf{U} = \xi\frac{\partial}{\partial z} + \tau\frac{\partial}{\partial t} + \Phi\frac{\partial}{\partial u} + \Phi^z\frac{\partial}{\partial u_z} + \Phi^t\frac{\partial}{\partial u_t} + \Phi^{zz}\frac{\partial}{\partial u_{zz}} + \Phi^{zt}\frac{\partial}{\partial u_{zt}} + \Phi^{tt}\frac{\partial}{\partial u_{tt}}.$$

◇

If we look at an arbitrary vector field (linear differential operator) $\mathbf{V}$ on $\mathcal{M}^{(n)}$, the components ξ_i, Φ^J could depend on all variables of this jet bundle. If $\mathbf{V}$ is a n-th order prolongation of $\mathbf{U}$, i.e. $\mathbf{V} = \mathrm{pr}^{(n)}\mathbf{U}$, all coefficients of this vector field will be determined by the coefficients of the original vector field $\mathbf{U}$, defined on the base space $\mathcal{M}$. Because of the natural projection $\mathcal{M}^{(0)} = \mathcal{M}$ and $\mathrm{pr}^{(0)}\mathbf{U} = \mathbf{U}$ we obtain

$$\xi_i = \xi_i(x_1, \ldots, x_p, u), \quad \Phi^0 = \Phi(x_1, \ldots, x_p, u).$$

We remember

$$\Pi_k^n \circ \mathrm{pr}^{(n)} g = \mathrm{pr}^{(k)} g, \qquad \forall g \in G,$$

and correspondingly, each coefficient of the vector field $\mathrm{pr}^{(k)}\mathbf{U}$ can depend on k-th and lower order derivatives of u, i.e., if $\#J = k$ we have $\Phi^J = \Phi^J(x, u^{(k)})$. We provide some simple examples to clarify the notations.

Example 8.1.10 Again let $p = 1$ and $k = 2$, i.e. $x \in \mathbb{R}$, $u(x) = y(x)$. We calculate the second prolongation of the vector field $\mathbf{U}$

$$\mathrm{pr}^{(2)}\mathbf{U} = \xi(x, y)\frac{\partial}{\partial x} + \eta(x, y)\frac{\partial}{\partial y} + \eta^{(1)}(x, y, y')\frac{\partial}{\partial y'} + \eta^{(2)}(x, y, y', y'')\frac{\partial}{\partial y''}.$$

Example 8.1.11 Let $p = 2$, $q = 1$, and $k = 2$. We denote the coordinates $(x_1, x_2) = (x, t)$, $u = u(x, t)$, then the second prolongation of $\mathbf{U}$ has the form

$$\begin{aligned}\mathrm{pr}^{(2)}\mathbf{U} &= \xi(x,t,u)\frac{\partial}{\partial x} + \tau(x,t,u)\frac{\partial}{\partial t} + \Phi(x,t,u)\frac{\partial}{\partial u} \\ &+ \underbrace{\Phi^{x}(x,t,u,u_x,u_t)}_{\Phi^{x}(x,u^{(1)})}\frac{\partial}{\partial u_x} + \underbrace{\Phi^{t}(x,t,u,u_x,u_t)}_{\Phi^{t}(x,u^{(1)})}\frac{\partial}{\partial u_t} \\ &+ \underbrace{\Phi^{xx}(x,t,u,u_x,u_t,u_{xx},u_{xt},u_{tt})}_{\Phi^{xx}(x,u^{(2)})}\frac{\partial}{\partial u_{xx}} + \Phi^{xt}(x,u^{(2)})\frac{\partial}{\partial u_{xt}} \\ &+ \Phi^{tt}(x,u^{(2)})\frac{\partial}{\partial u_{tt}}.\end{aligned}$$

◇

Theorem 8.1.2 *(Determining equations)*
Suppose $\Delta(x, u^{(n)}) = 0$, is a PDE defined on $\mathcal{M} \subset X \times V$. If G is a local continuous group of point transformations acting on $\mathcal{M}$ generated by Lie algebra L and

$$\mathrm{pr}^{(n)}\mathbf{U}\left(\Delta(x, u^{(n)})\right) = 0, \quad \forall \mathbf{U} \in L,$$

whenever $\Delta(x, u^{(n)}) = 0$, then G is a symmetry group of the system.

Example 8.1.12 Let $p = 1$ and let us consider the ordinary differential equation

$$y' = \frac{\mathcal{Y}(x,y)}{\mathcal{X}(x,y)}.$$

We know $\mathcal{A} = \mathcal{X}\frac{\partial}{\partial x} + \mathcal{Y}\frac{\partial}{\partial y}$ and $[\mathbf{U}, \mathcal{A}] = \lambda(x,y)\mathcal{A}$. The first prolongation of $\mathbf{U}$ is $\mathbf{U}^{(1)} = \mathbf{U} + \eta^{(1)}\frac{\partial}{\partial y'}$ and we obtain an equivalent determining equation

$$\mathbf{U}^{(1)}\left(y' - \frac{\mathcal{Y}}{\mathcal{X}}\right)\bigg|_{y'=\frac{\mathcal{Y}}{\mathcal{X}}} = 0,$$

or if we consider $\Delta(x, y^{(1)}) = \mathcal{X}(x,y)y' - \mathcal{Y}(x,y) = 0$, and apply the first prolongation $\mathbf{U}^{(1)}$ to this equation we get

$$\mathbf{U}^{(1)}\Delta(x, y^{(1)})\big|_{\Delta=0} = 0. \tag{8.1.17}$$

We see that we obtain determining equations in similar form as in Theorem 8.1.2. ◇

Definition 8.1.8 (Total differentiation) *Let us introduce a total differentiation given by the following formal infinite sums*

$$\begin{aligned} D_i &= \frac{\partial}{\partial x_i} + \sum_J u_{i,J}\frac{\partial}{\partial u_J} \\ &= \frac{\partial}{\partial x_i} + u_i\frac{\partial}{\partial u} + \sum_{i_1=1}^{p} u_{ii_1}\frac{\partial}{\partial u_{i_1}} + \ldots \\ &+ \sum_{i_1=1}^{p}\sum_{i_2=1}^{p}\cdots\sum_{i_r=1}^{p} u_{ii_1i_2\ldots i_r}\frac{\partial}{\partial u_{i_1i_2\ldots i_r}}\cdots, \quad i=1,\ldots,p, \quad 0 \le \#J \le \infty. \end{aligned}$$

The variables x_i are called independent variables and $u(x) = u(x_1, x_2, \ldots, x_p)$ is a dependent variable with derivatives denoted as $u_{i,J}$.

Example 8.1.13 Let $f(x, u, u^{(1)})$ be a smooth function defined on $\mathcal{M}^{(1)}$. The total derivatives D_i, $i = 1, \ldots, p$ act on a function $f(x, u, u^{(1)})$ involving a finite number of variables and define the unique smooth function on $\mathcal{M}^{(2)}$ as

$$D_i(f(x, u, u_{(1)}) = \frac{\partial f}{\partial x_i} + u_i\frac{\partial f}{\partial u} + \sum_{i_1=1}^{p} u_{ii_1}\frac{\partial f}{\partial u_{i_1}}.$$

In other words in this case we differentiate $f(x, u, u^{(1)})$ with respect to x_i while treating $u(x)$ and all derivatives $u_{i_1} = \frac{\partial u}{\partial x_{i_1}}$ as functions of $x = (x_1, x_2, \ldots, x_p)$. ◇

Example 8.1.14 If $f = u(x)$, we can calculate $u_i = D_i(u)$, or if $f = u_j$, we obtain $u_{ij} = D_i(u_j) = D_iD_j(u)$. It is easy to show that $D_iD_j(f) = D_jD_i(f)$. ◇

Example 8.1.15 We remark that total derivatives are not equivalent to partial derivatives. Let us look at a simple example. If $p = 1$ and we denote $x = x$, $u = y$, we get

$$D_x = \frac{\partial}{\partial x} + y'\frac{\partial}{\partial y} + y''\frac{\partial}{\partial y'} + \ldots + y^{(s+1)}\frac{\partial}{\partial y^{(s)}} + \ldots.$$

Let $f = xy'$, then

$$D_x(xy') = y' + xy'' \quad \not\Leftrightarrow \quad \frac{\partial}{\partial x}(xy') = y'.$$

The general differentiation rules are applied also to total derivatives. We formulate the most important of them.

Theorem 8.1.3 *(Faá di Bruno's formula, chain rule for a composite function)*
Let $p = 1$. If $u = f(y)$ and $y = \varphi(x)$, the k-th order derivative of the function $f(y)$ is given, in terms of $y', \dots, y^{(k)}$ and $f' = \frac{df}{dy}, \dots, f^{(k)} = \frac{d^k f}{dy^k}$, by the formula

$$D_x^k(f) = \sum_{\substack{l_1,\dots,l_k \geq 0 \\ l_1+2l_2+\dots+kl_k=k \\ p=l_1+\dots+l_k\geq 0}} \frac{k!}{l_1!\dots l_k!} f^{(p)} \left(\frac{y'}{1!}\right)^{l_1} \left(\frac{y''}{2!}\right)^{l_2} \dots \left(\frac{y^{(k)}}{k!}\right)^{l_k}.$$

Example 8.1.16 The Leibnitz formula for the product of functions $f\phi$ has the form

$$D_x^k(f\phi) = D_x^k(f)\cdot\phi + \sum_{s=1}^{k-1} \frac{k!}{(k-s)!s!} D_x^{k-s}(f) D_x^s(\phi) + f\cdot D_x^k(\phi).$$

◇

Example 8.1.17 Let us use the total derivatives in the case $p = 1$ then we get the prolongation formula in the form

$$\eta^{(n+1)} = D_x(\eta^{(n)}) - y^{(n+1)} D_x(\xi).$$

◇

Remark 8.1.3 *If we have one independent variable x and several dependent variables $u = (u_1, \dots, u_q)$, we obtain the prolongation formulas in a similar form*

$$\Phi_\alpha^1 = D_x(\Phi_\alpha^0) - u_1^\alpha D_x(\xi) \quad \Rightarrow \quad \Phi_\alpha^{n+1} = D_x(\Phi_\alpha^n) - u_{n+1}^\alpha D_x(\xi), \quad 0 \leq \alpha \leq q.$$

◇

Definition 8.1.9 (Total derivative, [59]) *Let $\Phi(x, u^{(n)})$ $(\dim X = p, \dim V = 1)$ be a smooth function of $x \in X$, $u \in V$ and derivatives of u up to the order n, defined on an open subset $\mathcal{M}^{(n)} = X \times V^{(n)}$. The total derivative of Φ with respect to x_i is a unique smooth function $D_i\Phi(x, u^{(n+1)})$, defined on $\mathcal{M}^{(n+1)}$, depending on derivatives of u up to the order $(n+1)$. It has the property that, if $u = f(x)$ is an arbitrary smooth function*

$$D_i\Phi(x, \mathrm{pr}^{(n+1)} f(x)) = \frac{\partial}{\partial x_i}\left(\Phi(x, \mathrm{pr}^{(n)} f(x))\right), \tag{8.1.18}$$

where $D_i\Phi$ is obtained by differentiating Φ with respect to x_i while treating $u(x)$ and its derivatives as functions of x,

$$D_i\Phi = \frac{\partial\Phi}{\partial x_i} + \sum_{\substack{J\\0\le\#J\le n}} u_{i,J}\frac{\partial\Phi}{\partial u_J}, \tag{8.1.19}$$

for $J = (j_1, \dots, j_k)$, $u_{i,J} = \frac{\partial u_J}{\partial x_i} = \frac{\partial^{k+1}u}{\partial x_i \partial x^{j_1}\partial x^{j_k}}$.

Example 8.1.18 Let $\Phi = xyu + y^2u_{xy}$, $(x, y) \in X$, $u \in V$. We get

$$\begin{aligned} D_x\Phi &= yu + xyu_x + y^2u_{xxy}, \\ D_y\Phi &= xu + xyu_y + 2yu_{xy} + y^2u_{xyy}, \\ D_xD_y\Phi = D_yD_x\Phi &= u + xu_x + yu_y + xyu_{xy} + 2yu_{xxy} + y^2u_{xxyy}. \end{aligned}$$

◇

Theorem 8.1.4 *Let*

$$\mathbf{V} = \sum_{i=1}^{p}\xi^i(x,u)\frac{\partial}{\partial x_i} + \Phi(x,u)\frac{\partial}{\partial u}$$

be a vector field (an infinitesimal generator), defined on an open subset $\mathcal{M} \subset X \times V$. The n-th prolongation of $\mathbf{V}$ is the vector field

$$\mathrm{pr}^{(n)}\mathbf{V} = \mathbf{V} + \sum_{\substack{J\\0\le\#J\le n\\1\le j_k\le p}} \Phi^J(x,u^{(n)})\frac{\partial}{\partial u_J}, \tag{8.1.20}$$

defined on the jet bundle $\mathcal{M}^{(n)} \subset X \times V^{(n)}$. The coefficient functions Φ^J of $\mathrm{pr}^{(n)}\mathbf{V}$ are given by

$$\Phi^J(x,u^{(n)}) = D_J\left(\Phi - \sum_{i=1}^{p}\xi^iu_i\right) + \sum_{i=1}^{p}\xi^iu_{i,J}, \tag{8.1.21}$$

where $u_i = \frac{\partial u}{\partial x_i}$, $u_{i,J} = \frac{\partial u_J}{\partial x_i}$.

8.2 The heat equation

Lie group analysis of the main types of PDEs of the second order: elliptical, parabolic and hyperbolic in the case of two independent variables was provided by Sophus Lie in [53].

We examine the parabolic equation which describes the heat conduction in a one-dimensional bar

$$u_t = u_{xx}, \tag{8.2.22}$$

where the thermal diffusivity has been normalized to one. The independent variables here are x and t, i.e. space and time variables. We have one dependent variable denoted by u. Hence, $\dim X = 2$, $\dim V = 1$. The heat equation is of second order and therefore, it can be identified with a linear subvariety in $\mathcal{M}^{(2)} = X \times V^{(2)}$, determined by

$$\Delta(x, t, u^{(2)}) = u_t - u_{xx} = 0.$$

We look for the symmetries admitted by this equation. Using the determining equations we define the basis of the admitted Lie algebra. Let

$$\mathbf{V} = \xi(x,t,u)\frac{\partial}{\partial x} + \tau(x,t,u)\frac{\partial}{\partial t} + \Phi(x,t,u)\frac{\partial}{\partial u}$$

be a vector field on $X \times V$. We use the second prolongation of $\mathbf{V}$

$$\begin{aligned}
&\mathrm{pr}^{(2)}\mathbf{V} = \mathbf{V}^{(2)}\\
&= \xi\frac{\partial}{\partial x} + \tau\frac{\partial}{\partial t} + \Phi\frac{\partial}{\partial u} + \Phi^x\frac{\partial}{\partial u_x} + \Phi^t\frac{\partial}{\partial u_t} + \Phi^{xx}\frac{\partial}{\partial u_{xx}} + \Phi^{xt}\frac{\partial}{\partial u_{xt}} + \Phi^{tt}\frac{\partial}{\partial u_{tt}}
\end{aligned}$$

in order to find the symmetries of our equation. We calculate

$$\mathrm{pr}^{(2)}\mathbf{V}(\Delta)\big|_{\Delta=0} = \Phi^t - \Phi^{xx}\big|_{u_t=u_{xx}} = 0. \tag{8.2.23}$$

Now we use the expressions for Φ^t and Φ^{xx} to get the system of determining equations on components of infinitesimal generators. With formula (8.1.21) we get

$$\begin{aligned}
\Phi^x &= D_x(\Phi - \xi u_x - \tau u_t) + \xi u_{xx} + \tau u_{xt}\\
&= D_x\Phi - u_x D_x\xi - u_t D_x\tau\\
&= \Phi_x + (\Phi_u - \xi_x)u_x - \tau_x u_t - \xi_u {u_x}^2 - \tau_u u_x u_t,
\end{aligned}$$

$$\begin{aligned}
\Phi^t &= D_t(\Phi - \xi u_x - \tau u_t) + \xi u_{xt} + \tau u_{tt}\\
&= \Phi_t - \xi_t u_x + (\Phi_u - \tau_t)u_t - \xi_u u_x u_t - \tau_u {u_t}^2,
\end{aligned}$$

$$\begin{aligned}
\Phi^{xx} &= D_x^2(\Phi - \xi u_x - \tau u_t) + \xi u_{xxx} + \tau u_{xxt}\\
&= \Phi_{xx} + (2\Phi_{xu} - \xi_{xx})u_x - \tau_{xx}u_t + (\Phi_{uu} - 2\xi_{xu}){u_x}^2 - 2\tau_{xu}u_x u_t\\
&\quad - \xi_{uu}{u_x}^3 - \tau_u u_t {u_x}^2 + (\Phi_u - 2\xi_x)u_{xx} - 2\tau_x u_{xt} - 3\xi_u u_x u_{xx}\\
&\quad - \tau_u u_t u_{xx} - 2\tau_u u_x u_{xt}.
\end{aligned}$$

After inserting these expressions into (8.2.23) and rearranging the terms according to the monomials, we obtain the system of determining equations as shown in Table 8.1. We now solve this system step by step. Equations (1) and (3) of Table 8.1 yield $\tau = \tau(x,t)$ and (4) additionally yields $\tau = \tau(t)$. If we insert this result into Equation (2) of Table 8.1, we obtain $\xi_u = 0$ and therefore, ξ is the function of two variables $\xi = \xi(x,t)$. Now we insert $\tau = \tau(t)$ and $\xi = \xi(x,t)$ in Equations (5) to (9) of Table 8.1.

As a result we obtain the modified system as shown in Table 8.2.

Because $\tau = \tau(t)$, Equation (5') in Table 8.2 yields $\xi(x,t) = \frac{1}{2}\tau_t(t)x + \gamma(t)$, i.e. $\xi(x,t)$ is a linear function of x with coefficients depending on t only. From Equation (7') of Table 8.2 we obtain $\Phi(x,t,u) = \alpha(x,t)u + \beta(x,t)$.

Table 8.1: The system of the determining equations for the heat equation.

	monomial	equation on the coefficient
	${u_{xx}}^2$	$\tau_u - \tau_u = 0$
(1)	${u_x}^2 u_{xx}$	$\tau_u = 0$
(2)	$u_x u_{xx}$	$2\xi_u + 2\tau_{xu} = 0$
(3)	$u_x u_{xt}$	$2\tau_u = 0$
(4)	u_{xt}	$2\tau_x = 0$
(5)	u_{xx}	$\tau_{xx} - \tau_t + 2\xi_x = 0$
(6)	${u_x}^3$	$\xi_{uu} = 0$
(7)	${u_x}^2$	$2\xi_{xu} - \Phi_{uu} = 0$
(8)	u_x	$\xi_{xx} - \xi_t - 2\Phi_{xu} = 0$
(9)	1	$\Phi_t - \Phi_{xx} = 0$

Table 8.2: The system of the modified determining equation where we used that τ depends only on t, i.e. $\tau = \tau(t)$, and correspondingly $\xi = \xi(x,t)$.

	monomial	equation on the coefficient
(5')	u_{xx}	$\tau_t = 2\xi_x$
(6')	${u_x}^3$	$0 = 0$
(7')	${u_x}^2$	$\Phi_{uu} = 0$
(8')	u_x	$\xi_{xx} - \xi_t = 2\Phi_{xu}$
(9')	1	$\Phi_t - \Phi_{xx} = 0$

Having in mind these results we can transform Equations (8') and (9') in Table 8.2 to

$$-\frac{1}{2}\tau_{tt}(t)x - \gamma_t(t) = 2\alpha_x(x,t), \tag{I}$$

respectively

$$(\alpha_t(x,t) - \alpha_{xx}(x,t))u + \beta_t(x,t) - \beta_{xx}(x,t) = 0. \tag{II}$$

Equation (I) leads to

$$\alpha(x,t) = -\frac{1}{8}\tau_{tt}(t)x^2 - \frac{1}{2}\gamma_t(t)x + \theta(t), \tag{I'}$$

and with Equation (II) we obtain the system of equations

$$\begin{aligned} \alpha_t(x,t) - \alpha_{xx}(x,t) &= 0, \\ \beta_t(x,t) - \beta_{xx}(x,t) &= 0. \end{aligned} \tag{II'}$$

We see that the function $\beta(x,t)$ itself is a solution of the heat equation. Equations (I') and (II') instead yield

$$\tau_{ttt} = 0, \quad \gamma_{tt} = 0, \quad \theta_t = -\frac{1}{4}\tau_{tt},$$

or correspondingly

$$\tau(t) = c_1t^2 + c_2t + c_3, \quad \gamma(t) = c_4t + c_5, \quad \theta(t) = -\frac{1}{2}c_1t + c_6,$$

where all c_i, $i = 1, \ldots, 6$, are constants.

Thus, we defined our vector field $\mathbf{V}$ completely. In other words we found the infinitesimal generator of the symmetry group of the heat equation

$$\xi(x,t,u) = c_1x + \frac{1}{2}c_2x + c_4t + c_5, \tag{8.2.24}$$

$$\tau(x,t,u) = c_1t^2 + c_2t + c_3, \tag{8.2.25}$$

$$\Phi(x,t,u) = -\frac{1}{4}c_1(x^2 + 2t)u - \frac{1}{2}c_4xu + c_6u + \beta(x,t). \tag{8.2.26}$$

The functions $\xi(x,t,u), \tau(x,t,u), \Phi(x,t,u)$ contain six arbitrary constants and the function $\beta(x,t)$ which represents an arbitrary solution of the heat equation. It means we obtain the algebra $L = \langle \mathbf{V}_1, \ldots, \mathbf{V}_6 \rangle$ spanned by generators $\mathbf{V}_1, \ldots, \mathbf{V}_6$. We obtain the generators $\mathbf{V}_1, \ldots, \mathbf{V}_6$ if we consequently put all constants in the expression for $\mathbf{V}$ except one to zero. The generator $\mathbf{V}_\beta$ gives rise to the infinite–dimensional subalgebra $\mathbf{V}_\beta = \beta(x,t)\frac{\partial}{\partial u}$. The group property described by $\mathbf{V}_\beta$ is nothing else than a well known superposition principle for linear equations. It means the sum of solutions is a solution as well.

Each of the generators $\mathbf{V}_1, \ldots, \mathbf{V}_6$ corresponds to a one-dimensional Lie group G_1, $i = 1, \ldots, 6$. Now let us list the generators and the corresponding groups

$$\begin{aligned}
\mathbf{V}_1 &= \tfrac{\partial}{\partial x} && \to G_1 : (x+\varepsilon, t, u),\\
\mathbf{V}_2 &= \tfrac{\partial}{\partial t} && \to G_2 : (x, t+\varepsilon, u),\\
\mathbf{V}_3 &= u\tfrac{\partial}{\partial u} && \to G_3 : (x, t, \mathrm{e}^{\varepsilon}u),\\
\mathbf{V}_4 &= x\tfrac{\partial}{\partial x} + 2t\tfrac{\partial}{\partial t} && \to G_4 : (\mathrm{e}^{\varepsilon}x, \mathrm{e}^{2\varepsilon}t, u),\\
\mathbf{V}_5 &= 2t\tfrac{\partial}{\partial x} - xu\tfrac{\partial}{\partial u} && \to G_5 : (x+2\varepsilon t, t, u\mathrm{e}^{-\varepsilon x-\varepsilon^2 t}),\\
\mathbf{V}_6 &= 4tx\tfrac{\partial}{\partial x} + 4t^2\tfrac{\partial}{\partial t} && \\
&\quad -(x^2+2t)u\tfrac{\partial}{\partial u} && \to G_6 : \left(\tfrac{x}{1-4\varepsilon t}, \tfrac{t}{1-4\varepsilon t}, u\sqrt{1-4\varepsilon t}\cdot \mathrm{e}^{\left(\frac{-\varepsilon x^2}{1-4\varepsilon t}\right)}\right),\\
\mathbf{V}_\beta &= \beta(x,t)\tfrac{\partial}{\partial u} && \to G_\beta : (x, t, u+\varepsilon\beta(x,t)).
\end{aligned} \tag{8.2.27}$$

The action of each of these one-dimensional groups is well defined. We demonstrated step-by-step which one-parameter Lie group G_i is generated by the corresponding operator $\mathbf{V}_i$. We take a solution of the heat equation and apply G_i to this solution, obviously, the result of transformation is still a solution of the same heat equation. If $u = f(x,t)$ is a solution, we can find the transformed solutions and collect the results in a table

$$\begin{aligned}
G_1 u &= f(x-\varepsilon, t),\\
G_2 u &= f(x, t-\varepsilon),\\
G_3 u &= \mathrm{e}^{\varepsilon} f(x,t),\\
G_4 u &= f(\mathrm{e}^{-\varepsilon}x, \mathrm{e}^{-2\varepsilon}t),\\
G_5 u &= \mathrm{e}^{-\varepsilon x+\varepsilon^2 t} f(x-2\varepsilon t, t),\\
G_6 u &= \frac{1}{\sqrt{1+4\varepsilon t}} \exp\left(-\frac{\varepsilon x^2}{1+4\varepsilon t}\right) f\left(\frac{x}{1+4\varepsilon t}, \frac{t}{1+4\varepsilon t}\right),\\
G_\beta u &= f(x,t) + \varepsilon\beta(x,t).
\end{aligned}$$

It is rather easy to give a simple intuitive explanation for some of these transformations.

Let us look at the first transformed solution $G_1 u$. The transformation is a shift of the x variable on a constant ε. It means that all our solutions are invariant under constant x-shifts or, in other words, are translation invariant in the x direction. We can imagine that this type of symmetry can be easily "seen" from the form of the heat equation. Really, all coefficients of (8.2.22) are independent of the variable x, any constant shift along this variable should leave all the solutions invariant.

The second one-parameter group G_2 describes the shifts on a constant value along the time variable t. Also this symmetry property can be easily "seen" from the form of the heat equation. All coefficients in the equation are time-independent, and each constant shift in t-direction should be admitted by the solution manifold. It means also that the solutions are translation invariant in t.

The transformation described by the group G_3 means that if we multiply any solution of the heat equation with an arbitrary constant it will still be a solution of the equation. It is a so-called homogeneity transformation.

We study now the transformation, described by G_4. Evidently that transformation describes a scaling transformation with different velocities in x- and t- directions. This type of scaling can also be "seen" from the form of the heat equation. We have the constant coefficients in the equation and the dependence on variables x, t controlled by derivatives only. The partial derivative on t is of the first order and the partial derivative on x is of the second order. Evidently, if we scale the x-variable with some constant coefficient, then to get the same effect on the coefficient of equation we should scale the t variable with a factor which is equal to the second power of the previous scaling coefficient.

The transformation which is described by the subgroup G_5 is not so trivial like previous ones, but is well-known. Physicists call it the *Galilean* transformation. This transformation was introduced by Galileo to describe in different coordinate systems the motion of a ball rolling down a ramp.

The transformation provided by G_6 is also non-trivial and gives us the information about the resolvent operator connected to the linear differential operator of the heat equation. A bit later we will show that with this transformation we obtain the fundamental solution of the given PDE.

The group G_β differs strongly from previous ones but the interpretation of this transformation is also trivial. The heat equation is a linear PDE. For such types of equations the superposition principle for solutions should be fulfilled. The transformed solution $G_\beta u$ represents exactly this principle: if we have a solution $f(x,t)$ and add any solution $\beta(x,t)$ of the same equation multiplied by any arbitrary constant, then we still have a solution of the heat equation.

How can we use the symmetry group? One of the possibilities is to generate more advanced solutions starting with rather simple ones. We take one of the solutions which we can get just by making an "educated guess". The heat equation contains just derivatives, it means any constant should be a solution of this equation. Now let $u = f(x,t) = c$, i.e. u be a constant, then we apply G_6 to this constant solution and obtain that

$$u(x,t) = \frac{c}{\sqrt{1+4\varepsilon t}} \exp\left(\frac{-\varepsilon x^2}{1+4\varepsilon t}\right)$$

is a solution. If $c = \sqrt{\frac{\varepsilon}{\pi}}$ we have the fundamental solution to the heat equation at the point $(x_0, t_0) = \left(0, -\frac{1}{4\varepsilon}\right)$. Using G_2 (translation in t-direction) we obtain

$$u(x,t) = \frac{1}{\sqrt{4\pi t}} \exp\left(-\frac{x^2}{4t}\right).$$

It is the classical fundamental solution of the heat equation. The fundamental solution is a very powerful tool for the investigations of different processes described by a linear equation because of the superposition principle. It is well known that the fundamental solution represents the "response" of a physical system (in this case the system describes the heat conduction in a one–dimensional bar) on a point excitation with the unit amplitude in the zero point. We interpret any initial condition for the heat equation as a set of corresponding point excitations with different amplitudes on different points of the x-axis. Then if we know the fundamental solution for a linear system we obtain any other solution by collecting all responses of the system on the excitation, in other words we just need to calculate the convolution of the fundamental solution with a function which describes the initial conditions [77].

We see that with very simple instruments (we just applied the symmetry transformations to a trivial constant solution), we got the fundament solution for the heat equation.

Our next question is how to find the most general transformation described by $G_1, \ldots, G_6, G_\beta$. We can certainly apply this most general transformation to any solution of the heat equation, not just to the trivial constant solution and in this way we get more advanced solutions. Such solution transformations can be very useful if we adapt the solution to the new boundary or initial conditions. If we apply each of the transformations described above then the group element g can be represented using the exponential map as

$$g = \exp(\mathbf{V}_\beta) \exp(\varepsilon_6 \mathbf{V}_6) \ldots \exp(\varepsilon_1 \mathbf{V}_1). \tag{8.2.28}$$

Applying this transformation g to the solution $f(x,t)$ we obtain a new solution, i.e. $u(x,t) = g \cdot f(x,t)$ and we obtain the expression

$$\begin{aligned} u(x,t) = {} & \frac{1}{\sqrt{1+4\varepsilon_6 t}} \exp\left(\varepsilon_3 - \frac{\varepsilon_5 x + \varepsilon_6 x^2 - {\varepsilon_5}^2 t}{1+4\varepsilon_6 t}\right) \\ & \times f\left(\frac{\mathrm{e}^{-\varepsilon_4}(x - 2\varepsilon_5 t)}{1+4\varepsilon_6 t} - \varepsilon_1, \frac{\mathrm{e}^{-2\varepsilon_4} t}{1+4\varepsilon_6 t} - \varepsilon_2\right) + \beta(x,t). \end{aligned} \tag{8.2.29}$$

Here $\varepsilon_1, \varepsilon_2, \ldots, \varepsilon_6$ are constants and $\beta(x,t)$ is an arbitrary solution of the heat equation.

Example 8.2.1 Let us consider a special solution of the heat equation which can be obtained using a one–dimensional subalgebra. Let

$$\mathbf{V} = 2a\mathbf{V}_1 + b^3\mathbf{V}_5 + \mathbf{V}_2,$$

where a, b are arbitrary fixed constants. This one–dimensional Lie subalgebra generates the transformations of t, x, u in the form

$$x_\varepsilon = x + 2b^3\varepsilon t + 2a\varepsilon + b^3\varepsilon^2, \tag{1}$$

$$t_\varepsilon = t + \varepsilon, \tag{2}$$

$$u_\varepsilon = u\exp\left(-b^3\varepsilon x - b^6\varepsilon^2 t - ab^3\varepsilon^2 - \frac{1}{3}b^6\varepsilon^3\right). \tag{3}$$

If we exclude the explicit dependence on the group parameter ε from the equations we obtain two invariants of this one–dimensional subgroup. From Equation (2) we can easily deduce $\varepsilon = t_\varepsilon - t$ and after substituting this result into (1) and (3) we obtain

$$\begin{aligned} x_\varepsilon &= 2at_\varepsilon + b^3t_\varepsilon^2 + x - 2at - b^3t^2, \\ u_\varepsilon &= u\exp\left(-b^3t_\varepsilon\left(x - b^3t^2 + at_\varepsilon - 2at + \frac{1}{3}b^3{t_\varepsilon}^2\right)\right. \\ &\quad \left. + b^3tx - \frac{2}{3}b^6t^3 - ab^3t^2\right). \end{aligned}$$

From this expression we can find the invariants

$$\begin{aligned} \mathrm{inv}_1 &= x - 2at - b^3t^2, \\ \mathrm{inv}_2 &= u\exp\left(b^3xt - ab^3t^2 - \frac{2}{3}b^6t^3\right). \end{aligned}$$

We introduce now the new invariant variables, i.e. instead of t, x, u we introduce just two variables, the first of them is an independent one, the other one is the dependent variable. In this way we reduce the given PDE to an ODE.

Let us denote $y = y(x,t) = \mathrm{inv}_1$ and $w = w(y) = \mathrm{inv}_2$. We replace independent and dependent variables x, t, u by invariants y, w and get

$$u(x,t) = w(y)\exp\left(-b^3xt + ab^3t^2 + \frac{2}{3}b^6t^3\right) = w(y)\zeta(x,t),$$

where $\zeta(x,t) = \exp\left(-b^3xt + ab_3t^2 + \frac{2}{3}b^6t^3\right)$. Differentiating yields

$$u_t = \zeta(x,t)\left(w(y)\left(-b^3x + 2ab^3t + 2b^6t^2\right) - 2(a+b^3t)w_y(y)\right),$$

$$u_x = \zeta(x,t)\left(-b^3tw(y) + w_y(y)\right),$$

$$u_{xx} = \zeta(x,t)\left(b^6t^2w(y) - 2b^3tw_y(y) + w_{yy}(y)\right),$$

and by inserting these differentials in the heat equation (8.2.22) we obtain the ordinary differential equation

$$w_{yy}(y) + 2aw_y(y) + b^3yw(y) = 0. \tag{8.2.30}$$

It is an ODE of the Laplace type and can be solved with the help of the Laplace transformation. The general solution of (8.2.30) is

$$w(y) = e^{-ay}\left(C_1\mathcal{A}i\left(\left(\frac{a}{b}\right)^2 - by\right) + C_2\mathcal{B}i\left(\left(\frac{a}{b}\right)^2 - by\right)\right),$$

where C_1, C_2 are arbitrary constants, and the functions $\mathcal{A}i(\cdot)$,$\mathcal{B}i(\cdot)$ are correspondingly the Airy functions of the first and second kind [1]. If we now replace the variables y, w with original variables x, t, u we get a special solution of the heat equation

$$\begin{aligned} u(x,t) &= w(y)\zeta(x,t) \\ &= \exp\left(-ax + 2a^2t - b^3xt + 2ab^3t^2 + \frac{2}{3}b^6t^3\right) \\ &\times \left(C_1\mathcal{A}i\left(\left(\frac{a}{b}\right)^2 - bx + 2abt + b^4t^2\right)\right. \\ &\left. + C_2\mathcal{B}i\left(\left(\frac{a}{b}\right)^2 - bx + 2abt + b^4t^2\right)\right). \end{aligned}$$

So we are able to find a non-trivial analytical solution of the heat equation in a closed form which includes arbitrary constants. We can say that we have found a family of solutions which depends on four constants a, b, C_1, C_2 . $\diamond$

In the example we used three symmetries combined in the special way within a one–dimensional group and reduced our partial differential equation to an ordinary one. If we take a different combination of symmetries we will get another solution and so on. It is evident that there are infinitely many possible combinations of infinitesimal generators and correspondingly we can find infinitely many such subgroup-invariant solutions. But it is also evident, that some of these solutions can be transformed one into another using the transformations from the same group. Such solutions build a class of solutions which are equivalent to each other. The other sets of solutions will be so different that we will not be able to find a transformation which reduces one of them to the other one. In this case we can say that we have found a set of the non-conjugate subalgebras of the corresponding dimension.

How many different non-equivalent invariant solutions exist? It depends on the internal structure of the underlying Lie algebra. The set of all non-conjugate subalgebras of each dimension less or equal to the dimension of the Lie algebra itself is called an optimal system of subalgebras as we described it in Chapter 6. In the next chapter we will use this idea to classify the invariant solutions.

Is it possible to solve a partial differential equation completely, using the optimal system of subalgebras? Can we describe all possible solutions of the given PDE knowing all non-equivalent invariant solutions? Unfortunately not. In the case of a linear ODE of second order we have two degrees of freedom, and if we are able to find a solution with two arbitrary constants we obtain a general solution. If we just fix the constants in some way we obtain all special solutions of this equation. If we have to do with linear second order PDE, then we have infinitely many degrees of freedom, the solution is defined up to arbitrary functions, not just up to two constants. For linear PDEs we are able to find with Lie group analysis the fundamental solutions and in this way provide at least theoretically the possibility of finding an analytical solution for given initial-boundary conditions. In the case of nonlinear PDEs the situation is much more complicated. In the general case we have no method to describe the complete set of all possible solutions. Certainly the invariant solutions for nonlinear PDEs are very helpful. We can understand the invariant solutions as some cuts in the much more complicated solution manifold of this equation. Still this way we get important information about the studied PDEs.

For the heat equation we obtained two types of symmetries - the infinitesimal generators $\mathbf{V}_1$ to $\mathbf{V}_6$ which spanned the six-dimensional Lie algebra $L_6 = \langle \mathbf{V}_1, \cdots, \mathbf{V}_6 \rangle$ and the infinite–dimensional symmetry described by $\mathbf{V}_\beta = \beta(x,t)\frac{\partial}{\partial u}$, where $\beta(x,t)$ is a solution of the heat equation. We can say that if we only use the finite-dimensional symmetries $\mathbf{V}_1, \ldots, \mathbf{V}_6$, we can obtain special classes, but we never obtain the complete set of solutions of the heat equation.

8.3 Some important types of group-invariant solutions of the heat equation

In this section we discuss some well-known types of invariant solutions of the heat equation. Each of these solutions is invariant under the action of some one–dimensional subgroup of the symmetry group G of the heat equation which we introduced in the previous section. Firstly such types of invariant solutions were introduced in physics during the investigation of physical phenomena. Physicists used "natural" symmetries of the systems to simplify the equation and to find the corresponding solutions. Because of that the corresponding symmetries were titled similar to the described properties of the phenomena for instance "a traveling wave" or some similar names.

Let now $\Delta(x, u^{(2)}) = 0$ be defined by the heat equation, i.e. $\Delta(x, u^{(2)}) = u_t - u_{xx} = 0$ and the six-dimensional Lie algebra $L_6 = \langle \mathbf{V}_1, \ldots, \mathbf{V}_6 \rangle$ be defined as in Section 8.2.9. The corresponding six-dimensional Lie group we denote as G.

Traveling wave solutions

First, we want to examine the so-called *traveling wave solutions.* We take a one–dimensional subgroup $H \subset G$, the corresponding one–dimensional Lie algebra we denote as $h = \langle \mathbf{U} \rangle$ where the infinitesimal generator has the form

$$\mathbf{U} = \mathbf{V}_2 + c\mathbf{V}_1 = \frac{\partial}{\partial t} + c\frac{\partial}{\partial x},$$

and c is a fixed constant. The characteristic equations define the change of all variables under the action of H and have the form

$$\frac{dt}{1} = \frac{dx}{c} = \frac{du}{0}.$$

From the first equation it follows that

$$x - ct = c_1, \text{ where } c_1 \text{ is a constant.}$$

From the second equation it follows that

$$du = 0 \cdot dt = 0 \ \text{ or } \ du = 0 \cdot dx = 0 \quad \Rightarrow \quad u \text{ is a constant.}$$

It means that $x - ct$ and u are invariants under transformations of this group H. The expression for $x - ct$ describes the orbits of the group, which are the planes $x - ct = \text{const}$. Using these two invariants we introduce now two new invariant variables instead of three variables x, t, u

$$z = \text{inv}_1 = x - ct, \quad v = v(z) = \text{inv}_2 = u(x,t).$$

Now we recalculate all derivatives in PDE in new variables

$$\frac{\partial}{\partial x} = \frac{\partial z}{\partial x} \cdot \frac{d}{dz} = \frac{d}{dz}, \quad \frac{\partial}{\partial t} = \frac{\partial z}{\partial t} \cdot \frac{d}{dz} = -c\frac{d}{dz}.$$

The original heat equation is now reduced to an ODE

$$-cv_z(z) = v_{zz}(z).$$

The solution of this ordinary differential equation can be calculated quickly. We substitute $w = v_z$ to get

$$w_z = -cw \quad \Rightarrow \quad \frac{dw}{w} = -c\,dz \quad \Rightarrow \quad w = \tilde{d}_1 \mathrm{e}^{-cz}.$$

By integrating once again we obtain

$$v(z) = -\frac{\tilde{d}_1}{c}\mathrm{e}^{-cz} + d_2$$

and therefore

$$u(x,t) = d_1 \mathrm{e}^{-c(x-ct)} + d_2,$$

where d_1 and d_2 are arbitrary constants. This solution represents the most general traveling wave solutions, these are translation-invariant solutions.

Scaling solutions

In this case the infinitesimal generator has the form

$$\mathbf{U} = \mathbf{V}_4 + a\mathbf{V}_3 = x\frac{\partial}{\partial x} + 2t\frac{\partial}{\partial t} + 2au\frac{\partial}{\partial u}, \quad \forall a \in \mathbb{R}.$$

The corresponding group transformation is

$$(x, t, u) \to (\varepsilon x, \varepsilon^2 t, \varepsilon^{2a} u).$$

Again, the invariants can be defined by the characteristic system of equations

$$\frac{dx}{x} = \frac{dt}{2t} = \frac{du}{2au}.$$

If we consider the first equation, we obtain

$$\ln x = \ln \sqrt{t} + \tilde{c}_1 \quad \Rightarrow \quad c_1\sqrt{t} = x \quad \Rightarrow \quad \text{inv}_1 = \frac{x}{\sqrt{t}}$$

and integration of the second equation yields

$$\ln t = \frac{1}{a}\ln u + \tilde{c}_2 \quad \Rightarrow \quad t^a = c_2 u \quad \Rightarrow \quad \mathrm{inv}_2 = t^{-a}u.$$

We get the most general *scaling-invariant solutions* if we use these invariants to reduce the heat equation to an ODE. We introduce new variables z, w

$$z = \mathrm{inv}_1 = \frac{x}{\sqrt{t}}, \quad w = w(z) = \mathrm{inv}_2 = t^{-a}u(x,t).$$

Then we recalculate all derivatives in the heat equation, using $u(x,t) = t^a w(z)$ and expression for z

$$\begin{aligned}
\frac{\partial u}{\partial t} &= \frac{\partial}{\partial t}(t^a w(z)) = a t^{a-1} w(z) + t^a w'(z)\frac{\partial z}{\partial t} \\
&= t^{a-1}\left(a w(z) - \frac{1}{2} z w'(z)\right), \\
\frac{\partial u}{\partial x} &= t^a w'(z)\frac{1}{\sqrt{t}}, \\
\frac{\partial^2 u}{\partial x^2} &= t^{a-1} w''(z).
\end{aligned}$$

Thus the original heat equation becomes an ODE of the following type

$$w''(z) + \frac{z}{2}w'(z) - a w(z) = 0. \tag{8.3.31}$$

The substitution $v = w\exp\left(-\frac{1}{8}z^2\right)$ reduces this equation to the parabolic cylinder equation

$$v_{zz} = \left(a + \frac{1}{4} + \frac{1}{16}z^2\right)v. \tag{8.3.32}$$

This linear equation has two independent solutions $\mathcal{U}$ and $\mathcal{V}$, which are parabolic cylinder functions [1], and v will be

$$v(z) = k\,\mathcal{U}\left(2a + \frac{1}{2}, \frac{z}{\sqrt{2}}\right) + \tilde{k}\,\mathcal{V}\left(2a + \frac{1}{2}, \frac{z}{\sqrt{2}}\right), \quad k, \tilde{k} \in \mathbb{R}$$

correspondingly

$$u(x,t) = t^a \exp\left(-\frac{x^2}{8t}\right)\left(k\,\mathcal{U}\left(2a + \frac{1}{2}, \frac{x}{\sqrt{2t}}\right) + \tilde{k}\,\mathcal{V}\left(2a + \frac{1}{2}, \frac{x}{\sqrt{2t}}\right)\right). \tag{8.3.33}$$

In the case $a = 0$, the solution has the form

$$u(x,t) = k^* \operatorname{erf}\left(\frac{x}{\sqrt{2t}}\right) + \tilde{k}^*.$$

where $k^*, \tilde{k}^* \in \mathbb{R}$ and $\operatorname{erf}(\cdot)$ is the Gaus error function.

Solutions invariant under Galilean transformation

In the last example we analyze solutions which are invariant under the action of *Galilean transformation.* The infinitesimal generator of this subgroup has the form

$$\mathbf{U} = \mathbf{V}_5 = 2t\frac{\partial}{\partial x} - xu\frac{\partial}{\partial u}.$$

The characteristic system of equations which describes the variable transformation is

$$\frac{dx}{2t} = \frac{du}{-xu} = \frac{dt}{0}.$$

It follows immediately from the system that t is an invariant and hence, $\text{inv}_1 = t$. Then the other equation yields

$$-\frac{x\,dx}{2t} = \frac{du}{u} \quad\Rightarrow\quad -\frac{x^2}{4t} = \ln u + \tilde{c} \quad\Rightarrow\quad \text{inv}_2 = \mathrm{e}^{\frac{x^2}{4t}} u(x,t).$$

We can choose new invariant variables z, w as

$$z = t, \quad w(z) = u\mathrm{e}^{\frac{x^2}{4t}},$$

in other words we take $u(x,t) = \exp\left(-\frac{x^2}{4t}\right) w(z)$. We recalculate the derivatives of $u(x,t)$ in new variables z, w and get

$$\begin{aligned}
\frac{\partial u}{\partial t} &= \frac{x^2}{4t^2}\mathrm{e}^{-\frac{x^2}{4t}} w(z) + \mathrm{e}^{-\frac{x^2}{4t}} w'(z),\\
\frac{\partial u}{\partial x} &= -\frac{x}{2t}\mathrm{e}^{-\frac{x^2}{4t}} w(z),\\
\frac{\partial^2 u}{\partial x^2} &= \left(-\frac{1}{2t} + \frac{x^2}{4t^2}\right)\mathrm{e}^{-\frac{x^2}{4t}} w(z).
\end{aligned}$$

The heat equation is reduced now to the first order ODE

$$2zw'(z) = w(z).$$

It can be easily solved. The solution has the form

$$w(z) = \frac{c}{\sqrt{z}}.$$

So in the original variables we obtain the following solution of the heat equation

$$u(x,t) = \frac{c}{\sqrt{t}} \exp\left(-\frac{x^2}{4t}\right), \quad c \in \mathbb{R}.$$

This invariant solution is the fundamental solution of the heat equation.

8.4 The Black-Scholes equation

In this section we want to examine the Black-Scholes equation. The Black-Scholes equation was introduced in the famous paper of Black and Scholes, 1973, [7] to calculate the fair price of a derivative product in financial markets (the short introduction in the finance terminology will be give in the next chapter).

If we denote the price of the derivative (European call or put, for instance) as u then this function satisfies the PDE

$$u_t + \frac{1}{2}\sigma^2 x^2 u_{xx} + rxu_x - ru = 0, \tag{8.4.34}$$

where x is the price of the asset, t time, σ the volatility, r the interest rate. The corresponding terminal and boundary conditions define the type of the derivative product of so-called European type. If we study American options or other derivative products of American type then instead of the equation we obtain an inequality. In other words we have to do with a parabolic partial differential inequality with free boundary. In this chapter we study just the most simple case - the symmetry properties of (8.4.34).

Lie group analysis of the Black-Scholes and similar linear equations was first done in [33]. Using the same procedure as in the case of the heat equation we find the infinitesimal generators of the symmetries

$$
\begin{aligned}
\mathbf{U}_1 &= \frac{\partial}{\partial t},\\
\mathbf{U}_2 &= x\frac{\partial}{\partial x},\\
\mathbf{U}_3 &= 2t\frac{\partial}{\partial t} + (\ln x + Dt)x\frac{\partial}{\partial x} + 2rtu\frac{\partial}{\partial u},\\
\mathbf{U}_4 &= \sigma^2 tx\frac{\partial}{\partial x} + (\ln x - Dt)u\frac{\partial}{\partial u},\\
\mathbf{U}_5 &= 2\sigma^2 t^2\frac{\partial}{\partial t} + 2\sigma^2 tx\ln x\frac{\partial}{\partial x} + \left((\ln x - DT)^2 + 2\sigma^2 rt^2 - \sigma^2 t\right)u\frac{\partial}{\partial u},\\
\mathbf{U}_6 &= u\frac{\partial}{\partial u},\\
\mathbf{U}_\phi &= \phi(t,x)\frac{\partial}{\partial u},
\end{aligned}
$$

with $D = r - \frac{\sigma^2}{2}$. Similar to the case of the heat equation the first six generators build a six-dimensional Lie algebra. We denote this six-dimensional Lie algebra spanned by operators $\mathbf{U}_1, \cdots, \mathbf{U}_6$ by $L_{BS} = \langle \mathbf{U}_1, \cdots, \mathbf{U}_6 \rangle$. The last generator $\mathbf{U}_\phi$ defines the infinite–dimensional subalgebra $\mathbf{U}_\phi = \langle \phi(t,x)\frac{\partial}{\partial u} \rangle$. The algebras, which are spanned by the six finite-dimensional generators of the heat equation L_6 and the six finite-dimensional generators of the Black-Scholes equation L_{BS} are isomorphic and similar. The corresponding variable substitutions y, τ, v to x, t, u to show this are

$$
\begin{aligned}
y &= \frac{\left(r - \sigma^2/2\right)}{\sigma^2/2}\ln\left(\frac{x}{x_0}\right) + \frac{\left(r - \sigma^2/2\right)^2}{\sigma^2/2}(T - t),\\
\tau &= \frac{\left(r - \sigma^2/2\right)^2}{\sigma^2/2}(T - t), \qquad (8.4.35)\\
v(\tau, y) &= \mathrm{e}^{r(T-t)} x^{-\frac{\left(r - \sigma^2/2\right)}{2\sigma^2}} u(x,t).
\end{aligned}
$$

Here we denoted with $u(x,t)$ a solution of the Black-Scholes equation and with $v(\tau, y)$ a solution of the heat equation $v_\tau = v_{yy}$.

The subgroup transformations which correspond to the generators of the algebra L_{BS} are

$$
\begin{aligned}
G_1: &\quad t_{\varepsilon_1} = t + \varepsilon_1,\ x_{\varepsilon_1} = x,\ u_{\varepsilon_1} = u, \quad \text{(translation invariance)},\\
G_2: &\quad t_{\varepsilon_2} = t, \quad x_{\varepsilon_2} = \mathrm{e}^{\varepsilon_2} x, \quad u_{\varepsilon_2} = u, \quad \text{(scaling invariance)},\\
G_3: &\quad t_{\varepsilon_3} = \mathrm{e}^{2\varepsilon_3} t, \quad x_{\varepsilon_3} = x^{\mathrm{e}^{\varepsilon_3}} \mathrm{e}^{D\mathrm{e}^{\varepsilon_3}(\mathrm{e}^{\varepsilon_3}-1)t}, \quad u_{\varepsilon_3} = u\mathrm{e}^{r(\mathrm{e}^{2\varepsilon_3}-1)},\\
G_4: &\quad t_{\varepsilon_4} = t, \quad x_{\varepsilon_4} = x\mathrm{e}^{\varepsilon_4 \sigma^2 t}, \quad u_{\varepsilon_4} = u x^{\varepsilon_4} \mathrm{e}^{\varepsilon_4(\frac{1}{2}\sigma^2 \varepsilon_4 - D)t},\\
G_5: &\quad t_{\varepsilon_5} = \tfrac{t}{1-2\varepsilon_5\sigma^2 t}, \quad x_{\varepsilon_5} = x^{\frac{1}{1-2\varepsilon_5\sigma^2 t}},\\
&\quad u_{\varepsilon_5} = u\sqrt{1-2\varepsilon_5\sigma^2 t}\,\exp\left(\tfrac{\varepsilon_5\left((\ln x - Dt)^2 + 2\sigma^2 r t^2\right)}{1-2\varepsilon_5\sigma^2 t}\right),\\
G_6: &\quad t_{\varepsilon_6} = t,\ x_{\varepsilon_6} = x,\ u_{\varepsilon_6} = \mathrm{e}^{\varepsilon_6} u, \quad \text{(homogeneity)},\\
G_\phi: &\quad t_\varepsilon = t,\ x_\varepsilon = x,\ u_\varepsilon = u + \varepsilon\phi(x,t). \quad \text{(linearity)}.
\end{aligned}
$$

We discuss the financial meaning of the Black-Scholes equation and its variations in the next chapter. Here let us just show how the substitutions (8.4.35) transform the Black-Scholes equation to the classical heat equation. With $x = x_0 \mathrm{e}^y$, $x_0 \in \mathbb{R}$, $t = T - \frac{\tau}{\sigma^2/2}$, $T\rangle 0$ and the terminal condition $u(x,t)|_{t=T} = \max(x - x_0, 0)$ for the European call option $u = x_0 \tilde{v}(y,\tau)$, the Black-Scholes Equation (8.4.34) becomes

$$\frac{\partial \tilde{v}}{\partial \tau} = \frac{\partial^2 \tilde{v}}{\partial y^2} + \left(\frac{r}{\sigma^2/2} - 1\right)\frac{\partial \tilde{v}}{\partial y} - \frac{r}{\sigma^2/2}\tilde{v}$$

and the terminal condition becomes the initial condition $\tilde{v}(y,0) = \max(\mathrm{e}^y - 1, 0)$ for the function $\tilde{v}(y,\tau)$. Further we substitute $\tilde{v}$ by v using

$$\tilde{v}(y,\tau) = \exp\left(-\left(\frac{r}{\sigma^2/2} - 1\right)y - \frac{1}{4}\left(\frac{r}{\sigma^2/2} + 1\right)^2 \tau\right) v(y,\tau).$$

After this substitution we obtain the classical heat equation

$$v_\tau = y_{yy}, \quad -\infty\langle y \langle \infty,\ \tau\rangle 0.$$

From the initial condition for $\tilde{v}$ we deduce the corresponding initial condition for v, i.e. from $\tilde{v}(y,0) = \max(\mathrm{e}^y - 1, 0)$ we get

$$v(y,0) = v_0(y) = \max\left(\exp\left(\left(\frac{r}{\sigma^2/2} + 1\right)\frac{y}{2}\right) - \exp\left(\left(\frac{r}{\sigma^2/2} - 1\right)\frac{y}{2}\right), 0\right).$$

Consequently, using the fundamental solution of the heat equation we obtain the solution $v(y,\tau)$ to the given initial condition $v_0(y)$ after integration

$$v(y,\tau) = \frac{1}{2\sqrt{\pi\tau}}\int_{-\infty}^{\infty} v_0(s)\exp\left(-\frac{(x-s)^2}{4\tau}\right) ds.$$

Using this result we obtain in original variables the solution of the Black-Scholes equation that describes the fair price of a call option

$$u(x,t) = c(x,t) = xN(d_1) - x_0 e^{-r(T-t)} N(d_2), \tag{8.4.36}$$

where $N(d) = \frac{1}{\sqrt{2\pi}} \int_{-\infty}^{d} \exp\left(-\frac{s^2}{2}\right) ds$ is the cumulative probability distribution function [1] for a standardized normal distribution and

$$d_1 = \frac{\ln\left(x/x_0\right) + \left(r + \sigma^2/2\right)(T-t)}{\sigma\sqrt{T-t}},$$
$$d_2 = \frac{\ln\left(x/x_0\right) + \left(r - \sigma^2/2\right)(T-t)}{\sigma\sqrt{T-t}}.$$

This exact solution was first presented in the famous paper [7]. The exact form of the solution and its simplicity made it extremely popular in the financial industry.

The invariant solutions for the call and put options are not unique invariant solutions of this equation.

Example 8.4.1 Let us finally calculate as an example another invariant solution of (8.4.34). Let us look for a scaling solution for instance. We use a linear combination of scaling operators

$$\mathbf{U} = \mathbf{U}_1 + \mathbf{U}_2 + \mathbf{U}_6 = \frac{\partial}{\partial t} + x\frac{\partial}{\partial x} + u\frac{\partial}{\partial u}.$$

For simplicity we take a special form of the linear combination of infinitesimal generators. It means we choose a one–dimensional subalgebra and look for solutions which are invariant under the action of the corresponding one–dimensional symmetry group.

We compute the invariants of this subgroup using the characteristic system

$$\frac{dt}{1} = \frac{dx}{x} = \frac{du}{u}$$

and get

$$\text{inv}_1 = -\ln x + t, \quad \text{inv}_2 = \frac{u(x,t)}{x}.$$

Those invariants are functionally independent, so they form a basis of invariants. We use these invariants as new independent and dependent variables

$$z = t - \ln x, \quad u(x,t) = xw(z).$$

Using the new variables we reduce the Black-Scholes equation to the second order ODE

$$\frac{\sigma^2}{2}w''(z) + (1 - D)w'(z) = 0. \tag{8.4.37}$$

By defining $v = w'$ and after simple integration we obtain the invariant solution

$$u(x,t) = c_1 x^{1-\tilde{r}} e^{\tilde{r}t} + c_2, \tag{8.4.38}$$

with $\tilde{r} = \frac{r-1}{\sigma^2/2} - 1$, and c_1, c_2 are constants. ◇

The variables substitution which is used to reduce the Black-Scholes equation to a classical heat equations is well-known in financial mathematics. In many textbooks this substitution is presented as some "cooking recipe". Usually these books do not even mention the Lie group theory. That is why it is quite difficult to understand what one can do with different variations of the BS equation, if it is possible to find a similar substitution and how one can find the substitution if it actually exists.

The tools of Lie group analysis give us algorithmic and constructive ways to find answers to such questions. In the next chapter we will study many examples of Black-Scholes model modifications (other examples are provided in [5]). Such modifications are necessary when we introduce some frictions typical of the real financial world. Such frictions are transaction costs, illiquidity problems, feedback effects, hedging costs of a portfolio and so on.

8.5 Problems with solutions

Usually solutions for problems connected with Lie group analysis of PDEs are very tedious and voluminous. Because of that we focus on examples which are useful to the reader but, at the same time, do not take too much space. The reader can try each problem on his own, or just follow the solution as given. An engaged reader can use the information provided in these problem settings in another way: since the formulations of all problems or their solutions contain the complete description of the admitted symmetry algebra, the reader can use this information to reformulate the problems by using a corresponding computer package and finding the admitted Lie algebra. Here one can use for instance the **Mathematica** based package **IntroToSymmetry** or **REDUCE** based package **ReLie** provided by [58]. One can compare the efficiency of different packages in this way. There exist about a dozen other packages developed for Lie group analysis which have their own strengths and weaknesses; we hope that the reader can find one which is most convenient for him or her.

In this section, the reader is asked to find typical reductions of the given equation and the most well known invariant solutions: traveling wave solutions or scaling invariant solutions. The reader can use for instance the **Mathematica** based package **SymboLie** [58] to find the optimal system of subalgebras for each of the problems. Using the optimal system of subalgebras one can provide a complete set of invariant solutions for the given equation. The reader can find a lot of examples of invariant solutions in [39].

Problem 8.5.1 *The nonlinear heat equation*

$$u_t = u_x^{\alpha-1} u_{xx}, \quad \alpha \geq 0, \alpha \neq 1$$

admits an infinitesimal generator

$$\mathbf{U} = (a + 2d\, t + e(1 - \alpha t))\frac{\partial}{\partial t} + (b + d\, x)\frac{\partial}{\partial x} + (c + d\, u + eu)\frac{\partial}{\partial u},$$

where $a, b, c, d, e \in \mathbb{R}$.

Is the corresponding symmetry algebra L a solvable algebra? If L is a solvable algebra then present the corresponding chain of ideals. Use a proper Lie subalgebra of L to obtain an ordinary differential equation determining the scaling invariant solutions.

Solution Since the generator possesses five independent constants, the corresponding symmetry algebra is of dimension 5 and L_5 is spanned by the following generators

$$\mathbf{U}_1 = \frac{\partial}{\partial t}, \ \mathbf{U}_2 = \frac{\partial}{\partial x}, \ \mathbf{U}_3 = \frac{\partial}{\partial u}, \ \mathbf{U}_4 = 2t\frac{\partial}{\partial t} + x\frac{\partial}{\partial x} + u\frac{\partial}{\partial u},$$
$$\mathbf{U}_5 = (1 - \alpha t)\frac{\partial}{\partial t} + u\frac{\partial}{\partial u}.$$

The commutator table is

	$\mathbf{U}_1$	$\mathbf{U}_2$	$\mathbf{U}_3$	$\mathbf{U}_4$	$\mathbf{U}_5$
$\mathbf{U}_1$	0	0	0	$2\mathbf{U}_1$	$-\alpha\mathbf{U}_1$
$\mathbf{U}_2$	0	0	0	$\mathbf{U}_2$	0
$\mathbf{U}_3$	0	0	0	$\mathbf{U}_3$	$\mathbf{U}_3$
$\mathbf{U}_4$	$-2\mathbf{U}_1$	$-\mathbf{U}_2$	$-\mathbf{U}_3$	0	$-2\mathbf{U}_1$
$\mathbf{U}_5$	$\alpha\mathbf{U}_1$	0	$-\mathbf{U}_3$	$2\mathbf{U}_1$	0

The first derived algebras of the algebra L_5 are three-dimensional $L^{(1)} = [L, L] = \langle \mathbf{U}_1, \mathbf{U}_2, \mathbf{U}_3 \rangle$, The second derived algebra $L^{(2)} = [L^{(1)}, L^{(1)}] = 0$

vanishes, therefore L_5 is a solvable algebra. We can present the chain of ideals

$$0 \subset \langle \mathbf{U}_1 \rangle \subset \langle \mathbf{U}_1, \mathbf{U}_2 \rangle \subset \langle \mathbf{U}_1, \mathbf{U}_2, \mathbf{U}_3 \rangle \\ \subset \langle \mathbf{U}_1, \mathbf{U}_2, \mathbf{U}_3, \mathbf{U}_4 \rangle \subset \langle \mathbf{U}_1, \mathbf{U}_2, \mathbf{U}_3, \mathbf{U}_4, \mathbf{U}_5 \rangle.$$

To reduce the nonlinear PDE to ODE, which determines scaling solutions, we should take a scaling generator $\mathbf{U}_4 = 2t\frac{\partial}{\partial t} + x\frac{\partial}{\partial x} + u\frac{\partial}{\partial u}$. The characteristic equations for invariants take the form

$$\frac{dt}{2t} = \frac{dx}{x} = \frac{du}{u} \quad \Rightarrow \quad \mathrm{inv}_1 = xt^{-1/2}, \ \mathrm{inv}_2 = u(x,t)t^{-1/2}.$$

We take inv_1 as an independent variable z and inv_2 as a dependent variable $V(z)$. Derivatives in the new variables can be rewritten

$$\begin{aligned}
\frac{\partial u(x,t)}{\partial t} &= \frac{\partial V(z)t^{1/2}}{\partial t} = \frac{1}{2}t^{1/2}V(z) + t^{1/2}\frac{\partial z}{\partial t}\frac{dV(z)}{dz} \\
&= \frac{1}{2}t^{-1/2}V(z) - \frac{1}{2}t^{-1}xV'(z) \\
&= \frac{1}{2}t^{-1/2}V(z) - \frac{1}{2}t^{-1/2}zV'(z), \\
\frac{\partial u(x,t)}{\partial x} &= \frac{\partial V(z)t^{1/2}}{\partial x} = t^{1/2}\frac{\partial z}{\partial x}\frac{dV(z)}{dz} = t^{-1/2}t^{1/2}V'(z) = V'(z), \\
\frac{\partial^2 u(x,t)}{\partial x^2} &= \frac{\partial}{\partial x}(V'(z)) = \frac{\partial z}{\partial x}V''(z) = t^{-1/2}V''(z).
\end{aligned}$$

Substituting the new variables in the nonlinear heat equation, we obtain the following equation

$$\frac{1}{2}t^{-1/2}V(z) - \frac{1}{2}t^{-1/2}zV'(z) = V'(z)^{\alpha-1}t^{-1/2}V''(z).$$

Dividing over $\frac{1}{2}t^{-1/2}$, we obtain the ODE

$$V(z) - zV'(z) = 2V'(z)^{\alpha-1}V''(z).$$

The solutions of this ODE are scaling invariant solutions of the original nonlinear heat equation.

Problem 8.5.2 *Using* **Mathematica** *find the Lie point symmetries of the sine-Gordon equation*

$$u_{xx} - u_{tt} = \sin(u).$$

The sine-Gordon equation describes for instance, a model for a vortex motion in a liquid. It is also a model equation for surfaces with constant negative curvature.

Solution Using the package **IntroToSymmetry** of **Mathematica** one can obtain the following symmetries of the sine–Gordon equation

$$\mathbf{U}_1 = \frac{\partial}{\partial x},\ \mathbf{U}_2 = \frac{\partial}{\partial t},\ \mathbf{U}_3 = t\frac{\partial}{\partial x} + x\frac{\partial}{\partial t},\ \mathbf{U}_4 = x\frac{\partial}{\partial x} + t\frac{\partial}{\partial t} - 2\tan u\frac{\partial}{\partial u},$$

$$\mathbf{U}_5 = (x^2+t^2)\frac{\partial}{\partial x} + 2xt\frac{\partial}{\partial t} - 4x\tan u\frac{\partial}{\partial u},$$

$$\mathbf{U}_6 = xt\frac{\partial}{\partial x} + \frac{x^2+t^2}{2}\frac{\partial}{\partial t} - 2t\tan u\frac{\partial}{\partial u}.$$

Problem 8.5.3 *Using the found symmetries for the sine-Gordon equation from the previous problem choose the subalgebra describing the translation invariant solutions (traveling wave solutions). Reduce the sine-Gordon equation to an ordinary differential equation in this case. Does this ordinary differential equation admit any Lie point symmetries?*

Solution To reduce this equation to ODE we choose the one-dimensional subalgebra describing translations $\mathbf{U} = \mathbf{U}_1 + a\mathbf{U}_2 = \frac{\partial}{\partial x} + a\frac{\partial}{\partial t}$ (in a general case we choose the linear combination of both operators $\mathbf{U} = a_1\mathbf{U}_1 + a_2\mathbf{U}_2$ and suppose that at least one of the coefficients a_1 or a_2 is not equal to zero. In this case we get simpler equations if we suppose that $a_1 \neq$ and divide by a_1). The characteristic system of equations for the invariants of this subgroup is

$$\frac{dx}{1} = \frac{dt}{a} = \frac{du}{0}$$

and we obtain

$$\mathrm{inv}_1 = ax - t,\ \mathrm{inv}_2 = u.$$

We take the first invariant as an independent variable z, while the second invariant is a dependent variable W. Derivatives over x, t can be recalculated into derivatives with respect to the new variables

$$\frac{\partial u}{\partial x} = \frac{\partial z}{\partial x}\frac{dW}{dz} = aW'(z), \qquad \frac{\partial u}{\partial t} = \frac{\partial z}{\partial t}\frac{dW}{dz} = -W'(z),$$

$$\frac{\partial^2 u}{\partial x^2} = a^2W''(z), \qquad \frac{\partial^2 u}{\partial t^2} = W''(z).$$

We insert the expressions in the sine-Gordon equation and obtain

$$(a^2-1)W''(z) = \sin W.$$

It can be proved using **Mathematica** that this ODE admits the following symmetries

$$\mathbf{U}_1 = \frac{\partial}{\partial x}, \qquad \mathbf{U}_2 = x\frac{\partial}{\partial x} - 2\tan y\frac{\partial}{\partial y},$$

$$\mathbf{U}_3 = (x^2 - 2/3(-1+a^2)y\csc y)\frac{\partial}{\partial x} - 4x\tan y\frac{\partial}{\partial y}.$$

One can use these symmetries to solve the ODE.

Problem 8.5.4 *The equation of a sub-sonic gas motion*

$$u_x u_{xx} + u_{yy} = 0$$

admits the 6–dimensional symmetry group. The corresponding infinitesimal generators of the Lie algebra are

$$\mathbf{U}_1 = \frac{\partial}{\partial x}, \quad \mathbf{U}_2 = \frac{\partial}{\partial y}, \quad \mathbf{U}_3 = \frac{\partial}{\partial u}, \quad \mathbf{U}_4 = y\frac{\partial}{\partial u}$$
$$\mathbf{U}_5 = x\frac{\partial}{\partial x} + 3u\frac{\partial}{\partial u}, \qquad \mathbf{U}_6 = y\frac{\partial}{\partial y} - 2u\frac{\partial}{\partial u}.$$

Describe the properties of this Lie algebra. Using the proper subalgebra find the ordinary differential equation for the traveling wave solutions of the sub-sonic gas motion equation.

Solution The commutator table of L_6 is

	$\mathbf{U}_1$	$\mathbf{U}_2$	$\mathbf{U}_3$	$\mathbf{U}_4$	$\mathbf{U}_5$	$\mathbf{U}_6$
$\mathbf{U}_1$	0	0	0	0	$\mathbf{U}_1$	0
$\mathbf{U}_2$	0	0	0	$\mathbf{U}_3$	0	$\mathbf{U}_2$
$\mathbf{U}_3$	0	0	0	0	$3\mathbf{U}_3$	$-2\mathbf{U}_3$
$\mathbf{U}_4$	0	$-\mathbf{U}_3$	0	0	$3\mathbf{U}_4$	$-3\mathbf{U}_4$
$\mathbf{U}_5$	$-\mathbf{U}_1$	0	$-3\mathbf{U}_3$	$-\mathbf{U}_4$	0	0
$\mathbf{U}_6$	0	$-\mathbf{U}_2$	$-\mathbf{U}_2$	$3\mathbf{U}_4$	0	0

There is no center in the algebra. It is easy to see that $\langle \mathbf{U}_1, \mathbf{U}_2, \mathbf{U}_3, \mathbf{U}_4 \rangle$ is an ideal.

Low central series is $L^{[2]} = \langle \mathbf{U}_1, \mathbf{U}_2, \mathbf{U}_3, \mathbf{U}_4 \rangle$, $L^{[3]} = L^{[1]}$.

We can provide some subalgebras $\langle \mathbf{U}_1 \rangle$, $\langle \mathbf{U}_1, \mathbf{U}_2 \rangle$, $\langle \mathbf{U}_1, \mathbf{U}_2, \mathbf{U}_3 \rangle$, $\langle \mathbf{U}_1, \mathbf{U}_2, \mathbf{U}_3, \mathbf{U}_4 \rangle$ as examples of different subalgebras.

The Lie algebra L_6 is solvable since the derived algebra $L^{(3)}$ vanishes

$$L^{(1)} = \langle \mathbf{U}_1, \mathbf{U}_2, \mathbf{U}_3, \mathbf{U}_4 \rangle, \; L^{(2)} = \langle \mathbf{U}_3 \rangle, \; L^{(3)} = 0.$$

Low central series $L^{[2]}$ coincides with the first derived algebra $L^{[2]} = L^{(1)}$.

For the reduction of PDE to an ODE which describes the traveling wave solutions we take the following subalgebra $\langle a_1\mathbf{U}_1 + a_2\mathbf{U}_2 \rangle$. We remark that like in previous problems we will suppose from the very beginning that $a_1 \neq 0$ and divide the infinitesimal generator of the subalgebra by a_1 to reduce the number of arbitrary constants. It means we look for the solution of a physical system which in any case depends on x. If we have no preferences we should not exclude the case $a_1 = 0$.

We introduce a new operator $\mathbf{U} = \mathbf{U}_1 + c\mathbf{U}_2 = \frac{\partial}{\partial x} + c\frac{\partial}{\partial y}$, where c is a constant. From the characteristic equations

$$\frac{dx}{1} = \frac{dy}{c} = \frac{du}{0}$$

it follows that the invariants are equal to $\text{inv}_1 = cx - y$ and $\text{inv}_2 = u(x, y)$. We take the first one as an independent variable z and the second as a dependent variable V. Rewriting the derivatives in new variables

$$\frac{\partial}{\partial x} = \frac{\partial z}{\partial x}\frac{d}{dz} = c\frac{d}{dz}, \quad \frac{\partial}{\partial y} = \frac{\partial z}{\partial y}\frac{d}{dz} = -\frac{d}{dz},$$

we obtain a second order ODE

$$V''(z)(c^3 V'(z) + 1) = 0.$$

Because of the special type of our equation we see that we obtain a solution if one of the multipliers in the equation is equal to zero. The first multiplier $V''(z) = 0$ provides a solution of the following type $V(z) = d_1 z + d_2$, $d_1, d_2 \in R$ which in the old variables takes the form

$$u(x, y) = d_1(cx - y) + d_2, \quad d_1, d_2, c \in \mathbb{R}.$$

The second multiplier $c^3 V'(z) + 1 = 0$ provides the solution of the form $V(z) = -c^{-3} z + d_3$, $d_3 \in \mathbb{R}$ which is included in the case displayed above.

8.6 Exercises

Here we present ten well known PDEs with rather rich admitted symmetry groups. They can be used to study different reductions or as in the problems listed before as some examples for computer aided calculations.

1. The nonlinear heat equation

$$u_t = u_x^{\alpha-1} u_{xx}, \quad \alpha \geq 0, \quad \alpha \neq 1$$

has an infinitesimal generator

$$\mathbf{U} = (a + 2dt + e(1 - \alpha t))\frac{\partial}{\partial t} + (b + dx)\frac{\partial}{\partial x} + (c + du + eu)\frac{\partial}{\partial u}$$

with five independent constants $a, b, c, d, e \in \mathbb{R}$. Use a proper subalgebra of L_5 to obtain an ordinary differential equation determining traveling wave solutions [48].

2. Provide the Lie group analysis of the Boussinesq equation (a model for shallow water waves or processes of a nonlinear filtration [75])

$$u_t = uu_{xx} + u_x^2$$

and study the corresponding symmetry algebra.

3. The Burgers equation

$$u_t = u_{xx} + uu_x$$

admits a 5–dimensional symmetry group. The corresponding infinitesimal generators of the Lie algebra are

$$\begin{aligned}
&\mathbf{U}_1 = \frac{\partial}{\partial x}, \quad \mathbf{U}_2 = \frac{\partial}{\partial t}, \quad \mathbf{U}_3 = t\frac{\partial}{\partial x} - \frac{\partial}{\partial u}, \\
&\mathbf{U}_4 = x\frac{\partial}{\partial x} + 2t\frac{\partial}{\partial t} - u\frac{\partial}{\partial u}, \\
&\mathbf{U}_5 = xt\frac{\partial}{\partial x} + t^2\frac{\partial}{\partial t} - (tu + x)\frac{\partial}{\partial u}.
\end{aligned}$$

Is the algebra L_5 a solvable algebra? If L_5 is a solvable algebra then present the corresponding chain of ideals.
Use a proper subalgebra of L_5 to obtain an ordinary differential equation determining traveling wave solutions.

Use a proper subalgebra to find the ordinary differential equation for the scaling invariant solutions of the Burgers equation.

4. The nonlinear heat equation

$$u_t = u_{xx} + \frac{1}{4t}(u_x)^2$$

admits the 4-dimensional symmetry group. The corresponding infinitesimal generators of the Lie algebra are

$$\begin{aligned}
&\mathbf{U}_1 = \frac{\partial}{\partial x}, \qquad \mathbf{U}_2 = \frac{\partial}{\partial u}, \\
&\mathbf{U}_3 = x\frac{\partial}{\partial u} - \frac{1}{2}\log|t|\frac{\partial}{\partial u}, \\
&\mathbf{U}_4 = x\frac{\partial}{\partial x} + 2t\frac{\partial}{\partial t} + 2u\frac{\partial}{\partial u}.
\end{aligned}$$

Using the proper subalgebra find the ordinary differential equation for the scale-invariant solutions of this equation.

5. The nonlinear heat equation

$$u_t = (u^4 u_x)_x$$

admits the 4–dimensional Lie symmetry group. The corresponding infinitesimal generators of the Lie algebra L_4 are

$$\mathbf{U}_1 = \frac{\partial}{\partial t}, \qquad \mathbf{U}_2 = \frac{\partial}{\partial x},$$

$$\mathbf{U}_3 = x\frac{\partial}{\partial x} + 2t\frac{\partial}{\partial t}, \qquad \mathbf{U}_4 = 2x\frac{\partial}{\partial x} + u\frac{\partial}{\partial u}.$$

Using the proper subalgebra find an ordinary differential equation determining scaling solutions.

6. The nonlinear heat equation

$$u_t + (e^u u_x)_x = 0$$

admits a 4–dimensional Lie algebra L_4 spanned by the infinitesimal generators

$$\mathbf{U}_1 = \frac{\partial}{\partial x}, \qquad \mathbf{U}_2 = \frac{\partial}{\partial t},$$

$$\mathbf{U}_3 = 2t\frac{\partial}{\partial t} + x\frac{\partial}{\partial x}, \qquad \mathbf{U}_4 = x\frac{\partial}{\partial x} + 2\frac{\partial}{\partial u}.$$

Is L_4 a solvable algebra? If L_4 is a solvable algebra then present the corresponding chain of ideals. Use a corresponding infinitesimal generator of L_4 to obtain an ordinary differential equation determining similarity (scaling invariant) solutions.

7. The nonlinear heat equation

$$u_t + \left(u^{-\frac{4}{3}} u_x\right)_x = 0$$

admits a 5–dimensional Lie algebra G_5 spanned by the infinitesimal generators

$$\mathbf{U}_1 = \frac{\partial}{\partial x}, \qquad \mathbf{U}_2 = \frac{\partial}{\partial t}, \qquad \mathbf{U}_3 = 2t\frac{\partial}{\partial t} + x\frac{\partial}{\partial x},$$

$$\mathbf{U}_4 = 2x\frac{\partial}{\partial x} - 3u\frac{\partial}{\partial u}, \qquad \mathbf{U}_5 = -x^2\frac{\partial}{\partial x} + 3xu\frac{\partial}{\partial u}.$$

Is L_5 a solvable algebra? If L_5 is a solvable algebra then present the corresponding chain of ideals. Use a corresponding infinitesimal generator of L_5 to obtain an ordinary differential equation determining similarity (scaling invariant) solutions of this equation.

8. The Harry Dym equation

$$u_t - u^3 u_{xxx} = 0$$

admits the 5–dimensional symmetry group. The corresponding infinitesimal generators of the Lie algebra are

$$\mathbf{U}_1 = \frac{\partial}{\partial x}, \qquad \mathbf{U}_2 = \frac{\partial}{\partial t}, \qquad \mathbf{U}_3 = x\frac{\partial}{\partial x} + u\frac{\partial}{\partial u},$$
$$\mathbf{U}_4 = -3t\frac{\partial}{\partial t} + u\frac{\partial}{\partial u}, \qquad \mathbf{U}_5 = x^2\frac{\partial}{\partial x} + 2xu\frac{\partial}{\partial u}.$$

Is L_5 a solvable algebra? If L_5 is a solvable algebra then present the corresponding chain of ideals. Using the proper subalgebra find the ordinary differential equation for the traveling wave (translation invariant) solutions of the Harry Dym equation.

9. The Korteweg-de-Vries (KdV) equation

$$u_t + uu_x + u_{xxx} = 0$$

which describes waves on shallow water admits a 4–dimensional Lie algebra L_4 spanned by the infinitesimal generators

$$\mathbf{U}_1 = \frac{\partial}{\partial x}, \; \mathbf{U}_2 = \frac{\partial}{\partial t}, \; \mathbf{U}_3 = 6t\frac{\partial}{\partial x} + \frac{\partial}{\partial u}, \; \mathbf{U}_4 = x\frac{\partial}{\partial x} + 3t\frac{\partial}{\partial t} - 2u\frac{\partial}{\partial u}.$$

Is L_4 a solvable algebra? If L_4 is a solvable algebra then present the corresponding chain of ideals. Use a proper subalgebra of L_4 to obtain an ordinary differential equation determining traveling wave solutions.

10. The equation of a sub-sonic gas motion

$$u_x u_{xx} + u_{yy} = 0$$

admits the 6–dimensional symmetry group. The corresponding infinitesimal generators of the Lie algebra are

$$\mathbf{U}_1 = \frac{\partial}{\partial x}, \qquad \mathbf{U}_2 = \frac{\partial}{\partial y}, \qquad \mathbf{U}_3 = \frac{\partial}{\partial u}, \qquad \mathbf{U}_4 = y\frac{\partial}{\partial u},$$
$$\mathbf{U}_5 = x\frac{\partial}{\partial x} + 3u\frac{\partial}{\partial u}, \qquad \mathbf{U}_6 = y\frac{\partial}{\partial y} - 2u\frac{\partial}{\partial u}.$$

Using the proper subalgebra of admitted symmetry algebra find an ordinary differential equation determining traveling wave solutions.

Chapter 9

Study of new models in financial mathematics

Since quite all models which we study using Lie group analysis in this chapter are modifications of the famous Black-Scholes model for pricing of derivative products [7], at first we describe the main features of the Black-Scholes model from the financial point of view. A more detailed description of this model can be found in the original paper of the authors and in textbooks and monographs devoted to this topic, for example in [70],[80].

Prices of financial products are often modeled as random variables or functions of random variables. If we look for mathematical models in continuous time then we naturally assume that some characteristic of an asset, for instance, its price, follows a stochastic process. We need some analog of the well known Taylor theorem for smooth functions of deterministic variables in order to describe small value changes of a function dependent on one or many random and deterministic variables. This analog is the Itô's Lemma.

We provide here one of the simplest versions of this Lemma to make discussions about financial models in this chapter more solid and clear. Consider first a Wiener process (also called Brownian motion) $\{dW_t, t \geq 0\}$. Here dW_t is a random variable, drawn from a normal distribution with the mean $E[dW_t] = 0$ and the variance $\mathrm{Var}[dW_t] = dt$.

We introduce now an Itô's process $\{X_t, t \geq 0\}$ with

$$dX_t = A(X_t, t)dt + B(X_t, t)dW_t, \tag{9.0.1}$$

where the mean (or drift) $A(X_t, t)$ and the variance $B^2(X_t, t)$ depend on a random variable X_t and a deterministic variable t. We consider now another stochastic process Y_t which depends explicitly on the process X_t and on time t, i.e. $Y_t = q(X_t, t)$. We suppose that the function $q(X_t, t)$ is differentiable at least one time with respect to both arguments and two times with respect to the first argument. The value changes dY_t of this process are defined by Itô's Lemma (see for instance [80], [44] or other textbooks).

Theorem 9.0.1 *(Itô's Lemma)*
Let X_t be a stochastic process of the type (9.0.1), the stochastic process $Y_t = q(X_t, t)$ explicitly depends on X_t and t, the function $q(\cdot,\cdot)$ is at least a twice differentiable function of the first argument and one times differentiable function of the second argument. Then the process Y_t satisfies the following stochastic differential equation

$$\begin{aligned} dY_t &= dq(X_t, t) \qquad (9.0.2) \\ &= \left(\frac{\partial q}{\partial t} + A(X_t, t)\frac{\partial q}{\partial X} + \frac{B^2(X_t, t)}{2}\frac{\partial^2 q}{\partial X^2} \right) dt + B(X_t, t)\frac{\partial q}{\partial X} dW_t. \end{aligned}$$

Here $\{dW_t, t \geq 0\}$ is a Wiener process.

In the formulation of Itô's Lemma we skipped the subindex t in the derivatives with respect to X_t to make the formula more transparent.

The BS model is based on many assumptions of the so-called ideal market which we formulate in Section 9.1. If one or other of these assumptions is not fulfilled then we have to do with a more complicated situation. The new models are usually nonlinear parabolic PDEs. In the next sections of this chapter we show different modifications of the BS model as well as application of Lie group analysis to these models.

9.1 Ideal financial market. Black-Scholes model for option pricing

In this section we introduce the main notations for the models used in modern financial mathematics. One of the most important assumptions of the Black-Scholes theory is the idea of ideal market. We can describe the ideal market as a market where the following assumptions are fulfilled:

1. All participants are price takers on the market. It means that the trading strategy of any actor on the market cannot move the prices on the corresponding assets. The condition is satisfied for usual market participants in the real financial market. For large traders as banks or hedging funds, it can be problematic because they have to trade big volumes of assets and their strategy can affect the prices of the assets in which they have a dominant position.

2. There are no arbitrage possibilities. It means nobody can get money without taking a risk.

3. The asset prices are log normally distributed and the asset volatility σ is a known constant, the asset pays no dividends during the lifetime of the option. The empirical study of asset price distributions shows that the distributions of liquid assets can be approximated with a log normal distribution quite well. The distributions of prices of the assets that are not so often traded or are strongly illiquid can differ from the log normal distribution dramatically. The condition that the assets do not pay any dividends is , in fact, not very restrictive. A lot of assets do not pay dividends at all, others pay them discreetly, for instance, once per year. This payment is irrelevant for the majority of the derivatives due to their short lifetimes. It is also rather easy to introduce the dividend payments in the Black-Scholes model, so we can develop the theory without them at first and then modify it as needed.

4. All assets are liquid and infinitely divisible, short selling is permitted, trading of the underlying asset can take place continuously. This assumption does not look very clear at first glance, but if we look at usual stocks presented in indexes then we see that they are very liquid. The typical volumes of the trading are so large, that one stock is just a small fraction of a traded bunch. Because of that we can look at assets as approximately infinitely divisible objects.

5. The relevant information on the market is spread instantaneously and is equally accessible for each participant. It means we exclude the inside trading completely and suppose that all market participants are able to inform themselves completely and instantaneously.

6. There are no frictions on the market. It means that we do not take into account any taxes, transaction costs or price spreads.

7. There exists a risk-free constant continuously compounded interest rate. The interest rates for savings and credit accounts are equal and are constant for the lifetime of the financial derivative product (e.g. several months). In this way we introduce the possibility of using absolutely liquid risk-free bonds or bank accounts with a fixed interest rate that is equal for all participants.

Using the concept of the ideal market Black and Scholes built a theory of the option pricing. They suppose that the value $\mathcal{V}(S_t, t)$ of an option depends on the price of the underlying asset S_t and time t only. The examples of the most popular options are call and put options. We define the European call (put) option in the following way: An European call $\mathcal{V}(S_t, t) = C(S_t, t)$ (or put $\mathcal{V}(S_t, t) = P(S_t, t)$) option is a contract which permits the owner (holder) of the option to buy (or sell) a prescribed stock S_T for a prescribed amount of money E (exercise price, strike price) in a prescribed time T in the future (expiry date, maturity) [80].

Remark 9.1.1 *The prescribed asset is often called an underlying asset or simply an underlying, the corresponding contract is called claim.*

The Black-Scholes model (BS model) allows us to define the price of such contracts using the assumptions listed above.

Corresponding to the assumption (3) in the BS model a stock price process S_t follows a geometric Brownian motion (in other words is a solution of a stochastic differential equations (SDE)),

$$dS_t = \mu S_t dt + \sigma S_t dW_t, \quad t \geq 0, \tag{9.1.3}$$

with constants $\mu \in \mathbb{R}$, $\sigma > 0$ and a standard Brownian motion $\{W_t, t \geq 0\}$ (also called Wiener process). Here S_t denotes the price of a stock at a time moment t (an index t will often be skipped to make the formulas more transparent), the constant μ is called a drift and σ a volatility of the stock prices. In other words the stock price dynamics is given by

$$S_t = S_0 \exp\left((\mu - \sigma^2/2)t + \sigma W_t\right), \quad t \geq 0. \tag{9.1.4}$$

The formulas (9.1.3) and (9.1.4) represent the stock price process in differential or integral form correspondingly.

Let us now assume that $\mathcal{V}(S_t, t)$ is a financial product related to the stock S_t. It can be an option written on the stock S_t or a portfolio containing this stock and a derivative product on this stock. Then the price of this product depends on the price of the stock and time. According to Itô's Lemma 9.0.2 the random walk that $\mathcal{V}(S_t, t)$ follows has the following form

$$d\mathcal{V} = \sigma S \frac{\partial \mathcal{V}}{\partial S} dW_t + \left(\mu S \frac{\partial \mathcal{V}}{\partial S} + \frac{1}{2}\sigma^2 S^2 \frac{\partial^2 \mathcal{V}}{\partial S^2} + \frac{\partial \mathcal{V}}{\partial t} \right) dt. \tag{9.1.5}$$

Let now $\mathcal{V}_t$ be a self-financing portfolio. The self-financing feature of a portfolio means that we neither remove nor invest some additional money in it, i.e. the change of the portfolio's value is completely defined by the price change of the stock. The portfolio is formed using some number of riskless bonds B_t and a number of risky assets S_t. We provide a dynamical hedging of our portfolio at any moment in time with a hedging strategy ϕ. It means that we sell (or buy) a number of stocks S_t and invest the money in the bond B_t (or vice versa) depending on the market price of the stock S_t to keep the portfolio riskless. Because of the ideal market assumption (6) the continuous trading will not lead us to bankruptcy. We just continuously change the amount of assets and bonds in our portfolio.

In every moment of time t the value of the dynamically hedged portfolio $\mathcal{V}_t$ is $\mathcal{V}_t^\phi = \delta_t S_t + \beta_t B_t$, where δ_t is a number of units of the stock (constant on each time interval Δt), B_t is the value of the bond and β_t is a number of units of the bond. We can put $B_0 = 1$ without loss of generality, it means we look at the relative prices $\tilde{\mathcal{V}}_t = \mathcal{V}_t/B_t$, $\tilde{S}_t = S_t/B_t$ and use the bond value as a numerator $\tilde{B}_t = B_t/B_0|_{t=0} = 1$, so we get a discounted price process $\tilde{S}_t$. To make all formulas transparent we skip further the tilde over discounted values. We rewrite the previous relation in the form $\mathcal{V}_t^\phi = \delta_t S_t + \beta_t e^{rt}$, here e^{rt} is the usual time discount factor with a risk free constant continuously compounded interest rate r. The value r corresponds to the riskless interest rate on the market for saving accounts.

The pair $\phi = (\delta_t, \beta_t)$ defines the self-financing hedging strategy that maintains the portfolio. We take the time intervals Δt between re-hedging infinitesimal small $\Delta t \to dt$ and use the arbitrage arguments correspondingly to the assumption (2) and Itô's Lemma 9.0.2, [7], [70], [55]. The arbitrage arguments mean that the mathematical expectation of the profit from the investment in the risky portfolio should coincide with the profit from a risk free account. In this way we get the famous Black - Scholes partial differential equation

$$u_t + \frac{1}{2}\sigma^2 S^2 u_{SS} + rSu_S - ru = 0, \tag{9.1.6}$$

where $S \in [0, \infty)$, $t \in [0, T]$. According to the BS model any financial product on the market is a solution of the BS Equation (9.1.6) under corresponding terminal conditions (payoff conditions) and boundary conditions. The value $u(S,t)$ can describe a price of the option, a value of a portfolio $u(S,t) = \mathcal{V}_t^\phi$ or a price of a hedging strategy. If we are looking for a price of a European call option $C(S,t) = u(S,t)$ we additionally demand that the solution of (9.1.6) equation satisfies the following conditions: payoff condition $u(S,T) = \max(S - E, 0)$, and boundary conditions $u(0,t) = 0$, $u(S,t) \to S - Ee^{-r(T-t)}$ by $S \to \infty$.

The symmetry properties of the Black-Scholes Equation (9.1.6) were studied in the previous chapter. Because the payoff structure of the call (and put) option does not destroy the scaling symmetry of the BS equation so we can get explicit formulas for these invariant solutions, which are known as Black-Scholes formulas (8.4.36)(see also [7], [80]). Existence of a simple explicit formula for a fair price of an option makes the BS model very popular. The model turned out to be very helpful and was quickly implemented in financial trading procedures.

All other models which we study in this chapter are various modifications of the Black-Scholes model. Many authors try to make better approximation of the real life processes by altering one or many of the ideal market assumptions (1) – (6) listed above. Usually such modified models are represented by nonlinear PDEs because feedback effects are implemented.

We provide a short discussion about the possibilities of altering one or more of the assumptions of the ideal market.

Let us first look at the case where assumption (1) does not hold. It means that we suppose that a large trader is present on the market and his trading strategy can move some asset prices. A lot of models regard this case. The development of adequate models for trading strategies of a large trader is extremely important for the banks, hedging and pensions funds, because it helps to avoid unnecessary losses.

Further models take into account frictions on the market (real existing taxes for example). It means that assumption (6) is not satisfied anymore. These models provide some hedging strategies in discrete time steps instead of the continuously instant hedging like in the Black-Scholes model.

The next important class of models are models where the liquidity assumption (4) does not hold. There models focus on liquidity problems, which are becoming more and more important for understanding market crisis phenomena (see for example [22]).

The main difference between all these new models and the well known Black-Scholes model lies in the practical applications. The Black-Scholes model in continuous or discrete form is widely used in financial practice, but all other models are still far from common acceptance. It means that in the nearest future new models will be developed and should be studied in detail. The financial market development in the modern world is more and more complicated, connected, and liquidity and feedback problems become more and more pressing issues. Attempts to take into account such phenomena lead to nonlinear partial differential equations. Because of that the study of all important features of these equation is so important now. Further in this chapter we study different financial models step-by-step using Lie group analysis and discuss the properties of the equations and the solutions.

Certainly the studied models are just a small part of the models that exist right now. It is due to the fact that new models appear rather often. It is not possible to list and study all of them, but it is possible to get an impression about the power of the presented method. In this chapter we present mostly our own results as examples of Lie group analysis applications, more detailed description of the mentioned results can be found in the corresponding papers.

9.2 Pricing options in illiquid markets

We look at the liquidity assumption (4) of traded assets in the ideal market. If the market is smooth, this assumption describes reality rather well. All main assets presented in corresponding stock indexes are more or less perfectly liquid. The small traders as well as banks or other large traders can liquidate their portfolios without any great losses. If we look not only at the main indexes but also at the assets that are less common, or at the extremely large portfolios of institutional investors, or assume that there is a crisis on the market (e.g. a liquidity shock) then the problem of liquidity becomes more and more important. In this situation the investors cannot be sure that they can liquidate their portfolios without considerable losses. The usual hedging strategies in such markets are not really successful and other modified strategies which take into account strong feedback effects from one's own trading should be used. The following subsections will be devoted to such models.

Remark 9.2.1 *In all examples of this chapter we study self-financing stock trading strategies Φ_t of the Markovian type, i.e. a trading strategy in the form $\Phi_t = \phi(S_t, t)$ for a smooth function ϕ. In this case the stock trading strategy Φ is a semimartingale. We would not go into details and explain these ideas but an interested reader can find the complete information in original papers and in textbooks like [70], [44], [69] etc.*

9.2.1 Frey, Frey-Patie, Frey-Stremme models of risk management for derivatives

In this section we present features of the models developed for pricing and trading strategies of a large trader. The price taker assumption (1) and the liquidity assumption (4) are not fulfilled for the large trader and should be omitted. First let us describe the general model setting then, consequently, the results of our investigation using Lie group analysis of corresponding nonlinear PDEs.

The models developed in [29], [30], [41], [54], [63] or [34] belong to the so called *reduced-form stochastic differential equation models*, the classification of the listed models is for instance presented in [15]. Under this modeling approach it is assumed that investors are *large traders* in the sense that their trading activity affects the equilibrium stock price. More precisely, given a liquidity parameter $\rho \geq 0$ and the stock trading strategy $\phi_t(S_t, t)$ representing this large trader, it is assumed that the stock price satisfies the SDE

$$dS_t = \sigma S_t dW_t + \rho S_t d\phi_t \,. \tag{9.2.7}$$

We interpret this equation as follows: in the case where the investor buys (sells) stock ($\Delta\phi_t > 0$) the stock price is pushed upward (downward) by $\rho S_{t-}\Delta\phi_t$; the strength of this price impact depends on the liquidity parameter ρ. It means also that we suppose that the price of the stock moves instantly with each rebalance of the portfolio. Note that for $\rho = 0$ the asset price simply follows the BS model with the reference volatility σ. The asset price process which results describing a large trader using a particular trading strategy ϕ is denoted by S^ϕ. We suppose additionally that the Markovian trading strategy $\phi(t, S_t)$ satisfies the constraint

$$1 - \rho S\phi_S(S,t) > 0 \text{ for all } (S,t).$$

This condition is mainly a condition on the permissible variations in the stock trading strategy of the large trader for a fixed liquidity parameter ρ. Applying Itô's formula (9.0.2) to (9.2.7) we see that S^ϕ is an Itô process with the following dynamics

$$dS_t^\phi = v^\phi(S_t^\phi, t)S_t^\phi dW_t + b^\phi(S_t^\phi, t)S_t^\phi dt$$

with *adjusted volatility* given by

$$v^\phi(S,t) = \frac{\sigma}{1 - \rho S\phi_S(S,t)}\,; \tag{9.2.8}$$

see for instance [29] for a detailed derivation. This adjusted volatility $v^\phi(S,t)$ increases (decreases) relatively to the reference (defined in the ideal market) volatility σ if $\phi_S > 0$ (respectively $\phi_S < 0$).

In the model (9.2.7), a portfolio with a stock trading strategy ϕ and a value $\mathcal{V}$ is called *self-financing*. The self-financing condition is satisfied if the equation $d\mathcal{V}_t = \phi_t dS_t^\phi$ is satisfied. Note that the form of the strategy ϕ affects the dynamics of S_t^ϕ instantly. This *feedback effect* gives rise to nonlinearities in the wealth dynamics. Suppose that $\mathcal{V}_t = u(S_t, t)$ and the functions $u(S,t)$ and $\phi(S,t)$ are smooth functions (we will often omit the indexes ϕ, t by S to make the formulas shorter). As before, applying Itô's Lemma (9.0.2) to the process $(u(S_t^\phi, t))_{t\geq 0}$ yields $\phi \equiv u_S$. Moreover, u must satisfy the equation $u_t + \frac{1}{2}(v^\phi(S,t))^2 S^2 u_{SS} = 0$. We use the expression (9.2.8) and the relation $\phi_S = u_{SS}$ to obtain the fully nonlinear PDE for $u(S,t)$. After the Feynman–Kac theorem [71] together with a fixed point argument [29] $u(S,t)$ satisfies the nonlinear partial differential equation

$$u_t + \frac{1}{2}\frac{\sigma^2}{(1-\rho S u_{SS})^2} S^2 u_{SS} = 0\,. \tag{9.2.9}$$

For pricing derivative securities a terminal condition corresponding to the particular payoff $u(S,T) = h(S)$ for all S where $S \geq 0$, $t \in [0,T]$, $T > 0$ should be added.

Note that (9.2.9) is a fully nonlinear equation in a sense that the coefficient of the highest derivative is a nonlinear function of this derivative. The value $1/(\rho S_t)$ is called the depth of the market at time t. If $\rho \to 0$ then simultaneously $d\phi_t \to 0$ in (9.2.7) and the Equation (9.2.9) is reduced to the Black–Scholes model.

In this model the parameter ρ is a characteristic of the market which defines the adjusted volatility (9.2.8) and does not depend on the payoff of the hedged derivatives. The value of ρ is fixed during the trading process and can be estimated in different ways. In the papers [28], [29] the value ρ was estimated in some special cases and ρ took the values equal to 0.1, 0.2, 0.3, 0.4. Later on, in [30] Frey and Patie improved the model (9.2.9). They introduced another liquidity coefficient, i.e. instead of one constant coefficient ρ , they introduced a coefficient $\rho\lambda(S_t)$ which depends on the current stock price S_t. It means the stochastic Equation (9.2.7) now takes the form

$$dS_t = \sigma S_t dW_t + \rho\lambda(S_t) S_t d\phi_t, \tag{9.2.10}$$

where $\lambda : \mathbb{R}^+ \to \mathbb{R}^+$ is a continuous bounded function for all $S \geq 0$, ρ is a constant parameter. In this model the depth of the market at time t is defined as $1/(\rho\lambda(S_t)S_t)$; $\lambda(S)$ is called a *level-depended liquidity profile*. The values of ρ and $\lambda(S)$ may be estimated from the observed option prices and depend on the payoff $h(S)$. The feedback-effect described in (9.2.10) leads to the other nonlinear version of the BS model,

$$u_t + \frac{\sigma^2 S^2}{2}\frac{u_{SS}}{(1-\rho S\lambda(S)u_{SS})^2} = 0, \tag{9.2.11}$$

with $S \in [0,\infty)$, $t \in [0,T]$.

From the analytical point of view the nonlinear model (9.2.9) is a special case of the more general model (9.2.11). We obtain the previous model from the Equation (9.2.11) if we put $\lambda(S) \equiv 1$. The Lie group properties of both equations were studied in [14], [20], [15]. Despite the different financial settings for both models (9.2.9) and (9.2.11) we look at the first model as a special case of the second one. We formulate the general results of the Lie group analysis for Equation (9.2.11). In this way we show with which type of the function $\lambda(S)$ the equation admits a non-trivial symmetry algebra.

Theorem 9.2.1 *[20] The differential Equation (9.2.11) with an arbitrary function $\lambda(S)$ possesses a trivial three-dimensional Lie algebra L_3 spanned by infinitesimal generators*

$$\mathbf{U}_1 = \frac{\partial}{\partial t}, \quad \mathbf{U}_2 = S\frac{\partial}{\partial u}, \quad \mathbf{U}_3 = \frac{\partial}{\partial u}.$$

Equation (9.2.11) admits a non-trivial four-dimensional Lie algebra L_4 spanned by generators

$$\mathbf{U}_1 = \frac{\partial}{\partial t}, \quad \mathbf{U}_2 = S\frac{\partial}{\partial u}, \quad \mathbf{U}_3 = \frac{\partial}{\partial u}, \mathbf{U}_4 = S\frac{\partial}{\partial S} + (1-k)u\frac{\partial}{\partial u},$$

with commutator relations

$$\begin{aligned} [\mathbf{U}_1, \mathbf{U}_2] = [\mathbf{U}_1, \mathbf{U}_3] = [\mathbf{U}_1, \mathbf{U}_4] = [\mathbf{U}_2, \mathbf{U}_3] = 0, \\ [\mathbf{U}_2, \mathbf{U}_4] = -k\mathbf{U}_2, \qquad [\mathbf{U}_3, \mathbf{U}_4] = (1-k)\mathbf{U}_3, \end{aligned} \tag{9.2.12}$$

only for the special form of the function

$$\lambda(S) \equiv \omega S^k, \omega, k \in \mathbb{R}. \tag{9.2.13}$$

Using these symmetry algebras one can provide symmetry reductions of the studied PDEs to some set of ODEs. The ODEs give rise to invariant solutions of the original equation.

It follows from (9.2.12) that for $k = 0, 1$ we can expect some simplification of the corresponding ODEs. The case $k = 0$ means that $\lambda(S) = \omega$ and Equation (9.2.11) is reduced exactly to (9.2.9). The case $k = 1$ has not been studied before in financial literature. Both cases $k = 0, 1$ are intrinsically connected to each other as it is shown in [20], where the special invariant solutions are also provided. Properties of the set of explicit invariant solutions are studied in detail in [14], [20], [15]. The invariant solutions reflect the strong nonlinearity of the given parabolic Equation (9.2.11) and cannot be obtained by the perturbation method because they "blow up" when $\rho \to 0$. It means they have no counterpart in the linear Black-Scholes model (9.1.6).

The symmetry algebra L_3 (or L_4) defines the corresponding symmetry group G_3 (or G_4) of the Equation (9.2.11) by the Lie equations (3.1.3)(see also [53]).

Theorem 9.2.2 *The action of the symmetry group G_3 of (9.2.11) with an arbitrary function $\lambda(S)$ is given by*

$$\tilde{S} = S, \tag{9.2.14}$$

$$\tilde{t} = t + a_2\epsilon, \tag{9.2.15}$$

$$\tilde{u} = u + a_3 S\epsilon + a_4\epsilon, \quad \epsilon \in (-\infty, \infty). \tag{9.2.16}$$

If the function $\lambda(S)$ has the special form (9.2.13) then the action of the symmetry group G_4 is represented by

$$\tilde{S} = Se^{a_1\epsilon}, \quad \epsilon \in (-\infty, \infty), \tag{9.2.17}$$

$$\tilde{t} = t + a_2\epsilon,$$

$$\tilde{u} = ue^{a_1(1-k)\epsilon} + \frac{a_3}{a_1 k} S\epsilon e^{a_1\epsilon}(1 - e^{-a_1 k\epsilon}) + \frac{a_4}{a_1(1-k)}(e^{a_1(1-k)\epsilon} - 1), \quad k \neq 0, \ k \neq 1 \tag{9.2.18}$$

$$\tilde{u} = ue^{a_1\epsilon} + a_3 S\epsilon e^{a_1\epsilon} + \frac{a_4}{a_1}(e^{a_1\epsilon} - 1), \ k = 0,$$

$$\tilde{u} = u + \frac{a_3}{a_1} S(e^{a_1\epsilon} - 1) + a_4\epsilon, \ k = 1, \tag{9.2.19}$$

where we assume that $a_1 \neq 0$ because the case $a_1 = 0$ coincides with the former case (9.2.14)-(9.2.16), $a_2, a_3, a_4 \in \mathbb{R}$ are arbitrary parameters.

It is possible to use the symmetry group to construct invariant solutions to the equation (9.2.11). In the first case, for an arbitrary form of $\lambda(S)$, the symmetry group G_3 is very poor and we can obtain just the following invariants

$$\text{inv}_1 = S, \tag{9.2.20}$$

$$\text{inv}_2 = u - (a_3 S + a_4)/a_2, \quad a_2 \neq 0.$$

Using these invariants we cannot obtain any non-trivial reduction of (9.2.11).

In the special case (9.2.13) the symmetry group admits two functionally independent invariants

$$\text{inv}_1 = \log S + at, \quad a = a_1/a_2, \ a_2 \neq 0 \tag{9.2.21}$$

$$\text{inv}_2 = u \ S^{(k-1)}. \tag{9.2.22}$$

The invariants can be used as new independent and dependent variables in order to reduce the partial differential Equation (9.2.11) with the special function $\lambda(S)$ defined by (9.2.13) to an ordinary differential equation.

The special case $\lambda(S) = \omega S^k$

Let us study a special case of (9.2.11) with $\lambda(S) = \omega S^k, \omega, k \in \mathbb{R}$. The equation under investigation is now

$$u_t + \frac{\sigma^2 S^2}{2} \frac{u_{SS}}{(1 - bS^{k+1}u_{SS})^2} = 0 \tag{9.2.23}$$

with the constant $b = \rho\omega$. As usual we assume that $\rho \in (0,1)$. The value of the constant ω depends on the corresponding option type and in our investigation it can be assumed that ω is an arbitrary constant, $\omega \neq 0$. The variables S, t are in the intervals

$$S > 0, \quad t \in [0,T], \quad T > 0. \tag{9.2.24}$$

Remark 9.2.2 *The case $b = 0$, i.e. $\rho = 0$ or $\omega = 0$ leads to the well known linear Black-Scholes model and we exclude this case from our investigations.*

We suppose that the denominator in Equation (9.2.11) (correspondingly (9.2.23)) is not equal to zero identically. Let us study the denominator in the second term of the Equation (9.2.23). It will be equal to zero if the function $u(S,t)$ satisfies the equation

$$1 - bS^{k+1}u_{SS} = 0. \tag{9.2.25}$$

The solution to this equation is a function $u_0(S,t)$

$$\begin{aligned}
u_0(S,t) &= \frac{1}{bk(k-1)}S^{1-k} + Sc_1(t) + c_2(t), \quad b \neq 0, k \neq 0, 1,\\
u_0(S,t) &= -\frac{1}{b}\log S + Sc_1(t) + c_2(t), \quad b \neq 0, k = 1,\\
u_0(S,t) &= \frac{1}{b}S\log S + Sc_1(t) + c_2(t), \quad b \neq 0, k = 0,
\end{aligned} \tag{9.2.26}$$

where the functions $c_1(t)$ and $c_2(t)$ are arbitrary functions of the variable t.

In the sequel we will assume that the denominator in the second term of the Equation (9.2.23) is not identically zero, i.e., a solution $u(S,t)$ is not equal to the function $u_0(S,t)$ (9.2.26) except in a discrete set of points.

Let us now introduce new invariant variables

$$\begin{aligned}
z &= \log S + at, \quad a \in \mathbb{R}, a \neq 0,\\
v &= u\ S^{(k-1)}.
\end{aligned} \tag{9.2.27}$$

After this substitution Equation (9.2.23) is reduced to an ordinary differential equation

$$av_z + \frac{\sigma^2}{2}\frac{v_{zz} + (1-2k)v_z - k(1-k)v}{(1 - b(v_{zz} + (1-2k)v_z - k(1-k)v))^2} = 0, \quad a, b \neq 0. \tag{9.2.28}$$

We obtain elementary solutions of this equation if we assume that $v =$ const. or $v_z =$ const. It is easy to prove that there exists a trivial solution $v = 0$ for any k and $v_z = 0$ if $k \neq 0, 1$, and the solutions $v = \text{const.} \neq 0$, $v_z = \text{const.} \neq 0$ if $k = 0, 1$ only. The condition that the denominator in (9.2.28) is not equal to zero, i.e.,

$$(1 - b(v_{zz} + (1-2k)v_z - k(1-k)v))^2 \neq 0 \tag{9.2.29}$$

corresponds to the Equation (9.2.25) in new variables z, v.

If the function $v(z)$ satisfies the inequality (9.2.29) then we can multiply both terms of Equation (9.2.28) with the denominator of the second term. In Equation (9.2.28) all coefficients are constants hence we can reduce the order of the equation. We assume that $v, v_z \neq$ const. We now choose v as a new independent variable and introduce as a new dependent variable $x(v) = v_z(z)$. This variable substitution reduces Equation (9.2.28) to a first order differential equation which is second order polynomial corresponding to the function $x(v)_v$. Under assumption (9.2.29) the set of solutions of the Equation (9.2.28) is equivalent to a union of solution sets for the following equations

$$\begin{aligned} x &= 0, \;\; k = 0, 1 \\ x_v &= -1 + 2k - \frac{\sigma^2}{4ab^2x^2} + \frac{1}{bx} + \frac{k(1-k)v}{x} - \frac{\sqrt{\sigma^2(\sigma^2 - 8abx)}}{4ab^2x^2}, \\ x_v &= -1 + 2k - \frac{\sigma^2}{4ab^2x^2} + \frac{1}{bx} + \frac{k(1-k)v}{x} + \frac{\sqrt{\sigma^2(\sigma^2 - 8abx)}}{4ab^2x^2}. \end{aligned} \tag{9.2.30}$$

Equations (9.2.30) are of an autonomous type if the parameter k is equal to $k = 0, 1$ only. We see that these are exactly the cases in which the corresponding Lie algebra (9.2.12) has a two-dimensional center. The case $k = 0$ was studied earlier in [14], [23] and in [20]. As an example in the next section we describe the case $k = 1$ studied in [20].

The special case $\lambda = \omega S$

If we put $k = 1$ in (9.2.13) then Equation (9.2.28) takes the form

$$v_z + q\frac{v_{zz} - v_z}{(1 - b(v_{zz} - v_z))^2} = 0, \tag{9.2.31}$$

where $q = \frac{\sigma^2}{2a}$, $a, b \neq 0$. It is an autonomous equation which possesses a simple structure. We use this structure and introduce a more simple substitution as described at the end of the previous section to reduce the order of equation.

One family of solutions to this equation is very easy to find. We just suppose that the value $v_z(z)$ is equal to a constant. The equation (9.2.31) admits as a solution the value $v_z = (-1 \pm \sqrt{q})/b$ consequently the corresponding solution $u(S,t)$ to (9.2.23) with $\lambda = \omega S$ can be represented by the formula

$$u(S,t) = \frac{1}{\rho\omega}(-1 \pm \sqrt{q})(\log S + at) + c, \ a > 0, \tag{9.2.32}$$

where c is an arbitrary constant.

To find other families of solutions we introduce a new dependent variable

$$y(z) = v_z(z) \tag{9.2.33}$$

and assume that the denominator of the Equation (9.2.31) is not equal to zero, i.e.

$$v(z) \neq -\frac{z}{b} + c_1\, e^z + c_2, \ \text{ i.e. } \ y(z) \neq -\frac{1}{b} + c_1\, e^z, \tag{9.2.34}$$

where c_1, c_2 are arbitrary constants.

We multiply both terms of Equation (9.2.31) by the denominator of the second term and obtain

$$yy_z^2 - 2\left(y^2 + \frac{1}{b}y - \frac{q}{2b^2}\right)y_z + \left(y^2 + \frac{2}{b}y + \left(\frac{1-q}{b^2}\right)\right)y = 0, \ \ b \neq 0. \tag{9.2.35}$$

We denote the left hand side of this equation as $F(y, y_z)$. The Equation (9.2.35) can possess exceptional solutions which are the solutions of a system

$$\frac{\partial F(y, y_z)}{\partial y_z} = 0, \quad F(y, y_z) = 0. \tag{9.2.36}$$

The first equation in this system defines a discriminant curve which has the form

$$y(z) = \frac{q}{4b}. \tag{9.2.37}$$

If this curve is also a solution of the original Equation (9.2.35) then we obtain an exceptional solution. We obtain an exceptional solution if $q = 4$, i.e. $a = \sigma^2/8$. It has the form

$$y(z) = \frac{1}{b}. \tag{9.2.38}$$

This solution belongs to the family of solutions (9.2.40) by the specified value of the parameter q. In all other cases the Equation (9.2.35) does not possess any exceptional solutions.

Hence the set of solutions to Equation (9.2.35) is a union of solution sets of the following equations

$$y = 0, \tag{9.2.39}$$

$$y = \left(-1 \pm \sqrt{q}\right)/b, \tag{9.2.40}$$

$$y_z = \left(y^2 + \frac{1}{b}y - \frac{q}{2b^2} - \sqrt{\frac{\sigma^2}{2ab^3}\left(\frac{q}{4b} - y\right)}\right)\frac{1}{y}, \ y \neq 0, \tag{9.2.41}$$

$$y_z = \left(y^2 + \frac{1}{b}y - \frac{q}{2b^2} + \sqrt{\frac{q}{b^3}\left(\frac{q}{4b} - y\right)}\right)\frac{1}{y}, \ y \neq 0, \tag{9.2.42}$$

where one of the solutions (9.2.40) is an exceptional solution (9.2.38) with $q = 4$. We denote the right hand side of equations (9.2.41), (9.2.42) as $f(y)$. The Lipschitz condition for equations of a type $y_z = f(y)$ is satisfied in all points where the derivative $\frac{\partial f}{\partial y}$ exists and is bounded. It is easy to see that this condition will not be satisfied by

$$y = 0, \quad y = \frac{q}{4b}, \quad y = \infty. \tag{9.2.43}$$

It means that on the lines (9.2.43) the uniqueness of the solutions of the equations (9.2.41), (9.2.42) can be lost. We study in detail the behavior of solutions in the neighborhood of the lines (9.2.43). For this purpose we look at the Equation (9.2.35) from another point of view. If we assume now that z, y, y_z are complex variables and denote

$$y(z) = \zeta, \quad y_z(z) = w, \quad \zeta, w \in \mathbb{C}, \tag{9.2.44}$$

then the Equation (9.2.35) takes the form

$$F(\zeta, w) = \zeta w^2 - 2\left(\zeta^2 + \frac{1}{b}\zeta - \frac{q}{2b^2}\right)w + \left(\zeta^2 + \frac{2}{b}\zeta + \frac{1-q}{b^2}\right)\zeta = 0, \tag{9.2.45}$$

where $b \neq 0$. The Equation (9.2.45) is an algebraic relation in $\mathbb{C}^2$ and defines a plane curve in this space. The polynomial $F(\zeta, w)$ is an irreducible polynomial if at all roots $w_r(z)$ of $F(\zeta, w_r)$ either the partial derivative $F_\zeta(\zeta, w_r)$ or $F_w(\zeta, w_r)$ are not equal to zero. It is easy to prove that the polynomial (9.2.45) is irreducible.

We can treat our Equation (9.2.45) as an algebraic relation which defines a Riemann surface $\Gamma \ : \ F(\zeta, w) = 0$ of $w = w(\zeta)$ as a compact manifold over the ζ-sphere. The function $w(\zeta)$ is uniquely analytically extended over the Riemann surface Γ of two sheets over the $\zeta-$sphere. We find all singular or branch points of $w(\zeta)$ if we study the roots of the first coefficient of the polynomial $F(\zeta, w)$, the common roots of equations

$$F(\zeta, w) = 0, \quad F_w(\zeta, w) = 0, \quad \zeta, w \in \mathbb{C} \cup \infty. \tag{9.2.46}$$

and the point $\zeta = \infty$. The set of singular or branch points consists of

$$\zeta_1 = 0, \quad \zeta_2 = \frac{q}{4b}, \quad \zeta_3 = \infty. \tag{9.2.47}$$

As expected we got the same set of points as in the real case (9.2.43) by the study of the Lipschitz condition but now the behavior of solutions at the points is more visible.

The points ζ_2, ζ_3 are the branch points at which two sheets of Γ are glued to each other. Let us remark that

$$w(\zeta_2) = \frac{1}{b}(q-4) + t\frac{1}{4\sqrt{-bq}} + \cdots, \quad t^2 = \zeta - \frac{q}{4b}, \tag{9.2.48}$$

where t is a local parameter in the neighborhood of ζ_2. For the special value of $q = 4$ the value $w(\zeta_2)$ is equal to zero.

At the point $\zeta_3 = \infty$ we have

$$w(\zeta) = \frac{1}{t^2} + \frac{1}{b} + t\sqrt{\frac{-q}{4b^3}}, \quad t^2 = \frac{1}{\zeta}, \quad \zeta \to \infty,$$

where t is a local parameter in the neighborhood of ζ_3. At the point $\zeta_1 = 0$ the function $w(\zeta)$ has the following behavior

$$w(\zeta) \sim -\frac{q}{b^2}\frac{1}{\zeta}, \quad \zeta \to \zeta_1 = 0, \qquad \text{on the principal sheet,} \tag{9.2.49}$$

$$w(\zeta) \sim (1-q)\,\zeta, \quad \zeta \to \zeta_1 = 0, \ q \neq 1, \qquad \text{on the second sheet,} \tag{9.2.50}$$

$$w(\zeta) \sim -2b^2\zeta^2, \quad \zeta \to \zeta_1 = 0, \ q = 1, \qquad \text{on the second sheet.} \tag{9.2.51}$$

Any solution $w(\zeta)$ to an irreducible algebraic Equation (9.2.45) is meromorphic on this compact Riemann surface Γ of the genus 0 and has a pole of the order one correspondingly (9.2.49) over the point $\zeta_1 = 0$ and the pole of the second order over $\zeta_3 = \infty$. It means also that the meromorphic function $w(\zeta)$ cannot be defined on a manifold of less than 2 sheets over the ζ sphere.

The solution of the differential equations (9.2.41) and (9.2.42) from this point of view is equivalent to the integration of a differential of a type $\frac{d\zeta}{w(\zeta)}$ on Γ and then the solution of the Abel's inverse problem of a degenerated type

$$\int \frac{d\zeta}{w(\zeta)} = z + \text{const.} \tag{9.2.52}$$

The integration can be done very easily because we can introduce a uniforming parameter on the Riemann surface Γ and represent the integral (9.2.52) in terms of rational functions merged possibly with logarithmic terms.

To do this we introduce a new variable (our uniforming parameter p)

$$\zeta = \frac{q(1-p^2)}{4b}, \tag{9.2.53}$$

$$w = \frac{(1-p)(q(1+p)^2-4)}{4b(p+1)}. \tag{9.2.54}$$

Then the equations (9.2.41) and (9.2.42) take the form

$$2q\int \frac{p(p+1)\mathrm{d}p}{(p-1)(q(p+1)^2-4)} = z + \text{const.}, \tag{9.2.55}$$

$$2q\int \frac{p(p-1)\mathrm{d}p}{(p+1)(q(p-1)^2-4)} = z + \text{const.} \tag{9.2.56}$$

The integration procedure of Equation (9.2.55) gives rise to the following relations

$$\begin{aligned} &2q\log(p-1) + (q-\sqrt{q}-2)\log((p+1)\sqrt{q}-2) \\ &+(q+\sqrt{q}-2)\log((p+1)\sqrt{q}+2) = 2(q-1)z + c, \quad q \neq 1, q > 0, \end{aligned} \tag{9.2.57}$$

$$\frac{1}{1-p} + \frac{1}{4}\log\frac{(p+3)^3}{(p-1)^5} = z + c, \quad q = 1. \tag{9.2.58}$$

$$\begin{aligned} &2\sqrt{(-q)}\arctan((p+1)\sqrt{(-q)}/2) - 2q\log(p-1) \\ &+(2-q)\log(4-q(p+1)^2) = 2(1-q)z + c, \quad q < 0, \end{aligned} \tag{9.2.59}$$

where c is an arbitrary constant. The Equation (9.2.56) leads to

$$\begin{aligned} &2q\log(p+1) + (q+\sqrt{q}-2)\log((p-1)\sqrt{q}-2) \\ &+(q-\sqrt{q}-2)\log((p-1)\sqrt{q}+2) = 2(q-1)z + c, \quad q \neq 1, q > 0, \end{aligned} \tag{9.2.60}$$

$$\frac{1}{p+1} + \frac{1}{4}\log\frac{(p-3)^3}{(p+1)^5} = z + c, \quad q = 1. \tag{9.2.61}$$

$$\begin{aligned} &-2\sqrt{(-q)}\arctan((p-1)\sqrt{(-q)}/2) - 2q\log(1+p) \\ &+(2-q)\log(4-q(p-1)^2) = 2(1-q)z + c, \quad q < 0. \end{aligned} \tag{9.2.62}$$

where c is an arbitrary constant.

The relations (9.2.57)-(9.2.62) are the first order ordinary differential equations and because of the substitutions (9.2.44) and (9.2.33) we have

$$p = \sqrt{1 - \frac{4b}{q}v_z}. \tag{9.2.63}$$

All these results are formulated in the following theorem.

Theorem 9.2.3 *The Equation (9.2.31) for arbitrary values of the parameters $q, b \neq 0$ can be reduced to the set of first order differential equations which consists of the equations*

$$v_z = 0, \quad v_z = (-1 \pm \sqrt{q})/b \tag{9.2.64}$$

and equations (9.2.57)-(9.2.62). The complete set of solutions of the Equation (9.2.31) coincides with the union of the solutions to these equations.

To solve equations (9.2.57)-(9.2.62) exactly we should first invert these formulas in order to obtain an exact representation of p as a function of z. If an exact formula for the function $p = p(z)$ is found we can use the substitution (9.2.63) to obtain an explicit ordinary differential equation of the type $v_z(z) = f(z)$ or another suitable type and integrate this equation if it is possible.

But even in the first step we would not be able to do this for an arbitrary value of the parameter q. It means we have just implicit representations for the solutions to the Equation (9.2.31) as solutions to the implicit first order differential equations (9.2.57)-(9.2.62).

Exact invariant solutions in the case of a fixed relation between variables S and t

For a special value of the parameter q we can invert the equations (9.2.57) and (9.2.60). Let us take $q = 4$, i.e., the relation between variables S, t is fixed in the form

$$z = \log S + \frac{\sigma^2}{8} t. \tag{9.2.65}$$

In this case the Equation (9.2.57) takes the form

$$(p-1)^2(p+2) = 2\ c\ \exp(3z/2) \tag{9.2.66}$$

and correspondingly the Equation (9.2.60) the form

$$(p+1)^2(p-2) = 2\ c\ \exp(3z/2), \tag{9.2.67}$$

where c is an arbitrary constant. It is easy to see that the equations (9.2.66) and (9.2.67) are connected by a transformation

$$p \to -p, \quad c \to -c. \tag{9.2.68}$$

This symmetry arises from the symmetry of the underlying Riemann surface Γ (9.2.45) and corresponds to a change of the sheets on Γ.

Theorem 9.2.4 *The second order differential equation*

$$v_z + 4\frac{v_{zz} - v_z}{(1 - b(v_{zz} - v_z))^2} = 0, \tag{9.2.69}$$

is exactly integrable for an arbitrary value of the parameter b. The complete set of the solutions for $b \neq 0$ is given by the union of the solutions (9.2.74), (9.2.75)-(9.2.78) and the solutions

$$v(z) = d, \quad v(z) = -\frac{3}{b}z + d, \quad v(z) = \frac{1}{b}z + d, \tag{9.2.70}$$

where d is an arbitrary constant. The last solution in (9.2.70) corresponds to the exceptional solution of the Equation (9.2.35).

For $b = 0$ Equation (9.2.69) is linear and its solutions are given by $v(z) = d_1 + d_2 \exp(3z/4)$, where d_1, d_2 are arbitrary constants.

Proof of Theorem 9.2.1

Because of the symmetry (9.2.68) it is sufficient to study either the equations (9.2.66) or (9.2.67) for $c \in \mathbb{R}$ or both these equations for $c > 0$. The value $c = 0$ can be excluded because it complies with the constant value of $p(z)$ and correspondingly constant value of $v_z(z)$, but all such cases were studied before and the solutions were given (9.2.70).

We study the Equation (9.2.67) in the case $c \in \mathbb{R} \setminus \{0\}$ and in this way obtain the complete class of exact solutions for equations (9.2.66)-(9.2.67).

Equation (9.2.67) for $c > 0$ has one real root only. It leads to an ordinary differential equation of the form

$$\begin{aligned} v_z(z) &= -\frac{1}{b}\left(1 + \left(1 + c\,e^{\frac{3z}{2}} + \sqrt{2\,c\,e^{\frac{3z}{2}} + c^2\,e^{3z}}\right)^{-\frac{2}{3}}\right. \\ &\quad \left. + \left(1 + c\,e^{\frac{3z}{2}} + \sqrt{2\,c\,e^{\frac{3z}{2}} + c^2\,e^{3z}}\right)^{\frac{2}{3}}\right), \quad c > 0. \end{aligned} \tag{9.2.71}$$

Equation (9.2.71) can be exactly integrated if we use an Euler substitution and introduce a new independent variable

$$\tau = 2\left(1 + c\,e^{\frac{3z}{2}} + \sqrt{2\,c\,e^{\frac{3z}{2}} + c^2\,e^{3z}}\right). \tag{9.2.72}$$

The corresponding solution is given by

$$
\begin{aligned}
v_r(z) = -\frac{1}{b}\Bigg(& \left(1 + c e^{\frac{3z}{2}} + \sqrt{2 c e^{\frac{3z}{2}} + c^2 e^{3z}}\right)^{-\frac{2}{3}} \\
& + \left(1 + c e^{\frac{3z}{2}} + \sqrt{2 c e^{\frac{3z}{2}} + c^2 e^{3z}}\right)^{\frac{2}{3}} \\
& + 2 \log\Bigg(\left(1 + c e^{\frac{3z}{2}} + \sqrt{2 c e^{\frac{3z}{2}} + c^2 e^{3z}}\right)^{-\frac{1}{3}} \\
& + \left(1 + c e^{\frac{3z!}{2}} + \sqrt{2 c e^{\frac{3z}{2}} + c^2 e^{3z}}\right)^{\frac{1}{3}} - 2\Bigg)\Bigg) + d,
\end{aligned} \tag{9.2.73}
$$

where $d \in \mathbb{R}$ is an arbitrary constant.

If on the right hand side of the Equation (9.2.67) the parameter c satisfies the inequality $c < 0$ and the variable z is chosen in the region

$$
z \in \left(-\infty, -\frac{4}{3} \ln |c|\right) \tag{9.2.74}
$$

then the equation on p possesses maximum three real roots.

These three roots of cubic equation (9.2.67) correspond to three differential equations of the type $v_z = (1 - p^2(z))/b$. The equations can be solved exactly and we find correspondingly three solutions $v_i(z),\ i = 1, 2, 3$.

The first solution is given by the expression

$$
\begin{aligned}
v_1(z) = & \frac{z}{b} - \frac{2}{b} \cos\left(\frac{2}{3} \arccos\left(1 - |c|\, e^{\frac{3z}{2}}\right)\right) \\
& - \frac{4}{3b} \log\left(1 + 2 \cos\left(\frac{1}{3} \arccos\left(1 - |c|\, e^{\frac{3z}{2}}\right)\right)\right) \\
& - \frac{16}{3b} \log\left(\sin\left(\frac{1}{6} \arccos\left(1 - |c|\, e^{\frac{3z}{2}}\right)\right)\right) + d,
\end{aligned} \tag{9.2.75}
$$

where $d \in \mathbb{R}$ is an arbitrary constant. The second solution is given by the formula

$$
\begin{aligned}
v_2(z) = & \frac{z}{b} - \frac{2}{b} \cos\left(\frac{2}{3}\pi + \frac{2}{3} \arccos\left(-1 + |c|\, e^{\frac{3z}{2}}\right)\right) \\
& - \frac{4}{3b} \log\left(1 + 2 \cos\left(\frac{1}{3}\pi + \frac{1}{3} \arccos\left(-1 + |c|\, e^{\frac{3z}{2}}\right)\right)\right) \\
& - \frac{16}{3b} \log\left(\sin\left(\frac{1}{6}\pi + \frac{1}{6} \arccos\left(-1 + |c|\, e^{\frac{3z}{2}}\right)\right)\right) + d,
\end{aligned} \tag{9.2.76}
$$

where $d \in \mathbb{R}$ is an arbitrary constant. The first and second solutions are defined up to the point $z = -\frac{4}{3}\ln|c|$ where they coincide (see Figure 9.1).

The third solution for $z < -\frac{4}{3}\ln|c|$ is given by the formula

$$\begin{aligned} v_{3,1}(z) = \frac{z}{b} - \frac{2}{b}\cos\left(\frac{2}{3}\arccos\left(-1 + |c|\, e^{\frac{3z}{2}}\right)\right) \\ - \frac{4}{3b}\log\left(-1 + 2\cos\left(\frac{1}{3}\arccos\left(-1 + |c|\, e^{\frac{3z}{2}}\right)\right)\right) \\ - \frac{16}{3b}\log\left(\cos\left(\frac{1}{6}\arccos\left(-1 + |c|\, e^{\frac{3z}{2}}\right)\right)\right) + d, \end{aligned} \quad (9.2.77)$$

where $d \in \mathbb{R}$ is an arbitrary constant. In the case $z > -\frac{4}{3}\ln|c|$ the polynomial (9.2.67) has one real root and the corresponding solution can be represented by the formula

$$\begin{aligned} v_{3,2}(z) = \frac{z}{b} - \frac{2}{b}\cosh\left(\frac{2}{3}\operatorname{arccosh}\left(-1 + |c|\, e^{\frac{3z}{2}}\right)\right) \\ - \frac{16}{3b}\log\left(\cosh\left(\frac{1}{6}\operatorname{arccosh}\left(-1 + |c|\, e^{\frac{3z}{2}}\right)\right)\right) \\ - \frac{4}{3b}\log\left(-1 + 2\cosh\left(\frac{1}{3}\operatorname{arccosh}\left(-1 + |c|\, e^{\frac{3z}{2}}\right)\right)\right) + d. \end{aligned} \quad (9.2.78)$$

The third solution is represented by formulas $v_{3,2}(z)$ and $v_{3,1}(z)$ for different values of the variable z. □

One of the sets of solutions (9.2.74), (9.2.75) -(9.2.78) for fixed parameters b, c, d is represented in Figure 9.1. The solution $v_r(z)$ (9.2.74) and the third solution given by both (9.2.77) and (9.2.78) are defined for any values of z. The solutions $v_1(z)$ and $v_2(z)$ cannot be continued after the point $z = -\frac{4}{3}\ln|c|$ where they coincide.

If we keep in mind that $z = \log S + \frac{\sigma^2}{8}t$ and $v(z) = u(S,t)$ we can represent exact invariant solutions of the Equation (9.2.23). The solution

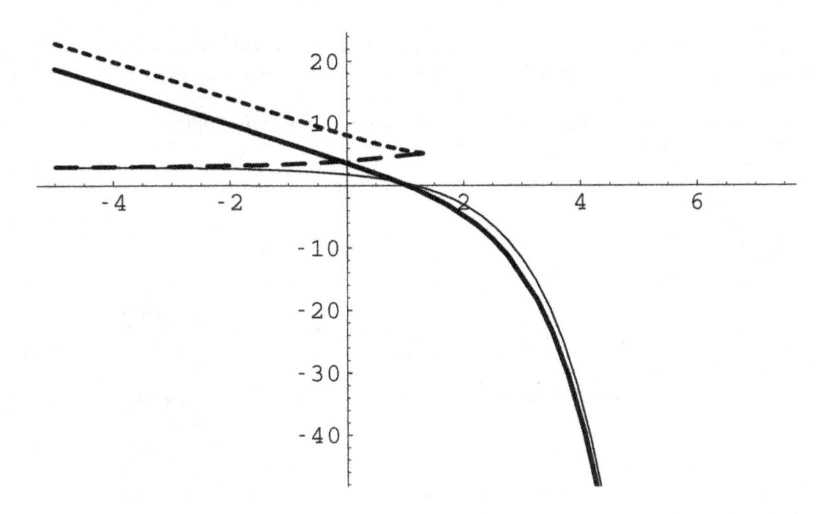

Figure 9.1: Plot of the solution $v_r(z)$ (9.2.74) (thick solid line), $v_1(z)$ (9.2.75) (short dashed line), $v_2(z)$ (9.2.76) (long dashed line) and the third solution $v_{3,1}(z), v_{3,2}(z)$ (9.2.77), (9.2.78), which is represented by the thin solid line. The parameters take values $|c| = 0.5, q = 4, d = 0, b = 1$ and the variable $z \in (-5, 4.5)$.

(9.2.74) provides an invariant solution $u_r(S,t)$ in the form

$$
\begin{aligned}
u_r(S,t) = & -\frac{1}{\omega\rho}\left(1 + c\,S^{\frac{3}{2}}e^{\frac{3\sigma^2}{16}t} + \sqrt{2\,c\,S^{\frac{3}{2}}e^{\frac{3\sigma^2}{16}t} + c^2\,S^3 e^{\frac{3\sigma^2}{8}t}}\right)^{-\frac{2}{3}} \\
& -\frac{1}{\omega\rho}\left(1 + c\,S^{\frac{3}{2}}e^{\frac{3\sigma^2}{16}t} + \sqrt{2\,c\,S^{\frac{3}{2}}e^{\frac{3\sigma^2}{16}t} + c^2\,S^3 e^{\frac{3\sigma^2}{8}t}}\right)^{\frac{2}{3}} \qquad (9.2.79) \\
& -\frac{2}{\omega\rho}\log\left(\left(1 + c\,S^{\frac{3}{2}}e^{\frac{3\sigma^2}{16}t} + \sqrt{2\,c\,S^{\frac{3}{2}}e^{\frac{3\sigma^2}{16}t} + c^2\,S^3 e^{\frac{3\sigma^2}{8}t}}\right)^{-\frac{1}{3}}\right. \\
& \left. +\left(1 + c\,S^{\frac{3}{2}}e^{\frac{3\sigma^2}{16}t} + \sqrt{2\,c\,S^{\frac{3}{2}}e^{\frac{3\sigma^2}{16}t} + c^2\,S^3 e^{\frac{3\sigma^2}{8}t}}\right)^{\frac{1}{3}} - 2\right) + d
\end{aligned}
$$

where $d \in \mathbb{R}$, $c > 0$.

In the case $c < 0$ we can obtain three solutions correspondingly if

$$
0 < S \leq |c|^{-\frac{4}{3}} \exp\left(-\frac{\sigma^2}{8}t\right). \qquad (9.2.80)
$$

The first solution is represented by

$$\begin{aligned} u_1(S,t) &= \frac{1}{\omega\rho}\left(\log S + \frac{\sigma^2}{8}t\right) \\ &-\frac{2}{\omega\rho}\cos\left(\frac{2}{3}\arccos\left(1-|c|\,S^{\frac{3}{2}}e^{\frac{3\sigma^2}{16}t}\right)\right) \\ &-\frac{4}{3\omega\rho}\log\left(1+2\cos\left(\frac{1}{3}\arccos\left(1-|c|\,S^{\frac{3}{2}}e^{\frac{3\sigma^2}{16}t}\right)\right)\right) \\ &-\frac{16}{3\omega\rho}\log\left(\sin\left(\frac{1}{6}\arccos\left(1-|c|\,S^{\frac{3}{2}}e^{\frac{3\sigma^2}{16}t}\right)\right)\right) \\ &+d, \end{aligned} \tag{9.2.81}$$

where $d \in \mathbb{R}$, $c < 0$. The second solution is given by the formula

$$\begin{aligned} u_2(S,t) &= \frac{1}{\omega\rho}\left(\log S + \frac{\sigma^2}{8}t\right) \\ &-\frac{2}{\omega\rho}\cos\left(\frac{2}{3}\pi+\frac{2}{3}\arccos\left(-1+|c|\,S^{\frac{3}{2}}e^{\frac{3\sigma^2}{16}t}\right)\right) \\ &-\frac{4}{3\omega\rho}\log\left(1+2\cos\left(\frac{1}{3}\pi+\frac{1}{3}\arccos\left(-1+|c|\,S^{\frac{3}{2}}e^{\frac{3\sigma^2}{16}t}\right)\right)\right) \\ &-\frac{16}{3\omega\rho}\log\left(\sin\left(\frac{1}{6}\pi+\frac{1}{6}\arccos\left(-1+|c|\,S^{\frac{3}{2}}e^{\frac{3\sigma^2}{16}t}\right)\right)\right) \\ &+d, \end{aligned} \tag{9.2.82}$$

where $d \in \mathbb{R}$, $c < 0$. The first and second solutions are defined for the variables under the conditions (9.2.80). They coincide along the curve

$$S = |c|^{-4/3}\exp\left(-\frac{\sigma^2}{8}t\right)$$

and cannot be continued further.

The third solution is defined by

$$\begin{aligned} u_{3,1}(S,t) = &\frac{1}{\omega\rho}\left(\log S + \frac{\sigma^2}{8}t\right) - \frac{2}{\omega\rho}\cos\left(\frac{2}{3}\arccos\left(-1+|c|\,S^{\frac{3}{2}}e^{\frac{3\sigma^2}{16}t}\right)\right) \\ &-\frac{4}{3\omega\rho}\log\left(-1+2\cos\left(\frac{1}{3}\arccos\left(-1+|c|\,S^{\frac{3}{2}}e^{\frac{3\sigma^2}{16}t}\right)\right)\right) \\ &-\frac{16}{3\omega\rho}\log\left(\cos\left(\frac{1}{6}\arccos\left(-1+|c|\,S^{\frac{3}{2}}e^{\frac{3\sigma^2}{16}t}\right)\right)\right) + d, \end{aligned} \tag{9.2.83}$$

where $d \in \mathbb{R}$ and S, t satisfied the condition (9.2.80).

In the case $\log S + \frac{\sigma^2}{8}t > -\frac{4}{3}\ln|c|$ the third solution can be represented by the formula

$$
\begin{aligned}
u_{3,2}(S,t) = {} & \frac{1}{\omega\rho}\left(\log S + \frac{\sigma^2}{8}t\right) - \frac{2}{\omega\rho}\cosh\left(\frac{2}{3}\text{arccosh}\left(-1+|c|\,S^{\frac{3}{2}}e^{\frac{3\sigma^2}{16}t}\right)\right) \\
& - \frac{16}{3\,\omega\rho}\log\left(\cosh\left(\frac{1}{6}\text{arccosh}\left(-1+|c|\,S^{\frac{3}{2}}e^{\frac{3\sigma^2}{16}t}\right)\right)\right) \qquad (9.2.84) \\
& - \frac{4}{3\,\omega\rho}\log\left(-1+2\cosh\left(\frac{1}{3}\text{arccosh}\left(-1+|c|\,S^{\frac{3}{2}}e^{\frac{3\sigma^2}{16}t}\right)\right)\right) + d.
\end{aligned}
$$

The solution $u_r(S,t)$ (9.2.79) and the third solution given by $u_{3,1}$, $u_{3,2}$ (9.2.83),(9.2.84) are defined for all values of variables t and $S > 0$. They have a common intersection curve of a type $S = \text{const.}\exp(-\frac{\sigma^2}{8}t)$. The typical behavior of all these invariant solutions is shown on Figure 9.2.

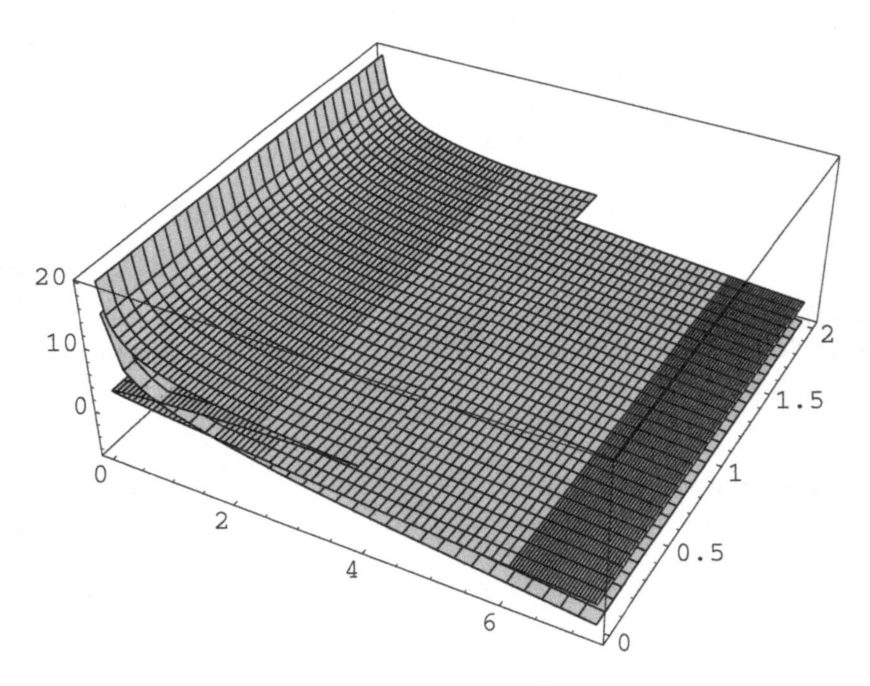

Figure 9.2: Plot of the solutions $u_r(S,t)$, $u_1(S,t)$, $u_2(S,t)$, $u_{3,1}(S,t)$, $u_{3,2}(S,t)$ for the parameters $|c| = 0.25$, $q = 4$, $b = 1.0$, $d = 0$. The variables t, S lie in intervals $S \in (0,9)$ and $t \in [0, 2.0]$. All invariant solutions change slowly in t-direction.

It should be noted that because of the symmetry properties (see Theorem 9.2.2) any solution remains a solution if we add a linear function of S to it,

$$u(S,t) \to u(S,t) + d_1 S + d_2, \tag{9.2.85}$$

with arbitrary constants d_1, d_2.

It means we have two additional constants to model boundary and terminal conditions.

We first study the non-trivial solutions, i.e., $u(S,t) \neq d_1 S + d_2$.

Previous results can be summed up in the following theorem describing the set of non-trivial invariant solutions of the Equation (9.2.11).

Theorem 9.2.5

1. *The Equation (9.2.11) possesses non-trivial invariant solutions only for the special form of the function $\lambda(S)$ given by (9.2.13).*

2. *In the case (9.2.13) the invariant solutions of the Equation (9.2.11) are defined by ordinary differential equations (9.2.30). In special cases $k = 0, 1$ equations (9.2.30) are of an autonomous type.*

3. *If $\lambda(S) = \omega S$, i.e. $k = 1$, then the invariant solutions of the Equation (9.2.11) can be defined by the set of first order ordinary differential equations (9.2.57)–(9.2.62) and Equation (9.2.64).*
   If additionally the parameter $q = 4$, or equivalent in the first invariant (9.2.21) we chose $a = \sigma^2/8$ then the complete set of invariant solutions (9.2.11) can be found exactly. This set of invariant solutions is given by formulas (9.2.79)–(9.2.84) and by solutions

   $$u(S,t) = d, \;\; u(S,t) = -3/b\,(\log S + \sigma^2 t/8), \;\; u(S,t) = 1/b\,(\log S + \sigma^2 t/8),$$

   where d is an arbitrary constant. This set of invariant solutions is unique up to the transformations of the symmetry group G_Δ given by Theorem 9.2.2

Properties of the invariant solutions

All solutions (9.2.79)-(9.2.84) have the form $u(S,t) = w(S,t)/(\omega\rho)$, where $w(S,t)$ is a smooth function of S, t. It means that the function $w(S,t)$ solves the Equation (9.2.23) with $b = \omega\rho = 1$, i.e.

$$w_t + \frac{\sigma^2 S^2}{2} \frac{w_{SS}}{(1 - S^2 w_{SS})^2} = 0. \tag{9.2.86}$$

In other words, if we find any solution to the equation (9.2.86) for any fixed boundary and terminal conditions, we can immediately obtain the corresponding solution of the Equation (9.2.23) if we just divide the function $w(S,t)$ by $b = \omega\rho$ (the solution $u(S,t)$ satisfies the boundary and terminal conditions which we obtain if we just divide the corresponding conditions on the function $w(S,t)$ by b). Therefore any ρ-dependence of a solution of (9.2.23) is trivial. It means as well that if the terminal conditions are fixed $u(S,T) = h(S)$ then the value $u(S,t)$ increases if the value of the parameter ρ increases. This dependence of hedge costs on the position of the large trader on the market is very natural.

If the parameter $\rho \to 0$ then the Equation (9.2.11) and correspondingly the Equation (9.2.23) are reduced to the linear Black-Scholes equation but solutions (9.2.79)-(9.2.84) which we obtained here are completely blown up by $\rho \to 0$ because of the factor $1/b = 1/(\omega\rho)$ in the formulas (9.2.79)-(9.2.84). It means that the solutions $u(S,t)$ (9.2.79), $u_1(S,t)$ (9.2.81), $u_2(S,t)$ (9.2.82), $u_{3,1}(S,t)$ (9.2.83), $u_{3,2}(S,t)$ (9.2.84), have no one counterpart in a linear case.

This phenomena is rather typical for nonlinear partial differential equations with singular perturbations and was described as well in [14], [23], [20] for the invariant solutions of Equation (9.2.23) with $k = 0$.

The families of exact solutions reflect the nonlinearity of this equation in an essential way. If we take a numerical method which was developed for the linear Black-Scholes equation or other types of parabolic equations and test it for a new type of a nonlinear equation we should take, if possible, the solutions which reflect this nonlinearity in the most complete way. The existence of non-trivial explicit solutions allows us to test different numerical methods usually used to calculate hedge-costs of derivatives (see [14]).

We obtain a typical terminal payoff function for the solutions (9.2.79)-(9.2.84) if we just fix $t = T$. All these payoffs are smooth functions. A typical payoff of any derivative is similar to a combination of calls and puts and is a continuous piecewise linear function. The smooth payoffs are more convenient for numerical calculations and usually one replaces the standard payoffs by the corresponding solution to the Black-Scholes model for a very small time interval. If in the case of smooth payoffs the numerical method does not work properly also, there is no hope that it works better in a worse case of piecewise linear functions. The fact that the payoffs are smooth does not affect its behavior anyhow.

In order to illustrate how we can model some typical payoffs we investigate asymptotic properties of the solutions (9.2.79)-(9.2.84) as $S \to 0$ and as $S \to \infty$.

Using the exact formulas for the solutions we retain the first two terms and obtain as $S \to 0$

$$u_1(S,t) \sim -\frac{1}{b}\left(2+\frac{4}{3}\log\left(|c|^2 2^{-2} 3^{-3}\right)+\frac{3}{8}\sigma^2 t+3\log S+\mathcal{O}(S^{3/2})\right), \quad (9.2.87)$$

$$u_2(S,t) \sim \frac{1}{b}\left(1+\frac{1}{3}\log\left(2^8 3^{-6}|c|^{-2}\right)+\frac{2^3\sqrt{2|c|}}{3^{4/3}}e^{\frac{3}{32}\sigma^2 t}S^{3/4}+\mathcal{O}(S^{3/2})\right), \quad (9.2.88)$$

$$u_{3,1}(S,t) \sim \frac{1}{b}\left(1+\frac{1}{3}\log\left(2^8 3^{-6}|c|^{-2}\right)-\frac{2^3\sqrt{2|c|}}{3^{4/3}}e^{\frac{3}{32}\sigma^2 t}S^{3/4}+\mathcal{O}(S^{3/2})\right), \quad (9.2.89)$$

$$u_r(S,t) \sim -\frac{1}{b}\left(2+2\log\left(2\ 3^{-2}\ c\right)+\frac{3}{8}\sigma^2 t+3\log S+\mathcal{O}(S^{5/4})\right). \quad (9.2.90)$$

In Figures 9.3 and 9.4 we present solutions (9.2.79)-(9.2.84) in the case where both additional constants $d_1 = d_2$ in (9.2.85) are equal to zero, i.e. there is no linear background. The parameters are $\sigma = 0.4$, $b = 1$, $|c| = 0.5$, $d_1 = 0$, $d_2 = 0$.

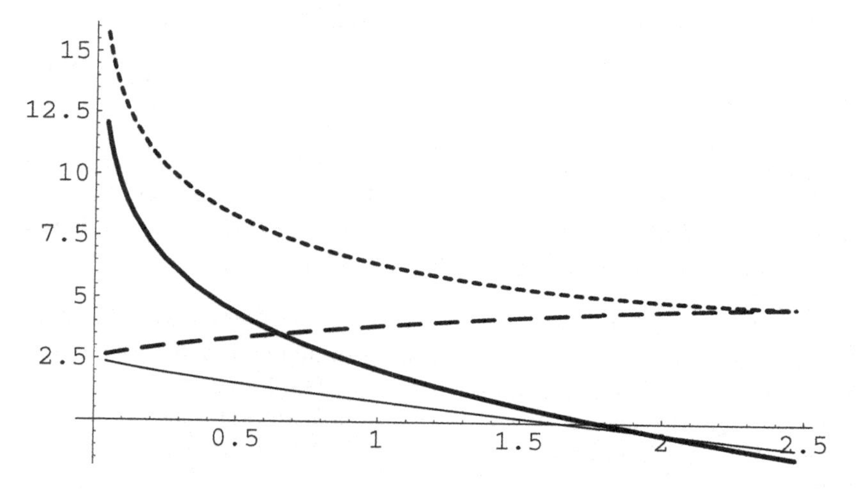

Figure 9.3: The behavior of the solutions $u_r(z)$ (9.2.79) (thick solid line), $u_1(z)$ (9.2.81) (short dashed line), $u_2(z)$ (9.2.82) (long dashed line) and $u_{3,1}(z)$ (9.2.83) (thin solid line) in the neighborhood of $S \sim 0$, for $t = 0$, where S lie in the interval (9.2.80).

If S is large enough we have just two solutions. The asymptotic behavior for both solutions $u_r(S,t)$,(9.2.79), and $u_{3,2}$,(9.2.84), coincides in the main terms as $S \to \infty$ and is given by the formula

$$u_r(S,t),\ u_{3,2}(S,t) \sim -\frac{1}{b}\left((2\,|c|)^{2/3} e^{\frac{3\sigma^2}{8}t} S + \log S + \mathcal{O}(1)\right), \quad S \to \infty. \tag{9.2.91}$$

The main term in the formulas (9.2.87)-(9.2.91) depends on the time and on the constant c.

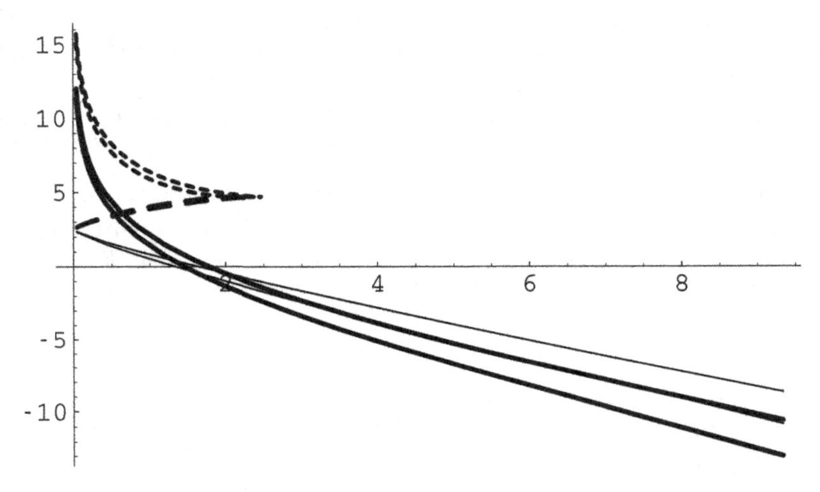

Figure 9.4: The same solutions as in Figure 9.3 but for times $t = 0$ and $t = 10$ and for $S \in (0.04, 9]$. The solutions are $u_r(z)$ (9.2.79) (thick solid line), $u_1(z)$ (9.2.81) (short dashed line), $u_2(z)$ (9.2.82) (long dashed line) and $u_{3,1}(z), u_{3,2}(z)$ (9.2.83), (9.2.84) (thin solid line).

The families of the exact solutions (9.2.79)-(9.2.84) have the following parameters ω, c, d_1, d_2 which can be used to match suitable boundary and terminal conditions with a desired accuracy. The formulas (9.2.87)-(9.2.91) can be useful for these purposes. In Figure 9.5 we represent as an example a long strip payoff and the solution $u_r(S,t)$, (9.2.79), which partly matches this payoff.

If we compare the price of the derivative product in the linear ideal world of the BS model and the price in the presence of the strong feedback effects like it is done in the models (9.2.9) and (9.2.11) we obtain different results. The prices in the presence of the feedback effect are expectedly higher, it means the feedback effect makes derivative products more expensive. In other words there exist two different fair prices on the market - one for the price takers and the other one for large traders, whose trading strategy changes the price of the traded asset.

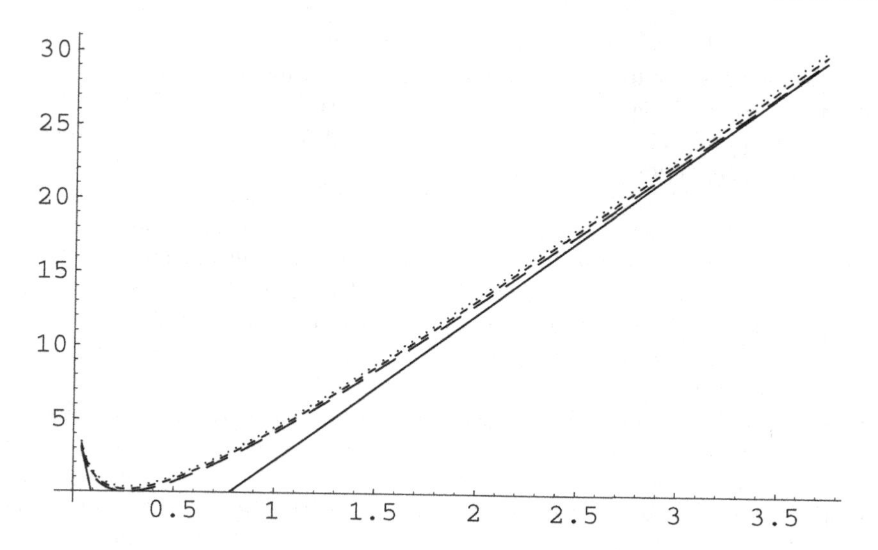

Figure 9.5: Plot of the solution $u_r(S,t)$ (9.2.79) for the parameters $|c| = 0.5$, $b = 1.0$, $\sigma = 0.3$, $d_1 = 11.5$, $d_2 = -9.0$. The variables S, t lie in intervals $S \in (0.04, 3.7)$ and $t = 0$ (dotted line), $t = 5$ (short dashed line), and $t = 10$ (long dashed line). Payoff for a long strip with 60 puts and 10 calls with exercise price 0.2 marked by the thin solid line.

9.2.2 Sircar–Papanicolaou model

Another modeling approach was suggested by Sircar and Papanicolaou in [72]. They proposed to take into account the feedback effect from the dynamic hedging strategies of the large trader on the price of traded asset. They present a class of nonlinear pricing models using the idea of a demand function $\mathcal{U}(\cdot)$ of the reference traders relative to the supply.

They obtain a nonlinear PDE of the following type

$$u_t + \frac{1}{2}\left[\frac{\mathcal{U}^{-1}(1-\rho u_S)\mathcal{U}'(\mathcal{U}^{-1}(1-\rho u_S))}{\mathcal{U}^{-1}(1-\rho u_S)\mathcal{U}'(\mathcal{U}^{-1}(1-\rho u_S)) - \rho S u_{SS}}\right]^2 \sigma^2 S^2 u_{SS} + r(Su_S - u) = 0. \tag{9.2.92}$$

Here as usual t denotes time, S and σ are the price and the volatility of the underlying asset respectively, ρ is a fixed constant liquidity parameter for the large trader, and the parameter r is the risk-free interest rate. The value $u(S,t)$ is the price of the derivative security and depends on the form of the demand function $\mathcal{U}(\cdot)$. The expression $\mathcal{U}^{-1}(\cdot)$ denotes the inverse function correspondingly. Because of the strong monotonicity of the demand function the existence of the inverse function $\mathcal{U}^{-1}(\cdot)$ is guaranteed. Depending on the priorities of the large trader the demand function can take different forms and correspondingly the nonlinear Equation (9.2.92) would have different analytical properties.

There exist different opinions on which of the demand functions is more relevant for really large traders. An unnecessary complicated demand function can make an analytical study of the nonlinear PDE extremely difficult. Because of that in the first steps of the model study one chooses the most simple type of the demand function to understand if a model is relevant to the feedback effect description. Sircar and Papanicolaou studied in [72] mainly the particular model arising from (9.2.92) taking $\mathcal{U}(\cdot)$ linear, i.e. they choose $\mathcal{U}(z) = \beta z,\ \beta > 0$. The authors focused their attention on the numerical solution and discussed the difference between this model and the classical Black-Scholes option pricing theory. In [12] the author studied a more general case of the demand function. Now the demand function of the type $\mathcal{U}(z) = \beta z^{\alpha},\ \alpha, \beta \neq 0$ is incorporated, the linear demand function used before in [72] is included here as a special case. To ensure that the model is self consistent and the demand function belongs to the class of the admitted demand functions we need the condition $\mathcal{U}'(z) = \beta\alpha z^{\alpha-1} > 0$.

In the case $\mathcal{U}(z) = \beta z^{\alpha}$ the model (9.2.92) takes the form

$$\frac{\partial u}{\partial t} + \frac{1}{2}\left[\frac{1 - \rho\frac{\partial u}{\partial S}}{1 - \rho\frac{\partial u}{\partial S} - \frac{\rho}{\alpha} S \frac{\partial^2 u}{\partial S^2}}\right]^2 \sigma^2 S^2 \frac{\partial^2 u}{\partial S^2} + r\left(S\frac{\partial u}{\partial S} - u\right) = 0. \tag{9.2.93}$$

Now the diffusion coefficient in the model Equation (9.2.93) depends on both, u_S and u_{SS} multiplied by the small perturbation parameter ρ. But in comparison to the previous models the diffusion coefficient depends also on the form of the demand function, to be specific on the parameter α which controlled the second derivative in it. The third fixed parameter in this model is the interest rate r. The presence of an additional parameter makes analytical studies of the Sircar–Papanicolaou model more demanding than the models discussed in the sections before.

In [12] the model was investigated with the Lie group method for both cases $r = 0$ and $r \neq 0$. As was to be expected, in the case $r = 0$ there exists a richer symmetry algebra than in the case $r \neq 0$. Two other parameters of the model ρ and α play different roles. The parameter α is not included in the generators of the algebra at all. It means that for any value of this parameter we have the same symmetry algebra. Later on if we introduce invariants and provide reductions then certainly the invariant solutions will depend on α. The parameter ρ has stronger influence on the form of the

Lie algebra generators. It is included explicitly and the form of generators changes if $\rho = 0$ or $\rho \neq 0$.

Here we formulate the main results concerning Lie symmetry algebra for the Sircar–Papanicolaou model. We set that $\rho \neq 0$, because the case $\rho = 0$ coincides with the original Black-Scholes model which was studied before.

Theorem 9.2.6 [12] *The Sircar–Papanicolaou model (9.2.93) for $r = 0$ admits a four-dimensional Lie algebra L_4 with the following infinitesimal generators*

$$\mathbf{U}_1 = -\frac{S}{2}\frac{\partial}{\partial S} + \left(\frac{S}{2\rho} - u\right)\frac{\partial}{\partial u}, \qquad \mathbf{U}_2 = \frac{\partial}{\partial u},$$
$$\mathbf{U}_3 = \frac{\partial}{\partial t}, \qquad \mathbf{U}_4 = \rho S\frac{\partial}{\partial S} + S\frac{\partial}{\partial u}. \tag{9.2.94}$$

The commutator relations are

$$[\mathbf{U}_1, \mathbf{U}_3] = [\mathbf{U}_1, \mathbf{U}_4] = [\mathbf{U}_2, \mathbf{U}_3] = [\mathbf{U}_2, \mathbf{U}_4] = 0,$$
$$[\mathbf{U}_3, \mathbf{U}_4] = 0, \ [\mathbf{U}_1, \mathbf{U}_2] = \mathbf{U}_2.$$

In the case $r \neq 0$ Equation (9.2.93) admits a three-dimensional Lie algebra L_3 with the following operators

$$\mathbf{U}_1 = \frac{\partial}{\partial t}, \quad \mathbf{U}_2 = (S - \rho)\frac{\partial}{\partial u}, \quad \mathbf{U}_3 = S\frac{\partial}{\partial S} + u\frac{\partial}{\partial u}. \tag{9.2.95}$$

The algebra L_3 is Abelian.

The Lie algebra L_4 (9.2.94) has a two–dimensional subalgebra $L_2 = \langle \mathbf{U}_1, \mathbf{U}_2 \rangle$ spanned by the generators $\mathbf{U}_1, \mathbf{U}_2$. The algebra L_4 is a decomposable Lie algebra and can be represented as a semi-direct sum $L_4 = L_2 \bigoplus \mathbf{U}_3 \bigoplus \mathbf{U}_4$. Usually one uses the symmetry algebra to get invariant solutions of the given equation. One looks for invariants and reduces the original PDE to some ODEs which give the invariant solutions of the studied PDE. Such a set of invariant solutions is not unique. After some simple transformations which for instance belong to the symmetry group of the studied equation we get other invariant solutions, which yet look very different. To avoid repeated calculations and study of equivalent solutions and yet not lose important invariant solutions which cannot be obtained from previous ones with simple transformations one introduces an optimal system of non-equivalent solutions. To get the optimal system of invariant solutions one needs an optimal system of subalgebras for the obtained symmetry algebra. The optimal system of subalgebras for L_3, L_4 in the general case was provided in Section 6.3, and can be found in [61]. In [12] in the cases $r = 0$ and $r \neq 0$ the optimal systems of subalgebras for respectively L_4 and L_3 were studied. We formulate here the results.

Theorem 9.2.7 *An optimal system of one–dimensional subalgebras for L_3 (9.2.95) consists of*

$$h_1 = \langle \mathbf{U}_3 \rangle, \quad h_2 = \langle \mathbf{U}_2 + x\mathbf{U}_3 \rangle, \quad h_3 = \langle \mathbf{U}_1 + x\mathbf{U}_2 + y\mathbf{U}_3 \rangle,$$

with parameters $x, y \in \mathbb{R}$. The optimal systems of one-, two- and three–dimensional subalgebras for L_4 (9.2.94) are listed in Table 9.1.

Table 9.1: [61] The optimal system of subalgebras h_i of the algebra L_4 with parameters are $x \in \mathbb{R}$, $\epsilon = \pm 1$, $\phi \in [0, \pi]$.

Dimen-sion	Subalgebras
1	$h_1 = \langle \mathbf{U}_2 \rangle$, $h_2 = \langle \mathbf{U}_3 \cos(\phi) + \mathbf{U}_4 \sin(\phi) \rangle$, $h_3 = \langle \mathbf{U}_1 + x(\mathbf{U}_3 \cos(\phi) + \mathbf{U}_4 \sin(\phi)) \rangle$, $h_4 = \langle \mathbf{U}_2 + \epsilon(\mathbf{U}_3 \cos(\phi) + \mathbf{U}_4 \sin(\phi)) \rangle$
2	$h_5 = \langle \mathbf{U}_1 + x(\mathbf{U}_3 \cos(\phi) + \mathbf{U}_4 \sin(\phi)), \mathbf{U}_2 \rangle$, $h_6 = \langle \mathbf{U}_3, \mathbf{U}_4 \rangle$, $h_7 = \langle \mathbf{U}_1 + x(\mathbf{U}_3 \cos(\phi) + \mathbf{U}_4 \sin(\phi)), \mathbf{U}_3 \sin(\phi) - \mathbf{U}_4 \cos(\phi) \rangle$, $h_8 = \langle \mathbf{U}_2 + \epsilon(\mathbf{U}_3 \cos(\phi) + \mathbf{U}_4 \sin(\phi)), \mathbf{U}_3 \sin(\phi) - \mathbf{U}_4 \cos(\phi) \rangle$, $h_9 = \langle \mathbf{U}_2, \mathbf{U}_3 \sin(\phi) - \mathbf{U}_4 \cos(\phi) \rangle$
3	$h_{10} = \langle \mathbf{U}_1, \mathbf{U}_3, \mathbf{U}_4 \rangle$, $h_{11} = \langle \mathbf{U}_2, \mathbf{U}_3, \mathbf{U}_4 \rangle$, $h_{12} = \langle \mathbf{U}_1 + x(\mathbf{U}_3 \cos(\phi) + \mathbf{U}_4 \sin(\phi)), \mathbf{U}_3 \sin(\phi) - \mathbf{U}_4 \cos(\phi), \mathbf{U}_2 \rangle$

Using this result we can easily provide an overview of the complete set of classes of non-equivalent invariant solutions. The corresponding reductions allowed us to get the complete set of invariant solutions for PDE. This set of invariant solutions is unique. Certainly each representative of one of the classes can be changed with an admitted transformation of the corresponding symmetry group, but with such transformations we cannot change the class. Using the optimal systems of subalgebras the author [12] provides the complete set of non-equivalent reductions of (9.2.92) to different ODEs. In most cases the reduced ODEs are solved explicitly. In both cases for $r = 0$ and $r \neq 0$ three different classes of non-trivial invariant solutions are described. We will not repeat complete descriptions of these classes of solutions, we will provide just one example of an invariant solution.

Let us take the case $r = 0$ and as new invariant variables let us introduce

$$z = S \exp(-t\rho\delta), \quad w = u - \frac{1}{\rho} S, \quad \delta \in \mathbb{R}. \tag{9.2.96}$$

Then the PDE (9.2.93) will be reduced to the ODE of the form

$$-\rho\delta z w_z + \frac{1}{2}\sigma^2 w_{zz} z^2 \left(\frac{\alpha w_z}{\alpha w_z + z w_{zz}} \right)^2 = 0. \tag{9.2.97}$$

This equation has two trivial solutions and one non-trivial of the form (in the original variables t, S, u)

$$u(S,t) = c_1 S^p \exp(-p\rho\delta\, t), \quad p = 1+\alpha-a\pm\sqrt{a(a-2\alpha)}, \; a = \frac{\sigma^2}{4\rho\delta}. \quad (9.2.98)$$

For each set of parameters $c_1, \delta, \rho, \sigma, \alpha$ we obtain two different solutions $u(S,t)$ because of the definition of the value p in (9.2.98). One set of such solutions is presented in Figure 9.6. From the financial point of view these solutions represent so called *power calls* and are used for speculation and hedging goals in financial markets (see also [76], [74]).

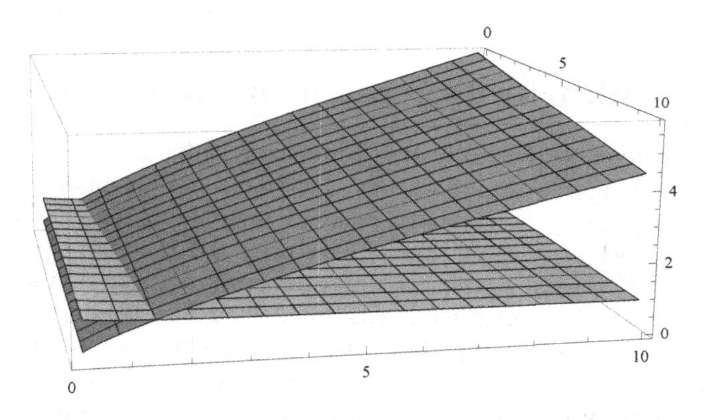

Figure 9.6: Plot of the solution (9.2.98). The parameters take the values $c_1 = 1$, $r = 0$, $\delta = 0.35$, $\sigma = 0.4$, $\alpha = 0.5; \rho = 0.1$ and the variables lie in intervals $S \in (0.2, 10)$, $t \in (0,1)$. The upper solution corresponds to the value $p = 0.76$ and the lower solution to $p = -0.047$.

We also get a similar set of invariant solutions for $r \neq 0$. Here we take the substitution

$$z = t, \quad w = -\frac{S^{\frac{\rho}{\delta}}}{\rho} + S^{\frac{\rho}{\delta}-1}u, \; \delta = \text{const.} \neq 0 \qquad (9.2.99)$$

and reduce Equation (9.2.93) to the ODE

$$w_z - \rho\left(\frac{\sigma^2\alpha^2(\delta-\rho)}{2(\alpha\delta-\rho)^2} + \frac{r}{\delta}\right) w = 0. \qquad (9.2.100)$$

The corresponding solution of (9.2.93) in variables t, S, u has the form

$$u(S,t) = c_1 S^{1-\frac{\rho}{\delta}} e^{\rho\gamma t} + \frac{1}{\rho} S, \qquad \gamma = \frac{\sigma^2\alpha^2(\delta-\rho)}{2(\alpha\delta-\rho)^2} + \frac{r}{\delta}, \qquad (9.2.101)$$

where $c_1, \delta \in \mathbb{R}, \delta \neq 0$ are arbitrary constants. In both cases (9.2.101) and (9.2.98) we obtain solutions which blow up when $\rho \to 0$ and do not converge to any solution of the linear BS model. It means we obtain solutions of a nonlinear PDE which have no counterpart in the linear case. Existence of such solutions is rather typical for nonlinear parabolic equations. Many authors try to get solutions of nonlinear parabolic PDEs using the idea of a small perturbation of the solutions of a linear parabolic equation. This is done in order to get approximation of the solutions of a nonlinear parabolic PDE. Evidently we cannot get all admissible solutions applying this method, especially not the ones described above. The complete description of all classes of non-equivalent invariant solutions of (9.2.93) is presented in [12].

9.3 Equilibrium or reaction-function models

Another approach to the study of self-financing hedging strategies of a large trader is used in the framework of a reaction-function model. The main subject of this model is a smooth *reaction function* ψ that determines the equilibrium stock price S_t at time t. This price is a function of some fundamental value F_t related to the stock and the stock position of a large trader. The ordinary investors or price takers using their own trading strategies do not move the stock price, but they act in a rational way on the price change in the market. The trading strategy of a large investor has a strong influence on this price because of his exposed stock position. Let the overall supply of the stock be normalized and equal to one. The normalized stock demand of the ordinary investors at the moment t is modeled as a function $D(F_t, S_t)$, where S_t is the proposed price of the stock. The function $D(F_t, S_t)$ represents the aggregated demand of all ordinary investors. The normalized stock demand of the large investor is written in the form $\rho\Phi_t$; where $\rho \geq 0$ is a parameter that measures the size of the trader's position relative to the total supply of the stock. According to this approach the mutual influence of the trading strategies leads to an equilibrium price for an asset. The equilibrium price S_t is then determined by the market clearing condition

$$D(F_t, S_t) + \rho\Phi_t = 1 . \tag{9.3.102}$$

The Equation (9.3.102) admits a unique solution S_t under suitable conditions on the demand function $D(F_t, S_t)$, for instance if the demand function is a monotonous function of S_t. Hence S_t can be expressed as a reaction function ψ of F_t and $\rho\Phi_t$, so that $S_t = \psi(F_t, \rho\Phi_t)$. Here the asset price is a function of the fundamental value F_t, trading strategy of the large investor and his position in the market. Throughout this section we assume that the fundamental value process F_t follows a geometric Brownian motion W_t with a constant volatility σ,

$$dF_t = \sigma F_t dW_t + \mu F_t dt, \tag{9.3.103}$$

where μ is a constant drift. Some authors assume that F_t represents a fundamental value of the company, others that this value represents the aggregated income of the ordinary investors. One typical assumption is that the price of the asset defined by $S_t = \psi(F_t, \rho\Phi_t)$ depends linearly on F_t, i.e. the reaction function $\psi(F_t, \rho\Phi_t)$ is of the form

$$S_t = \psi(F_t, \rho\Phi_t) = F_t\ g(\rho\Phi_t) \to \psi(f, \alpha) = f\ g(\alpha) \tag{9.3.104}$$

with some increasing function $g(\alpha)$, where $\alpha = \rho\Phi_t$. This form of the function $\psi(f, \alpha)$ appears for any model where

$$D(f, S) = \mathcal{U}(f/S) \tag{9.3.105}$$

for a strictly increasing function $\mathcal{U} : (0, \infty) \to \mathbb{R}$ with a suitable range. If we look at models described above from this point of view then we would see that in the case of the Frey-Stremme and Sircar-Papanicolaou models [31], [72] the reaction function has the form

$$\psi(f, \alpha) = C\frac{f}{1-\alpha}, \quad C = \text{const.}, C > 0.$$

In the paper Platen and Schweizer [62] the reaction function was used in the form

$$\psi(f, \alpha) = f \exp(\lambda\alpha), \quad \lambda\rangle 0.$$

We assume now that the demand function satisfies the relation (9.3.105) and call the function $\mathcal{U}$ a marginal utility function (definitions and properties of different utility functions are given in section 9.4). This function depends on the relative price f/S of a given asset. The fraction characterizes the relation between the fundamental value of the asset f and the market price S of the same asset. If the fundamental value f is fixed and the market price is low the asset seems to be cheap. The ordinary investors believe that the asset is undervalued and are encouraged to buy more, i.e. the demand grows up. If $f/S \to 0$, then the asset price is relatively high, the asset is too expensive and therefore the demand decreases.

As before we denote the normalized trading strategy of the large trader $\rho\Phi(S, t)$, where Φ is a smooth function. We apply Itô's formula (9.0.2) and obtain the market price of the asset following stochastic differential equation

$$dS_t = g(\rho\Phi(S_t, t))\, dF_t + \rho F_t g_\alpha(\rho\Phi(S_t, t))\Phi_S(S_t, t)\, dS_t + b(S_t, t)\, dt. \tag{9.3.106}$$

Here we used the fact that the price is given by $S_t = F_t g(\rho\Phi(S_t,t)))$, (9.3.104). The drift function $b(S_t,t)$ is irrelevant for further development of the model. We apply the restriction to the variations of the trading strategy of a large trader

$$\big(1 - \rho F_t g_\alpha(\rho\Phi(S_t,t))\Phi_(S_t,t))\big)\rangle 0 \quad \text{a.s.,} \tag{9.3.107}$$

to provide self consistency to the model. After some rearrangements of the formula (9.3.106) using (9.3.103) we get

$$dS_t = \frac{1}{1 - \rho\frac{g_\alpha(\rho\phi(S_t,t))}{g(\rho\phi(S_t,t))}S_t\phi_S(S_t,t)}\sigma S_t dW_t + \tilde{b}(S_t,t)dt\,, \tag{9.3.108}$$

where, again, the precise form of $\tilde{b}$ is irrelevant. If we apply the Itô-Wentzell formula to this Equation (we assume that the large trader uses the delta hedging strategy, i.e. $\phi(S,t) = u_S(S,t)$ (see [4], [72]) we obtain the following PDE

$$u_t + \frac{1}{2}\frac{\sigma^2 S^2 u_{SS}}{\left(1 - \rho\frac{g'(\rho u_S)}{g(\rho u_S)}Su_{SS}\right)^2} = 0 \tag{9.3.109}$$

for a self-financing hedging strategy $u(S,t)$ of a large trader. The SDE (9.3.108) and PDE (9.3.109) are main equations by the equilibrium or reaction-function approach. Depending on the form of the marginal utility function $\mathcal{U}(f/S)$ that we used, we would obtain different reaction functions $\psi(F_t, \rho\Phi_t)$ and correspondingly different functions $g(\rho\Phi(t,S_t))$. This function has a strong influence on the adjusted volatility in SDE (9.3.108) and on the main term in PDE (9.3.109). This PDE is a generalization of the Black-Scholes model and from the analytical point of view belongs to singular perturbed parabolic equations because the small parameter ρ is involved in the main term of the equation.

9.3.1 Schönbucher-Wilmott model for self-financing trading strategies

The detailed discussions devoted to the influence of a self-financing dynamic trading strategy of the large trader on a price process can be found in the paper of Schönbucher and Wilmott [69]. The authors analyze the stochastic process (9.3.108), and provide an option replication formula (they use the notation that is just slightly different from the one in (9.3.109) for illiquid markets and illustrate the price effects using the put-option replication trading). The nonlinear singular perturbed PDE (9.3.109) was studied using Lie group analysis in [16]. We present now the main results.

Theorem 9.3.1 *[16] The Equation (9.3.109), where $g(\alpha)$ is a two times differentiable function, admits a three-dimensional Lie algebra $L_3 = \langle \mathbf{U}_1, \mathbf{U}_2, \mathbf{U}_3 \rangle$ spanned by the following generators*

$$\mathbf{U}_1 = S\frac{\partial}{\partial S} + u\frac{\partial}{\partial u}, \quad \mathbf{U}_2 = \frac{\partial}{\partial u}, \quad \mathbf{U}_3 = \frac{\partial}{\partial t}. \tag{9.3.110}$$

The Lie algebra L_3 possesses a non-zero commutator relation $[\mathbf{U}_1, \mathbf{U}_2] = -\mathbf{U}_2$. It has a two-dimensional subalgebra $L_2 = \langle \mathbf{U}_1, \mathbf{U}_2 \rangle$ spanned by the generators $\mathbf{U}_1, \mathbf{U}_2$, while L_3 is a decomposable Lie algebra and can be represented as a semi-direct sum $L_3 = L_2 \bigoplus \mathbf{U}_3$.

An optimal system of subalgebras is listed in Table 9.2.

Table 9.2: The optimal system of one- and two–dimensional subalgebras of L_3, where parameters are $x \in \mathbb{R}$, $0 \leq \phi \leq \pi$, $\epsilon = \pm 1$.

Dimension of the subalgebra	Subalgebras
1	$h_1^g = \{\mathbf{U}_2\}$, $h_2^g = \{\mathbf{U}_1 \cos(\varphi) + \mathbf{U}_3 \sin(\varphi)\}$, $h_3^g = \{\mathbf{U}_2 + \epsilon\, \mathbf{U}_3\}$
2	$h_4^g = \big(\mathbf{U}_2, \mathbf{U}_3\big)$, $h_5^g = \big(\mathbf{U}_1, \mathbf{U}_3\big)$, $h_6^g = \big(\mathbf{U}_1 + x\, \mathbf{U}_3, \mathbf{U}_2\big)$

The Lie algebra L_3 (9.3.110) possesses three non-similar one–dimensional subalgebras denoted as h_1^g, h_2^g, h_3^g. For each one–dimensional subalgebra from this list there exists a corresponding one–dimensional subgroup of the symmetry group. Moreover all such subgroups are non-conjugate to each other.

We use this optimal system of subalgebras to find invariant reductions of the equation (9.3.109) to ODEs. To obtain the reduced form of the PDE we take invariant expressions as new dependent and independent variables. In the new variables PDE is reduced to ODEs. The results are presented in Table 9.3 (in the general case an optimal system for L_3 is presented in Table 9.2). The listed optimal system of subalgebras allowed us to get the corresponding system of subgroups in the symmetry group admitted by PDE (9.3.109).

Using invariants of these subgroups we obtain the set of non-equivalent invariant solutions. We denote the corresponding subgroups similarly to subalgebras, we just replace the small letter h with a capital letter H. Using the optimal system of subalgebras we get a complete set of invariant families

Table 9.3: In the first row we list one–dimensional subalgebras from the optimal system of L_3 (9.3.110). In the second row we list the corresponding invariants. In the last row we provide the transformations or reductions of the PDE (9.3.109) to ODEs. Here $\epsilon = \pm 1$, $0 \le \phi \le \pi$, $\sigma^2, \rho \in \mathbb{R}^+, c \in \mathbb{R}$ are parameters and Y', W' denote the differentiation of the corresponding function with respect to the invariant variable z.

Subalgebra	Invariants	Transformations/Reductions
h_1^g	$z = S,\ W(z) = t$	$u \to u + \text{const.}$
h_2^g	$z = S,\ W(z) = u - \epsilon t$	$2\epsilon + \frac{\sigma^2 z^2 Y'}{(1 - z(\ln g(\rho Y))')^2} = 0,\ Y = W',$ $g(\rho Y(z)) \neq cz.$
h_3^g	$z = \ln S - \gamma\, t\ ,\ \gamma = \text{cotan}(\phi)$ $W(z) = uS^{-1}$	$2\gamma W' - \frac{\sigma^2 (W' + W'')}{(1 - (\ln g(W' + W))')^2} = 0,$ $g(W + W') \neq ce^z.$

of solutions and do not lose any of them. Some of the invariant solution families are trivial, others are more interesting.

For example in the first line of Table 9.3 we see that the first subgroup H_1^g (we recall that in Table 9.3 we listed the corresponding subalgebras denoted with lower case h_1^g) describes translations with respect to the dependent variable u and cannot be used for any reduction. It can be used just to modify other solutions. In both other cases (H_2^g and H_3^g) the reductions of the given PDE to an ODE are possible. But the solutions invariant under action of H_2^g applied to the general model (9.3.109) are not really interesting because they have a trivial dependence on time (see the intersection of the second row and second line in Table 9.3).

If we look at the family of the solutions that are invariant under the action of the subgroup H_3^g we can see a different situation. All of them are the solutions of the second order ODE which is represented at the intersection of the last row and the last line of Table 9.3. The invariant solutions of (9.3.109) are represented in the form $u(S,t) = SW(\ln S - \gamma t)$. The ODE which determines these invariant solutions contains an arbitrary function $g(\alpha)$ so we cannot present the general solution of this nonlinear ODE, but if we study one special model with fixed function $g(\alpha)$ then it seems to be feasible to find explicit solutions.

We notice that the determining equations for the symmetry algebra admitted by (9.3.109) contain the arbitrary function $g(\alpha)$ [16]. If we do not specify this function we get the three–dimensional algebra presented in Theorem 9.3.1. For some special form of the function $g(\alpha)$ we can expect a richer symmetry group admitted by PDE. The determining system contains some differential conditions on $g(\alpha)$, if the function $g(\alpha)$ is a solution of these differential equations we obtain an additional symmetry. It is proved [16] that there exist just three cases for which (9.3.109) admits a four-dimensional symmetry algebra.

Theorem 9.3.2 *[16] Equation(9.3.109) admits a four-dimensional Lie algebra if and only if the function $g(\alpha)$ has one of the following forms:*

$$g(\alpha) = c_2 e^{c_1 \alpha}, \tag{9.3.111}$$

$$g(\alpha) = c_2 \alpha^{c_1}, \tag{9.3.112}$$

$$g(\alpha) = c_2(\rho + k\alpha)^{-\frac{1}{c_1}}, \quad c_1, c_2, k \text{ are constants.} \tag{9.3.113}$$

Because the function $g(\cdot)$ defines the main term in PDE (9.3.109) the properties of the corresponding financial models will depend also strongly on the form of this function. Let us list these equations with a short characteristic of the financial settings.

- If the function $g(\alpha)$ takes the form (9.3.111) then the PDE for the self-financing hedging strategy takes the form

$$u_t + \frac{1}{2}\frac{\sigma^2 S^2 u_{SS}}{(1 - \rho c_1 S u_{SS})^2} = 0. \tag{9.3.114}$$

  This equation was first introduced in [28]. The complete description of the symmetry group, invariant reductions and invariant solutions of this equation as well as of some generalization of this equation are given in [20], [15] and [14]. The short description of the results is presented in Section 9.2.1.

- If the function $g(\alpha)$ is given by (9.3.112) then the PDE which corresponds to this function $g(\alpha)$ has the form

$$u_t + \frac{1}{2}\frac{\sigma^2 S^2 u_{SS} u_S^2}{(u_S - c_1 S u_{SS})^2} = 0. \tag{9.3.115}$$

This equation is a new model for hedging strategies in the illiquid markets which was first studied in [16]. We describe the properties of this equation in this section.

- If similarly to the previous case we take the function $g(\alpha)$ in the form (9.3.113) then the PDE for a self-financing trading strategy takes the form

$$u_t + \frac{1}{2}\frac{\sigma^2(1+ku_S)^2}{c_1^2\left(1+ku_S+\frac{k}{c_1}Su_{SS}\right)^2}S^2u_{SS} = 0\,, c_1 \neq 0. \qquad (9.3.116)$$

This equation coincides with some of the models introduced in [72] depending on the value of the constant c_1. This model (9.3.116) was studied using Lie group analysis in [12] and [11]. We discussed the results in Section 9.2.2.

Let us first look at the analytical structure of the Equation (9.3.115). The second term has a non-trivial denominator. If the denominator is identically equal to zero then the equation is not well defined. To avoid singularities we exclude from further investigation the family of functions where the denominator is identically equal to zero, i.e. all the functions in the form

$$u(S,t) = d_1(t)S^\beta + d_2(t),\ \beta = 1+\frac{1}{c_1},\ c_1 \in \mathbb{R}\setminus\{0\}, \qquad (9.3.117)$$

where $d_1(t), d_2(t)$ are arbitrary functions of time.

Theorem 9.3.3 *Equation (9.3.115) admits a four-dimensional Lie algebra*

$$L_4 = \langle \mathbf{U}_1, \mathbf{U}_2, \mathbf{U}_3, \mathbf{U}_4\rangle$$

spanned by generators

$$\mathbf{U}_1 = S\frac{\partial}{\partial S},\quad \mathbf{U}_2 = u\frac{\partial}{\partial u},\quad \mathbf{U}_3 = \frac{\partial}{\partial u},\quad \mathbf{U}_4 = \frac{\partial}{\partial t}. \qquad (9.3.118)$$

The Lie algebra L_4 has the following non-zero commutator relation, $[\mathbf{U}_1, \mathbf{U}_3] = -\mathbf{U}_3$. The Lie algebra L_4 has a two-dimensional subalgebra $L_{4,2} = \langle \mathbf{U}_1, \mathbf{U}_3\rangle$ spanned by generators $\mathbf{U}_1, \mathbf{U}_3$. The algebra L_4 is a decomposable Lie algebra and can be represented as a semi-direct sum $L_4 = L_{4,2} \bigoplus \mathbf{U}_2 \bigoplus \mathbf{U}_4$.

The optimal system of one-dimensional subalgebras contains four subalgebras, the optimal system of two-dimensional subalgebras contains five subalgebras and the optimal system of three-dimensional subalgebras includes three subalgebras. All subalgebras from these systems are listed in Table 9.4.

Table 9.4: The optimal system of one-, two- and three–dimensional subalgebras of L_4 (9.3.118). Here $x \in \mathbb{R}$, $0 \le \phi \le \pi$, $\epsilon = \pm 1$ are parameters.

Dimension of the subalgebra	Subalgebras
1	$h_1 = \{\mathbf{U}_3\}$, $h_2 = \{\mathbf{U}_1 \cos\phi + \mathbf{U}_4 \sin(\phi)\}$, $h_3 = \{\mathbf{U}_2 + x(\mathbf{U}_1 \cos(\phi) + \mathbf{U}_4 \sin(\phi))\}$, $h_4 = \{\mathbf{U}_3 + \epsilon(\mathbf{U}_1 \cos(\phi) + \mathbf{U}_4 \sin(\phi))\}$
2	$h_5 = \{\mathbf{U}_2 + x(\mathbf{U}_1 \cos(\phi) + \mathbf{U}_4 \sin(\phi)), \mathbf{U}_3\}$, $h_6 = \{\mathbf{U}_2 + x(\mathbf{U}_1 \cos(\phi) + \mathbf{U}_4 \sin(\phi)), \mathbf{U}_1 \sin(\phi) - \mathbf{U}_4 \cos(\phi)\}$, $h_7 = \{\mathbf{U}_1, \mathbf{U}_4\}$, $h_8 = \{\mathbf{U}_3 + \epsilon(\mathbf{U}_1 \cos(\phi) + \mathbf{U}_4 \sin(\phi)), \mathbf{U}_1 \sin(\phi) - \mathbf{U}_4 \cos(\phi)\}$, $h_9 = \{\mathbf{U}_3, \mathbf{U}_1 \sin(\phi) - \mathbf{U}_4 \cos(\phi))\}$
3	$h_{10} = (\mathbf{U}_2, \mathbf{U}_1, \mathbf{U}_4)$, $h_{11}^2 = (\mathbf{U}_3, \mathbf{U}_1, \mathbf{U}_4)$, $h_{12} = (\mathbf{U}_2 + x(\mathbf{U}_1 \cos(\phi) + \mathbf{U}_4 \sin(\phi)), \mathbf{U}_1 \sin(\phi) - \mathbf{U}_4 \cos(\phi), \mathbf{U}_3)$

We use as before the symmetry algebra L_4 (9.3.118) to provide a complete set of reductions of Equation (9.3.115) to ODEs. In the cases where it is possible we provide also the exact solutions of these equations. The optimal system of one-, two- and three-dimensional subalgebras of the algebra (9.3.118) are listed in Table 9.4. Now we are interested in one-dimensional subalgebras because they can provide non-trivial invariant solutions of (9.3.115). From the four one-dimensional subalgebras listed in Table 9.4, the last three can be used for reductions.

The first subalgebra h_1 just means that an arbitrary constant can be added to any solution of (9.3.115). One may use this symmetry to modify existing solutions or to adjust a payoff. The second subalgebra h_2 in Table 9.4 corresponds to the symmetry subgroup H_2 with invariants

$$z = \ln S - \gamma\, t, \ W(z) = u, \quad \gamma \in \mathbb{R} \setminus \{0\}. \tag{9.3.119}$$

These invariants allow us to reduce (9.3.115) to ODE on the function $W(z)$

$$W' - \kappa \frac{(W'' - W')W'^2}{(W'' - \beta W')^2} = 0, \quad \kappa = \frac{\sigma^2}{2\gamma c_1^2}, \ \beta = 1 + \frac{1}{c_1}, \ c_1, \ \gamma \in \mathbb{R} \setminus \{0\}. \tag{9.3.120}$$

Remark 9.3.1 *The values of parameters $c_1, \gamma = 0$ lead to trivial demand function, invariants or equations so we exclude these values from our consideration.*

After the substitution $W' = Y$ the Equation (9.3.120) is reduced to a first order ODE

$$(Y')^2 - Y'Y(2\beta + \kappa) + Y^2(\beta^2 + \kappa) = 0. \qquad (9.3.121)$$

It has the following solutions

$$Y(z) = d_1 e^{k_{1,2}z}, \qquad k_{1,2} = \beta + \frac{\kappa}{2} \pm \frac{1}{2}\sqrt{\kappa(\kappa + 4(\beta - 1))}, \qquad (9.3.122)$$

or speaking in original variables, the Equation (9.3.115) possesses the solution in the form

$$u(S,t) = d_1\ S^{k_{1,2}} e^{-\gamma k_{1,2} t} + d_2, \qquad d_1, d_2 \in \mathbb{R}. \qquad (9.3.123)$$

Because of the parameter $\kappa = \frac{\sigma^2}{2\gamma c_1^2} \neq 0$ the solutions (9.3.123) do not coincide with the excluded set of functions (9.3.117).

The analytical structure of the reduced Equation (9.3.120) is defined by two parameters κ and β. The parameter β is defined by the demand function of the investor, and depends on the coefficient of the relative risk aversion. The parameter κ depends also on the volatility of the underlying asset and on the chosen symmetry reduction. It means the parameter κ describes the modified volatility of the asset in dependence on the utility preferences of the investor.

Solutions (9.3.123) represent power options (see also [76], [74]). The payoff of this kind of options is similar to one of usual options, but it includes some *gearing* of the asset price by the power $k_{1,2}$. For instance instead of the usual payoff function for a call, for payoff function at expiry day $t = T$ we have $\text{call}_{\text{Power}}(S,T) = \max[0, S^{k_{1,2}} - E]$, where E is the exercise price and $k_{1,2}$ is the power value. We present two typical solutions of the form (9.3.123) in Figure 9.7

The third one-dimensional subalgebra in Table 9.4 leads to a subgroup with the following invariants

$$\begin{aligned} &z = \ln S - \gamma t, \qquad W(z) = \ln u - \delta t, \\ &\delta = (x\cos(\phi))^{-1}, \gamma = \text{cotan}(\phi), \ \ \gamma, \delta \in \mathbb{R} \setminus \{0\}. \end{aligned} \qquad (9.3.124)$$

Here we assume that the group parameters x, ϕ are not equal to zero, because if these parameters are equal to zero the case becomes trivial.

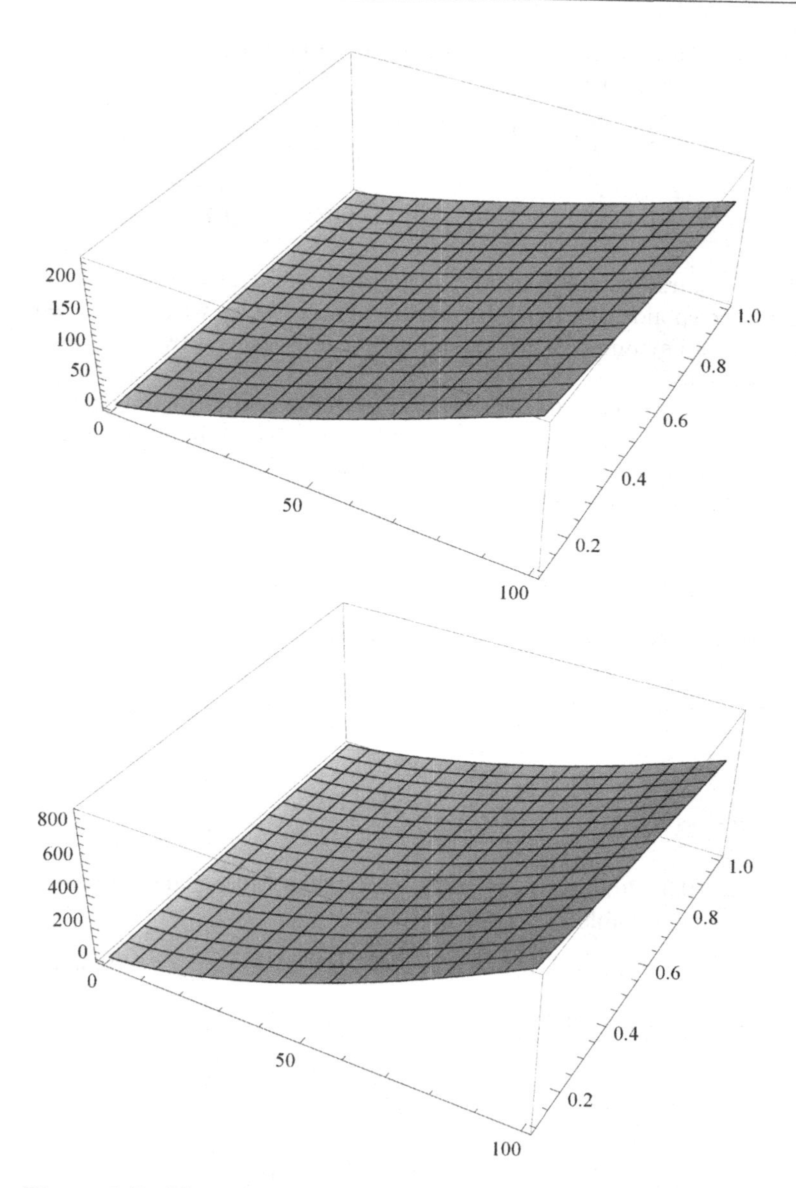

Figure 9.7: Plot of the explicit solutions $u(S,t)$ (9.3.123) with $k_1 = 1.298$ (left panel), and $k_2 = 1.762$ (right panel) for the parameters $\beta = 1.303$, $\kappa = 0.066$, $\gamma = 0.2776$, $\phi = 1.3$, $c_1 = 3.3$, $d_1 = 1$, $d_2 = 0$. The variables S, t are in intervals $S \in (0.1, 100)$ and $t \in (0.1, 1)$.

Using these invariants as new variables we reduce the PDE Equation (9.3.115) to a second order ODE

$$\delta - W' + \kappa \frac{W'^2(W'' + W'^2 - W')}{(W'' + W'^2 - \beta W')^2} = 0, \tag{9.3.125}$$

$$\beta = 1 + \frac{1}{c_1}, \quad \kappa = \frac{\sigma^2}{2\gamma c_1^2} \neq 0, \; c_1, \gamma, \delta \in \mathbb{R} \setminus \{0\}.$$

In (9.3.125) the parameter β reflects the form of the utility function, the parameter κ corresponds to adjusted volatility in accordance with the chosen utility function and symmetry, and the parameter δ is defined by the chosen symmetry subgroup.

We multiply (9.3.125) by the denominator of the second term and exclude from further study all the functions for which the denominator vanishes, i.e. the functions (9.3.117). After the substitution $W' = Y$, we reduce the Equation (9.3.125) to a first order ODE

$$\begin{aligned} &(Y')^2(\delta - Y) + Y'Y\left(2(\delta - Y)(Y - \beta) + \kappa Y\right) \\ &\quad + Y^2\left((\delta - Y)(Y - \beta)^2 + \kappa Y(Y - 1)\right) = 0. \end{aligned}$$

The set of solutions of this equation is equivalent to the unification of the sets of the solutions of the following equations

$$Y' = \frac{Y\left(2Y^2 - (\kappa + 2(\delta + \beta))Y + 2\beta\delta \pm \sqrt{\theta Y(Y - \zeta)}\right)}{2(\delta - Y)} \tag{9.3.126}$$

$$\theta = \kappa(4(\beta - 1) + \kappa), \quad \zeta = \frac{4(\beta - 1)\delta}{4(\beta - 1) + \kappa}, \quad \kappa = \frac{\sigma^2}{2\gamma c_1^2}.$$

We simplify (9.3.126) using the third Euler substitution and introduce instead of Y a new variable τ

$$Y = \theta\zeta(\theta - \tau^2)^{-1} = 4\kappa\delta(\beta - 1)(\theta - \tau^2)^{-1}. \tag{9.3.127}$$

After this transformation ODEs (9.3.126) take the form

$$\tau' = \frac{A \pm B\tau + \kappa(\beta - 2)\tau^2 \mp \beta\tau^3}{2\tau(\kappa \pm \tau)}, \tag{9.3.128}$$

$$\begin{aligned} A &= -\kappa^2(4(\beta - 1)^2 + 4(\beta - 1)(\delta - 1) + \kappa(\beta - 2)), \\ B &= \kappa(4(\beta - 1)(\beta - \delta) + \beta\kappa). \end{aligned}$$

The first integration defines the function $\tau(z)$ as

$$\int \frac{2\tau(\kappa \pm \tau)\mathrm{d}\tau}{A \pm B\tau + \kappa(\beta - 2)\tau^2 \mp \beta\tau^3} = z + d_1, d_1 \in \mathbb{R} \tag{9.3.129}$$

As the expressions for constants A and B in (9.3.129) contain an arbitrary symmetry parameter δ we can simplify this expression if we choose δ so that either $A = 0$ or $B = 0$. But also in these cases the integration procedure leads to voluminous expressions for $\tau(z)$. If the model parameters σ, c_1 are fixed then the integration can be performed. We obtain in this way $\tau(z)$ represented as a sum of logarithmic expressions of the type $\sum_{i=1}^{3} a_i \ln(\tau - b_i) = z + d_1$. It means that we have to find explicit expression for $\tau(z)$, insert in (9.3.127) to get $Y(z)$. Then we have to integrate one more to get $W(z)$ and to get the solution of (9.3.115) we have to replace z, W by S, t after (9.3.124). We skip this procedure in the general case because of the voluminous formulas. As we mentioned before one can perform this procedure for fixed financial parameters at least numerically.

We take the fourth one-dimensional subalgebra listed in Table 9.4. The invariants of the corresponding Lie subgroup can be chosen in the form

$$z = \ln S - \gamma t, \quad W(z) = u - \eta t,$$

here parameters are defined by

$$\gamma = \text{cotan}(\phi), \quad \eta = \frac{\epsilon}{\sin(\phi)}, \quad 0 < \phi < \pi, \epsilon = \pm 1. \tag{9.3.130}$$

Using the invariants (9.3.130) as new invariant variables we obtain a second order ODE as a reduction of (9.3.115)

$$\delta - W' + \kappa \frac{(W'' - W')W'^2}{(W'' - \beta W')^2} = 0, \quad \kappa = \frac{\sigma^2}{2\gamma c_1^2}, \quad \delta = \frac{\eta}{\gamma}. \tag{9.3.131}$$

We multiply both terms of the Equation (9.3.121) by the denominator of the second term (we excluded also in this case the set of functions (9.3.117)). After the substitution $W' = Y$ we obtain the first order ODE

$$(Y')^2(\delta - Y) + Y'Y(Y(2\beta + \kappa) - 2\beta\delta) - Y^2(Y(\beta^2 + \kappa) - \beta^2\delta) = 0. \tag{9.3.132}$$

This first order ODE is very similar to the Equation (9.3.126) obtained in the previous case and can be solved with a similar Euler transformation

$$Y = \theta\zeta(\theta - \tau^2)^{-1}, \quad \theta = \kappa(4(\beta - 1) + \kappa), \quad \zeta = \frac{4(\beta - 1)\delta}{(4(\beta - 1) + \kappa)}. \tag{9.3.133}$$

After the substitution (9.3.133) in Equation (9.3.132) we obtain the first order ODE for the function $\tau(z)$. Its solution takes the form

$$\frac{-4\delta}{\theta\zeta} \int \frac{\tau d\tau}{b_2\tau^2 \pm \tau + b_0} + 4 \int \frac{\tau d\tau}{(\theta - \tau^2)(b_2\tau^2 \pm \tau + b_0)} = z + d_1,$$

$$b_2 = \frac{\beta}{2\kappa\delta(\beta - 1)}, \quad b_0 = \kappa + 2\beta(1 - \zeta^{-1}), \quad d_1 \in \mathbb{R} \tag{9.3.134}$$

and can be easily integrated. For instance the solution of (9.3.132) with the parameter $\epsilon = 1$ has the form

$$
\begin{aligned}
&2\beta(\beta^2+\kappa)z + d_1 = 2(\beta^2+\kappa)\ln Y - \beta\sqrt{\theta}\ln\left(\theta Y - \frac{1}{2}a_1 + \sqrt{\theta Y(\theta Y - a_1)}\right) \\
&+ (\beta-2)\kappa\ln\big(Y(\beta^2(\kappa+2(\beta-1))) - 2\beta^2(\beta-1)\delta \\
&+\beta(\beta-2)\sqrt{Y(\theta Y - a_1)}\big), \qquad (9.3.135)
\end{aligned}
$$

where $a_1 = 4\kappa(\beta-1)\delta$, $\theta = 4\kappa(4(\beta-1)+\kappa^2)$, $d_1 \in \mathbb{R}$ (the solution of (9.3.132) with $\epsilon = -1$ has a similar form and we skip this expression). We present a typical solution of the form (9.3.135) in Figure 9.8.

To obtain an analytical solution of the Equation (9.3.131) we should first invert the function $z = f(Y)$ of the type (9.3.135) and then integrate $W(z) = \int Y(z)\mathrm{d}z$. In the general case neither the expression (9.3.135) nor the similar expression with $\epsilon = -1$ can be inverted explicitly. But there exists a possibility to solve this problem numerically and consequently solve Equation (9.3.115) in this way.

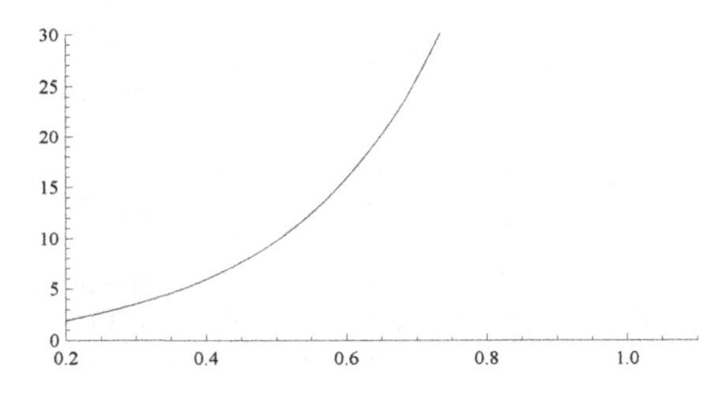

Figure 9.8: Plot of the solution to Equation (9.3.132), where $z \in (0.2, 1.1)$ and $Y \in (0.1, 30)$. The parameters are $c_1 = 0.5$, $\epsilon = 1$, $\phi = \pi/4$, $\sigma = 0.316$, $d_1 = 0$.

With an optimal system of subalgebras of the symmetry algebra admitted by (9.3.115) we can describe the complete set of non-equivalent reductions of this PDE to different ODEs. In all these cases we were able to solve the ODEs and obtained solutions in the exact form. In some cases we skipped the exact formulas because they are too voluminous.

It is interesting to remark that the models (9.3.114), (9.3.115) and (9.3.116) were introduced under different financial and economical assumptions. Now with the help of Lie group analysis we have clarified how deeply they are connected from the mathematical point of view.

Another interesting remark concerns the structure of the corresponding Lie algebras. The symmetry Lie algebras of the models (9.3.114), (9.3.115) and (9.3.116) are isomorphic. They are also isomorphic to the Lie algebra admitted by the risk-adjusted pricing methodology model described in Section 9.4.3.

9.4 Pricing and hedging in incomplete markets

In this section we discuss some models devoted to pricing and hedging in the incomplete markets. Details and a good historical overview of the development of the idea of incompleteness and price building under this condition can be found in [73]. Here we provide just a short introduction to the concept of incomplete markets and a connected idea of indifference prices.

The classic no-arbitrage theory assumes that the price for each asset is unique and each payoff can be replicated with other market-traded securities. In other words we can transfer risk perfectly. The situation changes dramatically if such risk transfer is not possible. The price of an asset becomes different for different investors. The uniqueness of the prices is lost in the incomplete markets and it makes the theory of pricing and hedging much more advanced. There are many causes of incompleteness in the market like transaction costs or constraints on portfolios. We can work with a non-traded asset like a labor income, property value, plant value and so on. An enterprise typically has a market value or market level (also called paper value) but as long as it is not sold we cannot speak about the price of this company. Yet we should be able to model the prices of the derivative products written on this enterprise at any moment of time.

To describe the behavior of a rational investor in an incomplete market the concept of utility was introduced in [78]. According to this idea an investor always prefers some asset A over another asset B if and only if $\mathsf{U}(A)\rangle\mathsf{U}(B)$, in other words a rational investor chooses the highest expected utility. From the economic point of view the utility function $\mathsf{U}(x)$ is the measure of the individual risk preferences of the investor and is defined by the risk aversion measure. It is expected that the utility function is a smooth increasing concave function. Later in [45] Kramkov and Schachermayer proved that for the correctness of the problem of maximizing the expected utility of the terminal wealth in incomplete markets one needs to pose the asymptotic conditions on the utility function U. It means that in addition to the standard

Inada conditions

$$\mathsf{U}'(0) = \lim_{x\to 0} \mathsf{U}'(x) = \infty, \tag{9.4.136}$$

$$\mathsf{U}'(\infty) = \lim_{x\to \infty} \mathsf{U}'(x) = 0, \tag{9.4.137}$$

it is necessary and sufficient that $\limsup_{x\to\infty} \frac{x\mathsf{U}'(x)}{\mathsf{U}(x)} < 1$. Nowadays a classification of utility functions described by Arrow-Pratt risk aversion measure introduced in [3], [65] is widely accepted. There are three types of utility functions widely used in practice:

1. The utility function of exponential growth $\mathsf{U}(x) = C - e^{-\gamma x}$, $\gamma > 0$, γ, C are constants. It is the so-called CARA-type *(Constant Absolute Risk Aversion)* utility function.

2. The utility function of power growth $\mathsf{U}(x) = C_1(x + C_2)^\gamma$, $0 < \gamma < 1, C_1 > 0$, C_1, C_2 are constants. It is the utility function of HARA-type *(Hyperbolic Absolute Risk Aversion)*.

3. The utility function of logarithmic growth $\mathsf{U}(x) = \log(x) + C$, C is a constant. It is the so-called DARA-type *(Decreasing Absolute Risk Aversion)* utility function.

The idea of an indifference price introduced in [36] was based on the idea of the utility function and an individual connected risk aversion of a market participant. The indifference price was the price which allowed a market participant to achieve the same level of the expected utility which he or she would achieve without trading this security with optimal trading otherwise (for more details see [35]). In incomplete markets the indifference prices typically arise by the investigation of optimal portfolio problems and are represented as solutions of nonlinear PDEs. The type of nonlinearity and corresponding properties of solutions depend on the type of utility function used. Most of the utility functions used are the exponential, power or logarithmic types, as listed above.

9.4.1 Musiela-Zariphopoulou model

Musiela and Zariphopoulou in [56] introduced an example of a price valuation for a derivative product using the indifference price valuation theory. They regard a market with two risky assets, a riskless bond B_t with interest rate equal to zero and an European claim with maturity T written on the non-traded risky asset. One of the risky assets is a liquid stock S_t with the price process given by

$$dS_t = \mu S_t dt + \sigma S_t dW_t^S, \quad S_{t=0} = S\rangle 0 \tag{9.4.138}$$

(similar to the Black-Scholes model, (9.1.3)). The other risky asset Y_t is a non-traded asset. The level or paper value Y_t of this asset is given by

$$dY_t = b(Y_t, t)dt + a(Y_t, t)dW_t^Y, \quad Y_{t=0} = y \in \mathbb{R}. \tag{9.4.139}$$

The standard Brownian motions W_t^S and dW_t^Y are correlated with the correlation coefficient $\rho \in (-1, 1)$. The coefficients $a(\cdot,\cdot)$ and $b(\cdot,\cdot)$ satisfy all necessary conditions that are needed for the SDE (9.4.139) to possess a unique strong solution.

The problem can be formulated as follows: find the price for an European claim written on Y_t with the payoff at T defined by $Q = q(Y_T)$ where the function q is bounded.

By the used valuation method the maximal expected utilities will be compared in both cases with and without involving the derivative product Q. The investor can trade the liquid assets S_t and B_t for all $t \leq T$ in both cases. The price for the derivative product Q will be fair if the investor according to his risk preferences (type of the utility function) is indifferent towards the choice of one of these two scenarios.

The authors in [56] chose the exponential utility function as the investor utility. It describes the individual risk preferences of the CARA-type as follows

$$\mathsf{U}(x) = -e^{-\gamma x}, \quad \gamma\rangle 0. \tag{9.4.140}$$

For the indifference price $u(y, t)$ of the European claim with the payoff $Q = q(Y_T)$ they obtained the PDE of the following type

$$u_t + \frac{a^2(y,t)}{2} u_{yy} + \left(b(y,t) - \rho\frac{\mu}{\sigma}a(y,t)\right) u_y + \frac{\gamma}{2}(1-\rho^2)a^2(y,t)u_y^2 = 0. \tag{9.4.141}$$

Here $a(y,t), b(y,t)$ are smooth functions of the spatial variable y and time t; parameters $\sigma, \mu, \gamma, \rho$ are fixed constants. It is a quasi-linear PDE of the Burger's type. We can find the corresponding symmetry algebra and use the invariants to find symmetry reductions of this Equation (9.4.141), see [15].

Let us now provide one example of such invariant solutions. We use the following invariants as the independent and dependent variables. The substitutions

$$z = \log y - \kappa^2/2t = \log(ye^{-\kappa^2/2t}), \quad u(y,t) = \gamma^{-1}(1-\rho^2)^{-1}w(z),$$

where κ is a constant reduce (9.4.141) to an ODE on $w(z)$ if the functions $a(y,t)$ and $b(y,t)$ have special forms. Let $a(y,t) = \kappa y p(z)$, and $b(S,t) =$

$\kappa y p(z)\left(\rho\mu\sigma^{-1} - \kappa p(z) f(z)\right)$, where $p(z)$, $f(z)$ are arbitrary functions and $p(z) \neq 0$. Under this substitution the Equation (9.4.141) is reduced to the ordinary differential equation

$$w_{zz} + (f(z) - 1 - p(z)^{-2})w_z + w_z^2 = 0. \tag{9.4.142}$$

After standard transformations we obtain the solution of ODE and correspondingly the Equation (9.4.141) in the form

$$u(y,t) = \frac{1}{\gamma(1-\rho^2)} \log\left(C_1 + \int e^{-\Phi(z)} \mathrm{d}z\right) + C_2,$$
$$\Phi(z) = \int \left(f(z) - 1 - p(z)^{-2}\right) \mathrm{d}z, \tag{9.4.143}$$

where C_1, C_2 are arbitrary constants. The functions $p(z), f(z)$ can be defined after the payoff $u(y,T) = Q = q(y)$. The boundedness of the function $g(y)$ can be reformulated as the following condition on the function $\Phi(z)$ at $t = T$

$$\frac{e^{-\Phi(z)}}{C_1 + \int e^{-\Phi(z)} \mathrm{d}z} \leq 0.$$

Using the solution $u(y,t)$ of (9.4.141) we can also explicitly represent the price process which provides this indifference price (similar to [56]). We introduce

$$H_t = u(Y_t, t), \quad 0 \leq t \leq T, \quad F(z) = \frac{e^{-\Phi(z)}}{C_1 + \int e^{-\Phi(z)} \mathrm{d}z}$$

then applying stochastic analysis techniques we obtain that H_t satisfies the following SDE

$$dH_t = p(z)F(z)\left(\left(-\frac{\kappa}{2}p(z)F(z) + \rho\frac{\mu}{\sigma}\right)dt + dW_t^Y\right).$$

The invariant solution (9.4.143) has the similar structure as the solution which the authors of [56] got using the Hopf-Cole substitution

$$u(y,t) = \gamma^{-1}(1-\rho^2)^{-1}\log(v(y,t)).$$

They got the linear equation for $v(y,t)$ of the following form

$$v_t + \frac{1}{2}a^2(y,t)v_{yy} + (b(y,t) - \rho\frac{\mu}{\sigma}a(y,t))v_y = 0.$$

We additionally applied the second invariant and reduced this equation to ODE and solved it. This procedure is very similar to that which was used by Black and Scholes [7] to get the explicit solution for the European call option (8.4.36).

9.4.2 Problems of optimal consumption with random income

Our goal in this section is to present the admitted Lie algebras for a PDE describing a value function and investment and consumption strategies for a portfolio with illiquid asset that is sold at a random moment in time with a prescribed distribution.

Management of a portfolio that includes an illiquid asset is an important problem of modern financial mathematics. One of the ways to model illiquidity among others is to build an optimization problem and assume that one of the assets in a portfolio cannot be sold until a certain finite, infinite or random moment in time. This framework usually leads to a three–dimensional PDE on the value function. The nonlinear PDE is then studied with different analytic and numeric methods. Usually one tries to find an inner symmetry of the equation to reduce the number of variables at least to two or if possible to one. Low-dimensional problems are better studied and are easier to handle. In all the papers devoted to three-dimensional Hamilton–Jacobi–Bellman (HJB) equations, the authors provide some variable substitution without any remark on how to get similar substitution in other cases or why they use this or that substitution. It is possible to carry out a complete Lie group analysis of PDEs that arise for a portfolio optimization problem, find the inner algebraic structure behind the studied PDEs and use it to obtain reductions. Yet practical application of these procedures is connected with rather tedious and voluminous calculations which can be sometimes facilitated by modern computer packages. For example in this case we used the package **ReLie** kindly provided by the author [58].

In [18], [19] the authors described in detail an optimization problem that corresponds to the following situation: an investor has an illiquid asset that has some paper value and cannot be sold until some moment in time that is random with a prescribed distribution. The investor tries to maximize an average consumption investing into a risky asset that is partly correlated with an illiquid one, and into a riskless asset that has a constant dividend rate.

The investor's portfolio includes a riskless bond B_t, a risky asset S_t and a non-traded asset H_t that generates stochastic income, i.e., dividends or costs of maintaining the asset. The liquidation time of the portfolio τ is a random-distributed continuous variable. The risk-free bank account B_t, with the interest rate r, follows

$$dB_t = rB_t\,dt, \qquad 0 < t \leq \tau, \tag{9.4.144}$$

where r is taken to be constant. The stock price S_t follows the geometrical

Brownian motion

$$dS_t = S_t(\alpha\, dt + \sigma\, dW^1), \quad 0 < t \le \tau, \tag{9.4.145}$$

with the continuously compounded rate of return $\alpha > r$ and the standard deviation σ. The lower case index t denotes the spot value of the asset at the moment t. The illiquid asset H_t cannot be traded up to the random liquidation time τ. It has a paper value correlated with the stock price (9.4.145) and governed by

$$\frac{dH_t}{H_t} = (\mu - \delta)\, dt + \eta(\rho\, dW^1 + \sqrt{1-\rho^2}\, dW^2), \quad 0 < t \le \tau, \quad H_t = h > 0. \tag{9.4.146}$$

Here μ is the expected rate of return of the illiquid risky asset, δ is the rate of dividend paid by the illiquid asset, η is the standard deviation of the rate of return, (W^1, W^2) are two independent standard Brownian motions, and $\rho \in (-1, 1)$ is the correlation coefficient between the stock index and the illiquid risky asset. The parameters μ, δ, η, ρ are all assumed to be constant. The stochastically distributed time τ does not depend on the Brownian motions (W^1, W^2). The probability density function of the τ distribution is denoted by $\phi(t)$, whereas $\Phi(t)$ denotes the cumulative distribution function, and $\overline{\Phi}(t)$, the survival function, also known as *a reliability function*, $\overline{\Phi}(t) = 1 - \Phi(t)$. We omit here the explicit notion of the possible parameters of the distribution in order to make the formulas shorter.

We assume that the investor consumes at rate c from the liquid wealth and the allocation-consumption plan (π, c) consists of the allocation of the portfolio with the cash amount $\pi = \pi(t)$ invested in stocks, the consumption stream $c = c(t)$ and the rest of the capital kept in bonds. Let us note that further on we sometimes omit the dependence on t in some of the equations for the sake of clarity of the formulas. The consumption stream c is admissible if and only if it is positive and there exists a strategy that finances it. All the income is derived from the capital gains and the investor must be solvent. In other words, the liquid wealth process L_t must cover the consumption stream. The wealth process L_t is the sum of cash holdings in bonds, stocks and *random* dividends from the non-traded asset minus the consumption stream, i.e. it must satisfy the balance equation

$$dL_t = \big(rL_t + \delta H_t + \pi_t(\alpha - r) - c_t\big)\, dt + \pi_t\sigma\, dW^1, \quad 0 < t \le \tau \quad L_t = l > 0.$$

The investor wants to maximize the overall utility consumed up to the random time of liquidation τ, given by

$$\mathcal{U}(c) := E\left[\int_0^\infty \overline{\Phi}(t)\mathsf{U}(c(t))\, dt\right]. \tag{9.4.147}$$

It means we work with the problem (9.4.147) that corresponds to the *value function* $V(l,h,t)$, which is defined as

$$V(l,h,t) = \max_{(\pi,c)} E\left[\int_t^\infty \overline{\Phi}(t)\mathsf{U}(c(t))\,dt \mid L(t)=l, H(t)=h\right]. \quad (9.4.148)$$

The value function $V(l,h,t)$ satisfies the HJB equation for the value function, in terms of l, h and t

$$\begin{aligned} V_t(l,h,t) + \frac{1}{2}\eta^2 h^2\, V_{hh}(l,h,t) + (rl+h)\,V_l(l,h,t) + \\ (\mu-\delta)h\,V_h(l,h,t) + \max_{\pi} G[\pi] + \max_{c\geq 0} H[c] &= 0, \quad (9.4.149) \\ G[\pi] = \frac{1}{2}V_{ll}(l,h,t)\,\pi^2\sigma^2 + V_{hl}(l,h,t)\,\eta\rho\pi\sigma h + \pi(\alpha-r)\,V_l(l,h,t), \\ H[c] = -c\,V_l(l,h,t) + \overline{\Phi}(t)\mathsf{U}(c), \quad (9.4.150) \end{aligned}$$

with the boundary condition

$$V(l,h,t) \to 0, \quad as \;\; t \to \infty.$$

For this problem authors proved in [18], [19] the following theorem for the value function $V(l,h,t)$.

Theorem 9.4.1 *There exists a unique viscosity solution of the corresponding HJB Equation (9.4.148) if*

1. $\mathsf{U(c)}$ *is strictly increasing, concave and twice differentiable in* c,
2. $\lim_{\tau\to\infty}\overline{\Phi}(\tau)E[\mathsf{U(c(\tau))}] = 0$, $\overline{\Phi}(\tau) \sim e^{-\kappa\tau}$ *or faster as* $\tau\to\infty$,
3. $\mathsf{U(c)} \leq \mathsf{M(1+c)^\gamma}$ *with* $0\langle\gamma\langle 1$ *and* $M\rangle 0$,
4. $\lim_{c\to 0}\mathsf{U'(c)} = +\infty$, $lim_{c\to+\infty}\mathsf{U'(c)} = 0$.

We now look at the Equation (9.4.149) in the case of exponentially distributed liquidation time i.e. $\overline{\Phi}(t) = -e^{-\kappa t}$ and logarithmic utility function $U(c) = \log c$ and just for this case we provide here some results of Lie group analysis. Further cases of different liquidation time distributions and utility functions and details of the Lie group analysis can be found in [17].

For this particular case authors proved in [19], [18] that the value function $V(l,h,t)$ is a concave monotonous increasing function. Using these properties of the value function we find both maxima in (9.4.149). We formally maximize both functions $G[\pi], H[c]$ with respect to π and c. It means we insert in HJB Equation (9.4.149)

$$\pi(t) = -\frac{\eta\rho\sigma h V_{hl}(l,h,t) + (\alpha - r)V_l(l,h,t)}{\sigma^2 V_{ll}(l,h,t)}, \quad c(t) = \frac{e^{-\kappa t}}{V_l(l,h,t)}$$

and obtain the following three-dimensional PDE

$$V_t(t,l,h) + \frac{1}{2}\eta^2 h^2 V_{hh}(t,l,h) + (rl + \delta h)V_l(t,l,h) + (\mu - \delta)hV_h(t,l,h)$$
$$-\frac{(\alpha - r)^2 V_l^2(t,l,h) + 2(\alpha - r)\eta\rho h V_l(t,l,h)V_{lh}(t,l,h) + \eta^2\rho^2\sigma^2 h^2 {V_{hl}}^2(t,l,h)}{2\sigma^2 V_{ll}(t,l,h)}$$
$$-(1 + \kappa t)e^{-\kappa t} - e^{-\kappa t}\ln V_l(t,l,h) = 0. \quad (9.4.151)$$

Using the **ReLie** software [58] (package for **REDUCE**) one can find three-dimensional Lie algebra L_3 admitted by this equation. The Lie algebra $L_3 = \langle \mathbf{U}_1, \mathbf{U}_2, \mathbf{U}_3 \rangle$ is spanned by three infinitesimal generators

$$\mathbf{U}_1 = \frac{1}{\kappa}\frac{\partial}{\partial t} - V\frac{\partial}{\partial V}, \quad \mathbf{U}_2 = \frac{\partial}{\partial V},$$
$$\mathbf{U}_3 = l\frac{\partial}{\partial l} + h\frac{\partial}{\partial h} + \frac{1}{\kappa}e^{-\kappa t}\frac{\partial}{\partial V}. \quad (9.4.152)$$

The following theorem can therefore be proved with the help of the **ReLie** software:

Theorem 9.4.2 *The Equation (9.4.151) admits a three-dimensional Lie algebra L_3 spanned by generators (9.4.152). The Lie algebra L_3 has the following non-zero commutator relation $[\mathbf{U}_1, \mathbf{U}_2] = \mathbf{U}_2$. It means that L_3 is a decomposable Lie algebra and can be represented as a semi-direct sum $L_3 = L_2 \bigoplus \mathbf{U}_3$ where $L_2 = \langle \mathbf{U}_1, \mathbf{U}_2 \rangle$. The optimal system of subalgebras for the Lie algebra L_3 (9.4.152) is listed in the Table 9.5*

Table 9.5: The optimal system of one- and two-dimensional subalgebras of L_3, where parameters are $\xi \in [-1, 1]$, $\epsilon = \pm 1$, $\zeta \in \mathbb{R}$.

Dimension of the subalgebra	Subalgebras
1	$h_1 = \langle \mathbf{U}_2 \rangle$, $h_2 = \left\langle \xi \mathbf{U}_1 + \sqrt{1-\xi^2}\mathbf{U}_3 \right\rangle$, $h_3 = \langle \epsilon\ \mathbf{U}_2 + \mathbf{U}_3 \rangle$
2	$h_4 = \langle \mathbf{U}_2, \mathbf{U}_3 \rangle$, $h_5 = \langle \mathbf{U}_1, \mathbf{U}_3 \rangle$, $h_6 = \langle \mathbf{U}_1 + \zeta\ \mathbf{U}_3, \mathbf{U}_2 \rangle$

To describe all non-similar invariant solutions to (9.4.151) we need to find an optimal system of subalgebras of L_3 listed in the Table 9.5. The construction of such systems for all real three-dimensional Lie algebras is given in [61] or one can use the software package **SymboLie** [58] (a supplement package for **Mathematica**). We use further the notation h_i for the subalgebras of the Lie algebra L_3 and H_i for the corresponding subgroups of the group G_3.

The choice of an optimal system of subalgebras is non-unique; likewise the choice of invariant expressions which we use as new variables to simplify the equation or to reduce the dimension of the studied equation. Sometimes one can take a favorable subalgebra and an invariant expression in this way so as to obtain a more convenient form of the reduced equation. Certainly the reduced equations are similar under the group of point transformations admitted by the original equation, but it is often rather difficult to find this similarity point transformation.

We provide here just two examples of possible reductions of the original three-dimensional Equation (9.4.151). First let us use a one-dimensional subalgebra h_3 from Table 9.5 and obtain a two-dimensional PDE after this reduction, in the second example we use the two-dimensional subalgebra h_5 and a reduced equation becomes an ODE.

Case $H_3(h_3)$. Under h_3 we denote the conjugacy class of subalgebras which have one discrete parameter $\epsilon = \pm 1$ and are spanned by the following generator $h_3 = \langle \pm \mathbf{U}_2 + \mathbf{U}_3 \rangle$. To find invariants of the subgroup H_3 we solve the characteristic system of the equations

$$\frac{dV}{\pm 1 + \frac{1}{\kappa} e^{-\kappa t}} = \frac{dl}{l} = \frac{dh}{h} = \frac{dt}{0}. \tag{9.4.153}$$

From the last expression we can immediately conclude that t is invariant in this case, i.e. $\text{inv}_1 = t$ and t is unaltered under the action of the subgroup H_3. This variable t will also be one of the independent variables in the reduced equation. The second equation in the system, i.e. $\frac{dl}{l} = \frac{dh}{h}$ immediately gives us the invariant expression $\text{inv}_2 = \frac{l}{h}$. We denote this invariant expression as $z = \frac{l}{h}$ and use it later as a new invariant independent variable to reduce the dimension of the studied PDE (9.4.151). The first and the third expressions in (9.4.153) give us the next equation to solve. The solution is an invariant expression $\text{inv}_3 = (\pm 1 + \frac{1}{\kappa} e^{-\kappa t}) \ln h - V(t, l.h)$. We use this expression to introduce a new dependent variable W

$$V(t, l, h) = (\pm 1 + \frac{1}{\kappa} e^{-\kappa t}) \ln h + W(z, t). \tag{9.4.154}$$

Using the new invariant variables t, z, W we reduce the three-dimensional

PDE (9.4.151) to a two-dimensional one

$$W_t - e^{-\kappa t}\ln W_z - \kappa t e^{-\kappa t} + \frac{\mu - \delta - \frac{\eta^2}{2} - \kappa}{\kappa} e^{-\kappa t} \pm (\mu - \delta - \frac{\eta^2}{2})$$
$$+ \frac{\eta^2}{2}(2zW_z + z^2 W_{zz}) + ((r - \mu - \delta)z + \delta)W_z + \frac{1}{2\sigma^2 W_{zz}}$$
$$\cdot (-(\alpha - r)^2 {W_z}^2 - 2(\alpha - r)\eta\rho\sigma(-{W_z}^2 + zW_z W_{zz}) - \eta^2\rho^2\sigma^2(W_z - zW_{zz})^2)$$
$$= 0.$$

The optimal policies (π, c) in new variables are

$$\pi = h\frac{\eta\rho\sigma\frac{\partial}{\partial z}(z\frac{\partial W}{\partial z}) - (\alpha - r)\frac{\partial W}{\partial z}}{\sigma^2 \frac{\partial^2 W}{\partial z^2}}, \qquad c = h\frac{1}{\frac{\partial W}{\partial z}}. \tag{9.4.155}$$

Case $H_5(h_5)$. The two-dimensional Lie algebra h_5 is spanned on two generators $\mathbf{U}_1$ and $\mathbf{U}_3$. To find the corresponding invariants of the two-parameter group H_5 we should solve the system of equations

$$\kappa dt = -\frac{dV}{V}, \tag{9.4.156}$$
$$\frac{dl}{l} = \frac{dh}{h} = \kappa e^{\kappa t} dV.$$

We obtain two invariants $\text{inv}_1 = \frac{l}{h}$ and $\text{inv}_2 = \kappa e^{\kappa t} V(t, l, h) - \ln h$. We choose one of them as an independent variable and the other as a dependent variable

$$z = \frac{l}{h}, \quad v(z) = \kappa e^{\kappa t} V(t, l, h) - \ln h. \tag{9.4.157}$$

We replace in (9.4.151) the variables l, h, t, V with new invariant variables z, v and obtain an ODE. All solutions of this ODE describe the solutions of (9.4.151) which are invariant under the action of the two-parameter subgroup H_5. After the variable substitution (9.4.157) the PDE (9.4.151) takes the form

$$z^2(v'')^2 + 2(c_1 z + c_2)v'v'' + c_3 \ln(v')v'' + c_3 v v'' + (c_4 z + c_5)v'' + c_6(v')^2 + c_7 v' = 0, \tag{9.4.158}$$

where the constants c_1, c_2, c_3, c_4 are defined by

$$c_1 = \left(1 - \frac{\mu - \delta}{\eta^2} + \frac{r}{\eta^2} + \frac{(\alpha - r)\rho}{\eta\sigma^2}\right), c_2 = \frac{2\mu}{\eta^2}, \quad c_3 = -\frac{2\kappa}{\eta^2}, \quad c_4 = -\rho^2,$$
$$c_5 = \left(-1 - 2\frac{\kappa}{\eta^2} + 2\frac{\kappa}{\eta^2}\ln\kappa\right), \quad c_6 = \frac{2(\alpha - r)\rho}{\eta\sigma^2}, c_7 = -\left(\frac{(\alpha - r)^2}{\eta^2\sigma^2} + \rho^2\right).$$

The optimal policies, therefore, in these new variables are

$$\pi = h\frac{\eta\rho\sigma(v' - zv'') - (\alpha - r)v'}{\sigma^2 v''}, \quad c = \kappa h\frac{1}{v'}. \tag{9.4.159}$$

As an example of another possible reduction which corresponds to the same subalgebra, we can look at a point transformation which we used for (9.4.151) in [18].

In [17] the four cases of the HJB equations (9.4.149) are studied with Lie group analysis. One takes two different utility functions: a log-utility and a HARA utility function as well as two different distributions for the liquidation time for an illiquid asset: an exponential and Weibull distributions. For all these four cases a detailed analysis of admitted Lie algebras, optimal system of subalgebras and possible reductions is provided in [17]. After corresponding reductions one obtains instead of a three-dimensional PDE, two-dimensional PDEs, or in some cases even ODEs. Such reduced equations are more convenient for numerical calculations.

9.4.3 Transaction costs models

In general, markets with transaction costs are incomplete because the uniqueness of the prices is lost and any participant has to pay the price modified by individual transaction costs and taxes. The incompleteness of the market is in this case not so evident as in previous cases but it still exists.

We listed in the very beginning of this chapter important properties of the ideal market which form a foundation of the Black-Scholes-Merton theory. We pointed out that the ideal market is frictionless, i.e. the assumption (6) is fulfilled. It means we do not have to pay any transaction costs or taxes of any type. This condition allows the instantaneous trading and continuous dynamic hedging. Under this condition one obtained the Black-Scholes linear parabolic equation. If one tries to apply this strategy in the real world without any adaptation then in a very short time period he would become bankrupt. The transaction costs and trading frictions are small but if somebody trades every tick they add up to a tremendous amount. Every portfolio manager has a dilemma: either you hedge every infinitesimally small tendency on the market and in short time become bankrupt due to the transaction costs or you ignore the market developments and avoid trading to save the transaction costs but then your portfolio is exposed to extremely high risk such as the possibility of losing the complete portfolio. The portfolio managers look for the compromise - they usually use the dynamic hedging strategy but they do not trade at each tick. The main problem in the modeling of hedging strategies is to include at least the main transaction costs and to describe the optimal hedging strategy for a self-financing portfolio.

The first work on the problem of option pricing in the presence of transaction costs was developed by Leland [49]. He introduced the idea of a periodic revision of a hedged portfolio. In his work Leland assumed that the level of transaction costs is constant, i.e. we have a market with proportional transaction costs. Such a type of transaction costs exists in the financial markets. They are not the only transaction costs but they are very typical ones. Leland reduced this problem to a nonlinear partial differential equation with an adjusted volatility. It is a general feature that any feedback effects as well as transaction costs lead to a nonlinearity in the corresponding PDE. Leland claimed that the terminal value of the portfolio approximates the payoff as the length of a revision interval tends to zero, i.e. if the revision is extremely frequent. Later, Kabanov and Safarian [42] proved that Leland's conjecture based on an approximate replication fails. It means that his model has a non-trivial limiting hedging error relative to simulated market prices (see as well the detailed discussion in [43]). From the mathematical point of view the problems arise in the limiting cases as revisions become unboundedly frequent. From the practical point of view the extremely frequent revisions are not desirable and the average errors are less than one-half of one per cent of the price suggested by Leland's formula [50].

The nonlinear PDE obtained in the Leland model is relativly simple and gives the possibility to approximate the influence of the transaction costs. But in comparison to the perfect portfolio within the Black-Scholes theory in Leland's framework the portfolio is unprotected between the revisions.

Kratka [46] suggested improving the Leland model for pricing derivative securities in the presence of proportional transaction costs. He additionally took into account the risk of the exposure of the unprotected portfolio in between the revisions. Later Jandačka and Ševčovič [41] modified Kratka's approach in order to derive a scale-invariant model. Let us describe the setting of the model introduced in [41] in the next section.

The risk-adjusted pricing methodology model

We give now a brief description of a so-called risk-adjusted pricing methodology (RAPM) model introduced in [41]. The authors describe the risk from the volatile portfolio by the average value of the variance of the synthesized portfolio. In the model setting the transaction costs and the unprotected portfolio risk both depend on the time interval between two revisions. The minimizing of the total risk leads to the RAPM model. The full description of the model and some numerical solutions to European and American options are presented in [79].

The authors of [41] assume similarly to the Black-Scholes model that the stock price dynamics is given by the geometric Brownian motion (9.1.4). Further it is assumed that the risk-free bond earns at a continuous compound constant rate r.

The time-steps Δt at which the portfolio can be revised and hedged against the price change of the underlying asset S_t are non-infinitesimal and fixed. The authors [41] introduce the idea of a switching time t^* for the last revision of the portfolio before the maturity time. In this way the time interval $(0, T)$ is divided in two parts, in the first part $(0, t^*)$ the revisions of the portfolio will be done regularly with the same period. In the second part of the time interval (t^*, T) there are no revisions and correspondingly no transaction costs, the portfolio is unprotected. Because of that it is assumed that the time interval (t^*, T) is very small. It is also assumed that the price of the contingent claim $u(S, t), \ \ t \in [t^*, T]$ is defined as in the classical Black-Scholes formula (here T is the maturity time). Further it is assumed that the model (similar to Leland's model) does not include the cost of establishing the initial investor's portfolio composition.

The value of the dynamically hedged portfolio V_t depends on chosen hedging strategy ϕ and is equal to $V_t^{\phi} = \delta_t S_t + \beta_t B_t$ at time t. Here δ_t is a number of units of the stock (a constant on each time interval Δt), B_t is the value of the bond and β_t is the number of units of the bond. As before we put $B_0 = 1$ and without loss of generality can rewrite the previous relation in the form $V_t^{\phi} = \delta_t S_t + \beta_t e^{rt}$. The self-financing hedging strategy that maintains the portfolio is the pair $\phi = (\delta_t, \beta_t)$.

In the model the total risk premium r_R contains two parts $r_R = r_{TC} + r_{VP}$, here TC denotes the part of the risk premium connected with transaction costs and VP the part connected with volatile portfolio. In any time-step Δt the change of portfolio V_t^{ϕ} is represented by $\Delta V_t^{\phi} = V_{t+\Delta t}^{\phi} - V_t^{\phi} = \beta_t e^{rt}(e^{r\Delta t} - 1) + \delta_t(S_{t+\Delta t} - S_t) - r_R S_t \Delta t$. The transaction costs part of the risk premium r_{TC} is modeled by the expression

$$r_{TC} = \frac{C\sigma S|u_{SS}|}{\sqrt{2\pi}}, \quad C = \frac{S_{ask} - S_{bid}}{S}, \tag{9.4.160}$$

where C is the round trip transaction costs per unit of transaction in dollars (for details see [49], [37] or [47]) and the value $u(S, t)$ is the value function of the contingent claim with respect to the asset price S and time t. The risk r_{VP} is connected with a volatility and during the time-step Δt unprotected portfolio is modeled by

$$r_{VP} = \frac{1}{2} R\sigma^4 S^2 (u_{SS})^2 \Delta t, \tag{9.4.161}$$

where R is a risk premium coefficient introduced in [46] and [41]. It represents the marginal value of the investor's exposure to a risk.

The total risk premium after [79] depends on the time-lag Δt and is a strongly convex function between two consecutive portfolio revisions. To obtain a risk-adjusted Black-Scholes equation the authors [41] minimize the total risk premium $r_R = r_{TC} + r_{VP}$. They then obtain the following value for the optimal time-lag

$$\Delta t_{opt} = \frac{C^{2/3}}{\sigma^2 (R\sqrt{2\pi}|Su_{SS}|)^{2/3}}.$$

Finally the authors use Itô's Lemma to obtain the RAPM model in the form of a nonlinear partial differential equation

$$u_t + \frac{1}{2}\sigma^2 S^2 u_{SS}(1 - \mu(Su_{SS})^{\frac{1}{3}}) - ru + rSu_S = 0, \quad \mu = 3\left(\frac{C^2 R}{2\pi}\right)^{1/3}, \tag{9.4.162}$$

where $t \in (0, t^*)$ and the value t^* is determined by the implicit equation $T - t^* = \min_{S>0} \Delta t_{opt}(S, t^*)$. The nonlinear term in the Equation (9.4.162) has the highest derivative u_{SS}, it means that this equation possesses a non-trivial singular perturbed algebraic structure. To represent a well-posed parabolic problem the following condition should be satisfied

$$Su_{SS}(S, t) < \left(\frac{3}{4\mu}\right)^3. \tag{9.4.163}$$

The condition (9.4.163) for instance will not be fulfilled for usual call and put options at $S = E$ and $t \to T^-$, where E is the strike price of the corresponding option.

To avoid the problems with the singularities in the model the authors introduce as previously mentioned the switching time t^*. They demand that the condition (9.4.163) is satisfied when $t = t^*$. In other words the authors assume that in the very small time interval $T - t^*$ the value $u(S, t)$ is near to the payoff function and one can use the Black-Scholes linear parabolic equation. It means one replaces the piecewise linear payoff function at T with the smooth Black-Scholes solution at t^*. The *new* payoff function is smooth and the condition (9.4.163) can be satisfied. The equation for the switching time t^* can be reduced to the form $T - t^* = CR^{-1}\sigma^{-2}$ (for European call and put options). It has a positive solution and the condition (9.4.163) is satisfied for these options if

$$\frac{C}{R} < \sigma^2 T, \quad CR < \frac{\pi}{8}. \tag{9.4.164}$$

We show in the next section an application of Lie group analysis to the RAPM model (9.4.162). Using symmetries of the equation one can describe the complete set of invariant solutions as was done in [13]. The analytical solutions can be used as a benchmark for numerical methods, for example. We also show that the RAPM model possesses four–dimensional symmetry algebras for the cases when $r = 0$ and when $r \neq 0$; both algebras are isomorphic.

Symmetry properties of the RAPM model

The symmetry analysis of the RAPM model was provided in [13], and the following theorem was proved.

Theorem 9.4.3 *The Equation (9.4.162) admits a four-dimensional Lie algebra L_4 with the following infinitesimal generators*

$$\mathbf{U}_1 = S\frac{\partial}{\partial S} + u\frac{\partial}{\partial u}, \quad \mathbf{U}_2 = e^{rt}\frac{\partial}{\partial u}, \quad \mathbf{U}_3 = \frac{\partial}{\partial t}, \quad \mathbf{U}_4 = S\frac{\partial}{\partial u}. \tag{9.4.165}$$

The commutator relations are

$$[\mathbf{U}_1, \mathbf{U}_2] = -\mathbf{U}_2, \ [\mathbf{U}_2, \mathbf{U}_3] = -r\mathbf{U}_2, \tag{9.4.166}$$

$$[\mathbf{U}_1, \mathbf{U}_3] = [\mathbf{U}_1, \mathbf{U}_4] = [\mathbf{U}_2, \mathbf{U}_4] = [\mathbf{U}_3, \mathbf{U}_4] = 0. \tag{9.4.167}$$

The algebra L_4 is a decomposable Lie algebra and can be written as a semidirect sum

$$L_4 = L_2 \bigoplus e_3 \bigoplus e_4, \quad L_2 = \langle\ e_1, e_2\ \rangle, \quad [e_1, e_2] = e_2. \tag{9.4.168}$$

The generators in the case $r \neq 0$ takes the form

$$e_1 = (r-1)\mathbf{U}_1 + \mathbf{U}_3 = (r-1)S\frac{\partial}{\partial S} + (r-1)u\frac{\partial}{\partial u} + \frac{\partial}{\partial t}, \qquad e_2 = \mathbf{U}_2 = e^{rt}\frac{\partial}{\partial u},$$
$$e_3 = r\mathbf{U}_1 + \mathbf{U}_3 = rS\frac{\partial}{\partial S} + ru\frac{\partial}{\partial u} + \frac{\partial}{\partial t}, \qquad e_4 = \mathbf{U}_4 = S\frac{\partial}{\partial u}. \tag{9.4.169}$$

For $r = 0$ the generators of L_4 have the form

$$e_1 = -\mathbf{U}_1 = -S\frac{\partial}{\partial S} - u\frac{\partial}{\partial u}, \qquad e_2 = \mathbf{U}_2 = \frac{\partial}{\partial u},$$
$$e_3 = \mathbf{U}_3 = \frac{\partial}{\partial t}, \qquad e_4 = \mathbf{U}_4 = S\frac{\partial}{\partial u}. \tag{9.4.170}$$

Table 9.6: [61] The optimal system of subalgebras h_i of the algebra L_4, where $a \in \mathbb{R}$, $\epsilon = \pm 1$, $\phi \in [0, \pi]$.

Dimen-sion	Subalgebras
1	$h_1 = \langle e_2 \rangle$, $h_2 = \langle e_3 \cos(\phi) + e_4 \sin(\phi) \rangle$, $h_3 = \langle e_1 + a(e_3 \cos(\phi) + e_4 \sin(\phi)) \rangle$, $h_4 = \langle e_2 + \epsilon(e_3 \cos(\phi) + e_4 \sin(\phi)) \rangle$
2	$h_5 = \langle e_1 + a(e_3 \cos(\phi) + e_4 \sin(\phi)), e_2 \rangle$, $h_6 = \langle e_3, e_4 \rangle$, $h_7 = \langle e_1 + a(e_3 \cos(\phi) + e_4 \sin(\phi)), e_3 \sin(\phi) - e_4 \cos(\phi) \rangle$, $h_8 = \langle e_2 + \epsilon(e_3 \cos(\phi) + e_4 \sin(\phi)), e_3 \sin(\phi) - e_4 \cos(\phi) \rangle$, $h_9 = \langle e_2, e_3 \sin(\phi) - e_4 \cos(\phi) \rangle$
3	$h_{10} = \langle e_1, e_3, e_4 \rangle$, $h_{11} = \langle e_2, e_3, e_4 \rangle$, $h_{12} = \langle e_1 + a(e_3 \cos(\phi) + e_4 \sin(\phi)), e_3 \sin(\phi) - e_4 \cos(\phi), e_2 \rangle$

An optimal system of subalgebras of algebra L_4 (9.4.168) of dimensions one, two and three is represented in Table 9.6

We notice that the commutator relations (9.4.166) depend on the parameter r, i.e. on the interest rate included in the model. Depending on whether $r = 0$ or $r \neq 0$, we obtain different commutation relations if we use $\mathbf{U}_1, \ldots, \mathbf{U}_4$ as basis generators for the algebra L_4. It is easy to prove that in both cases the algebras L_4 for $r = 0$ and L_4 for $r \neq 0$ are isomorphic. The proper choice of generators $e_1, \ldots, e_4$ provides a direct proof of the isomorphism. We provide in Theorem 9.4.3 the choice of generators $e_1, \ldots, e_4$ in both cases.

Using the generators $e_1, \ldots, e_4$ we display an optimal system of subalgebras of the algebra L_4 (for both cases $r = 0$ and $r \neq 0$) in Table 9.6. It is possible to list the complete set of symmetry reductions of Equation (9.4.162) in both cases. It is also possible to provide exact solutions to all reduced equations in an explicit or parametric form (see [13]). Due to the exact form of the solutions it is possible to compare different structures of these solutions in both cases (where the interest rate is $r = 0$ and $r \neq 0$).

We provide a short overview of these results and some examples of solutions. First we look at the case where the interest rate for riskless account $r \neq 0$. We can get an invariant solution of (9.4.162) if we choose as new variables

$$z = Se^{-(r-1)t}, \quad w = u(S,t)e^{-(r-1)t} - \alpha e^t. \tag{9.4.171}$$

It is easy to see that PDE (9.4.162) after this substitution will be reduced to a second order ODE

$$-w + zw_z + \frac{1}{2}\sigma^2 z^2 w_{zz}\left(1 - \mu(zw_{zz})^{1/3}\right) = 0. \tag{9.4.172}$$

This equation admits a further reduction to a first order ODE

$$v = -\frac{\sigma^2}{2} z(v_z - \mu v_z^{4/3}) \tag{9.4.173}$$

and after the substitution

$$v(z, w) = zw_z - w. \tag{9.4.174}$$

We get a solution of (9.4.173) in a parametric form. We denote $v_z = \theta$ then the Equation (9.4.173) has the following parametric solution

$$\begin{aligned} v(\theta) &= -\frac{\sigma^2}{2} z(\theta)(\theta - \mu\theta^{4/3}) \\ &= -c_1 \frac{\sigma^2}{2} \theta^{\beta}(1 - \mu\theta^{1/3}) \left(\frac{1}{\beta} - \mu\frac{\sigma^2}{2}\theta^{1/3}\right)^{-1-3\beta}, \\ z(\theta) &= c_1 \left(1 + \frac{\sigma^2}{2}\left(1 - \mu\theta^{\frac{1}{3}}\right)\right)^{-1-3\beta} \theta^{-1+\beta} \\ &= c_1 \left(\frac{1}{\beta} - \mu\frac{\sigma^2}{2}\theta^{\frac{1}{3}}\right)^{-1-3\beta} \theta^{-1+\beta} \end{aligned} \tag{9.4.175}$$

where $\beta = \left(1 + \frac{\sigma^2}{2}\right)^{-1}$ and $c_1 = \text{const}$. We obtain the parametric solution of (9.4.172) using the substitution (9.4.174) which now takes the form

$$v(\theta) = z(\theta)w_z - w = (\ln z(\theta))_\theta^{-1} w_\theta - w \tag{9.4.176}$$

in this representation

$$w(\theta) = z(\theta)\left(c_2 + g(\theta)\right), \quad c_2 = \text{const}, \tag{9.4.177}$$

where the function $g(\theta)$ is given by

$$\begin{aligned} g(\theta) = &-\left(\frac{\mu\sigma^2}{2}\right)^{-2} \frac{3\beta+1}{\beta^2}\theta^{\frac{1}{3}} - \frac{3\beta+1}{\mu\sigma^2\beta}\theta^{\frac{2}{3}} - \frac{4}{3}\theta + \frac{\sigma^2}{2}\theta(1 - \mu\theta^{\frac{1}{3}}) \\ &- \left(\frac{\mu\sigma^2}{2}\right)^{-3} \frac{1}{\beta^2(3+\beta)} \ln\left(1 + \frac{\sigma^2}{2}\left(1 - \mu\theta^{\frac{1}{3}}\right)\right). \end{aligned} \tag{9.4.178}$$

Expressions (9.4.177) and (9.4.175) give a parametric representation of a solution $w(z)$ of the equation (9.4.172).

Using the optimal system of subalgebras of the symmetry algebra L_4 for (9.4.162) we can systematically describe all equivalence classes of invariant solutions.

In Table 9.6 we have four one-dimensional subalgebras. The first one describes the invariance of the solutions of (9.4.162) by $u(S,t) \rightarrow u(S,t) + \alpha e^{rt}$, where α is an arbitrary constant.

The second one produces a one–dimensional subgroup with invariants

$$z = Se^{-rt}, \quad w = \frac{u}{S} - \frac{\tau}{r}\ln S, \quad r \neq 0, \quad \tau = \tan(\phi), \ \phi \in [0,\pi], \ \phi \neq \pi/2. \tag{9.4.179}$$

Using these invariants we obtain the equation in the reduced form

$$(\tau + rz(zw_{zz} + 2w_z))\left(1 - \mu r^{-\frac{1}{3}}\left(\tau + rz(zw_{zz} + 2w_z)\right)^{\frac{1}{3}}\right) + \frac{2r\tau}{\sigma^2} = 0, \tag{9.4.180}$$

$$r \neq 0, \qquad \tau = \tan(\phi), \quad \phi \in [0,\pi], \ \phi \neq \pi/2$$

which after substitution $w_z(z) = v(z)$ has the form of an ODE of the first order

$$\left(\tau + r(z^2v)_z\right)\left(1 - \mu r^{-\frac{1}{3}}\left(\tau + r(z^2v)_z\right)^{\frac{1}{3}}\right) + \frac{2r\tau}{\sigma^2} = 0. \tag{9.4.181}$$

We denote $(\tau + r(z^2v)_z)^{1/3} = p(z)$, then we get an algebraic equation of the fourth order on the value $p(z)$

$$p^3\left(1 - \mu r^{-\frac{1}{3}}p\right) + \frac{2r\tau}{\sigma^2} = 0. \tag{9.4.182}$$

We denote the real roots of (9.4.182) by k_i then we have just to integrate two simple ODEs

$$\tau + r(z^2v)_z = k_i^3, \qquad w_z(z) = v(z). \tag{9.4.183}$$

to get solutions of the equation (9.4.180) and correspondingly of (9.4.162)

$$u(S,t) = \frac{k_i^3}{r}S\ln S - (k_i^3 - \tau)tS + c_1S + c_2e^{rt}, \tag{9.4.184}$$

where $c_1, \ c_2 \in \mathbb{R}, r \neq 0, \tau = \tan(\phi), \phi \in [0,\pi], \quad \phi \neq \pi/2$.

From the financial point of view we can describe the solution as a portfolio with some number of assets S and bonds which is hedged against price changes in the presence of transaction costs. We see that the price of the hedged portfolio increases proportionally to $S \ln S$ due to transaction costs, and it is more expensive than in the BS model.

Another non-equivalent class of the invariant solutions of (9.4.162) can be obtained using the third one–dimensional subalgebra in Table 9.6. We use the invariants

$$z = Se^{-(r+\gamma)t}, \qquad u(S,t) = Sw(z) + \zeta S \log S, \tag{9.4.185}$$

$$\gamma = (1 + a\cos(\phi))^{-1}, \;\; \zeta = \frac{a\sin(\phi)}{r(1 + a\cos(\phi)) - 1}, a \in \mathbb{R}, \;\; \phi \in [0,\pi] \tag{9.4.186}$$

as new variables in the RAPM Equation (9.4.162) to obtain ODE

$$\frac{\sigma^2}{2}\left(z(zw)_{zz} + \zeta\right)\left(1 - \mu\left(z(zw)_{zz} + \zeta\right)^{\frac{1}{3}}\right) + r\zeta - \gamma z w_z = 0. \tag{9.4.187}$$

The solutions of this equation can be given in the parametric form

$$\begin{aligned} z(\theta) &= \exp\left(\int \frac{d\theta}{k_i(\theta)^3 - \theta - \zeta}\right), \\ w(\theta) &= \int \frac{\theta d\theta}{k_i(\theta)^3 - \theta - \zeta}, \quad \theta \in \mathbb{R} \end{aligned} \tag{9.4.188}$$

where $k_i(\theta)$ is one of the real roots of the fourth order algebraic equation

$$\frac{\sigma^2}{2} k_i(\theta)^3 (1 - \mu k_i(\theta)) + r\zeta - \gamma\theta = 0. \tag{9.4.189}$$

The investigation of the analytical properties of these solutions is rather complicated because of the parametric form. We can use these expressions to evaluate numerically the graphical representation of the solutions or for other properties of different group parameters.

The last class of non-equivalent invariant solutions to (9.4.162) with $r \neq 0$ can be obtained using the fourth subalgebra in Table 9.6. Here the invariant variables have the form

$$z = Se^{-rt}, \;\; u(S,t) = Sw(z) + \left(\frac{\tau}{r} + \frac{\epsilon}{r\cos(\phi)} z^{-1}\right) S \log S,$$
$$\tau = \tan(\phi), \phi \in [0,\pi], \;\; \phi \neq \pi/2, \epsilon = \pm 1$$

and reduced Equation (9.4.162) takes the following form

$$\frac{\sigma^2}{2}\left(z(zw)_{zz} + \frac{\tau}{r} + \frac{\epsilon}{rz\cos(\phi)}\right)\left(1 - \mu\left(z(zw)_{zz} + \frac{\tau}{r} + \frac{\epsilon}{rz\cos(\phi)}\right)^{\frac{1}{3}}\right)$$
$$+\tau + \frac{\epsilon}{z\cos(\phi)} = 0. \tag{9.4.190}$$

We notice that the expression $p(z) = \left(z(zw)_{zz} + \frac{\tau}{r} + \frac{\epsilon}{rz\cos(\phi)}\right)^{\frac{1}{3}}$ is constant and the value $p(z)$ satisfies an algebraic equation of the fourth order

$$p^3(z)\left(1 - \mu p(z)\right) + \frac{2\tau}{\sigma^2} + \frac{2\epsilon}{z\sigma^2\cos(\phi)} = 0. \tag{9.4.191}$$

We denote the real roots by $k_i(z)$ then for each root $k_i(z)$ we obtain the solutions of the equation (9.4.162) which are given as two-parametric families of functions

$$\begin{aligned} u(S,t) &= e^{rt}\int\left(\int \frac{k_i(z)^3}{z}\mathrm{d}z\right)\mathrm{d}z + S\left(\tau t + c_1\right) \\ &+ e^{rt}\left(\frac{\epsilon}{\cos(\phi)}t + c_2\right), \\ z &= Se^{-rt}, \quad \tau = \tan(\phi), \phi \in [0,\pi], \ \phi \neq \pi/2, c_1, c_2 \in \mathbb{R}, \epsilon = \pm 1. \end{aligned}$$

To obtain the invariant solutions of (9.4.162) with $r = 0$ we repeat the procedure described above. The general structure of the optimal system of subalgebras is the same in both cases but the forms of infinitesimal generators differ. The invariants and the reductions therefore take other forms as well.

The first subalgebra in the optimal system of subalgebras in Table 9.6 does not provide any new reduction like in the case $r \neq 0$. The generator of the subalgebra h_1^0 has a very simple form $e_2 = \frac{\partial}{\partial u}$ in the case $r = 0$. This means that we are dealing with a subgroup of translations in the u-direction. Hence, to each solution of the Equation (9.4.162) with $r = 0$, we can add an arbitrary constant without destroying the property of the function to be a solution.

The second subalgebra h_2^0 has the form $h_2^0 = \langle\ e_3\cos(\phi) + e_4\sin(\phi)\ \rangle$, it means that in terms of the variables (t, S, u) we have the subalgebra of the following type

$$h_2^0 = \langle\ \cos(\phi)\frac{\partial}{\partial t} + \sin(\phi)S\frac{\partial}{\partial u}\ \rangle. \tag{9.4.192}$$

The invariants of the corresponding subgroup are given by

$$z = S, \quad w = u(S,t) - \tau tS, \quad \tau = \tan(\phi), \quad \phi \in [0,\pi], \ \phi \neq \pi/2. \tag{9.4.193}$$

If we use the variables z, w as new independent and dependent variables we obtain the following reduction of the RAPM model (9.4.162) with $r = 0$

$$\frac{\sigma^2}{2}zw_{zz}\left(1 - \mu\left(z\,w_{zz}\right)^{\frac{1}{3}}\right) + \tau = 0, \quad \tau = \tan(\phi), \quad \phi \in [0,\pi], \ \phi \neq \pi/2. \tag{9.4.194}$$

We denote $(z\ w_{zz})^{\frac{1}{3}} = p(z)$ and obtain an algebraic fourth order equation for the value $p(z)$

$$p^3\,(1-\mu p) + \frac{2\tau}{\sigma^2} = 0. \tag{9.4.195}$$

As before we denote the real roots of this equation as k_i. To find solutions to the ODE (9.4.194) we just have to integrate it twice

$$z\ w_{zz} = k_i^3. \tag{9.4.196}$$

Then the corresponding solutions of the Equation (9.4.194) are given by

$$u(S,t) = k_i^3 S\,(\ln S - 1) + \ \tau t S + c_1 S + c_2, \tag{9.4.197}$$

where $\tau = \tan(\phi)$, $c_1,\ c_2 \in \mathbb{R}, \phi \in [0,\pi], \phi \neq \pi/2$.

This solution is very similar to the solution (9.4.184) for the case $r \neq 0$, we should remark that the roots k_i are denoted in the same way but are defined by different equations in these cases.

The third subalgebra h_3^0 for $r = 0$ has the form

$$h_3^0 = \langle\ a\cos(\phi)\frac{\partial}{\partial t} - S\frac{\partial}{\partial S} + (a\sin(\phi)S - u)\frac{\partial}{\partial u}\ \rangle, \tag{9.4.198}$$

where $a \in \mathbb{R}$, $\phi \in [0,\pi]$ are parameters. The invariants z, w of the corresponding group are given by the expressions

$$z = Se^{\delta t}, \quad u(S,t) = Sw(z) + \zeta S\log S,$$

where the parameters are defined as

$$\delta = (a\cos(\phi))^{-1}, \quad \zeta = a\sin(\phi), \ a \in \mathbb{R}, \ a \neq 0, \ \phi \in [0,\pi], \ \phi \neq \pi/2, \tag{9.4.199}$$

and the reduced equation takes the form

$$\frac{\sigma^2}{2}\left(z(zw)_{zz} + \zeta\right)\left(1 - \mu\left(z(zw)_{zz} + \zeta\right)^{\frac{1}{3}}\right) + \delta z w_z = 0. \tag{9.4.200}$$

The solutions to this equation can be represented in the parametric form (9.4.188), where $k_i(v)$ is one of the real roots of the equation

$$\frac{\sigma^2}{2}k_i(v)^3(1 - \mu k_i(v)) + \delta v = 0, \tag{9.4.201}$$

and the parameter δ is defined in (9.4.199).

The fourth subalgebra h_4^0 for $r = 0$ in Table 9.6 has the form

$$h_4^0 = \langle \ \epsilon \cos(\phi)\frac{\partial}{\partial t} + (1 + \epsilon \sin(\phi) S)\frac{\partial}{\partial u} \ \rangle, \tag{9.4.202}$$

where $\epsilon = \pm 1$, $\phi \in [0, \pi]$ are parameters.

The invariants z, w of this subgroup H_4^0 are given by the expressions

$$z = S, \qquad w(z) = u(S,t) - \tau t S - \frac{\epsilon\, t}{\cos(\phi)}, \quad \tau = \tan(\phi), \tag{9.4.203}$$

and the RAPM model is reduced to the ODE of the form

$$\frac{\sigma^2}{2} z^2 w_{zz}\left(1 - \mu\,(z w_{zz})^{\frac{1}{3}}\right) + \tau z + \frac{\epsilon}{\cos(\phi)} = 0, \tag{9.4.204}$$

where $\tau = \tan(\phi)$, $\phi \in [0, \pi]$, $\phi \neq \pi/2$, $\epsilon = \pm 1$.

The structure of Equation (9.4.204) is very similar to previous cases and we can use similar tools to solve it. We first substitute $(z w_{zz})^{1/3} = p(z)$. Then for the function $p(z)$ we obtain a fourth order algebraic equation but now its coefficients depend on the variable z

$$p(z)^3\,(1 - \mu p(z)) + \frac{2\tau}{\sigma^2} + \frac{2\epsilon}{z\sigma^2 \cos(\phi)} = 0, \tag{9.4.205}$$

where $\tau = \tan(\phi)$, $\phi \in [0, \pi]$, $\phi \neq \pi/2$, $\epsilon = \pm 1$. For each real root $k_i(z)$ of this equation we have then to solve a linear ODE

$$z\; w_{zz} = k_i(z)^3. \tag{9.4.206}$$

The corresponding invariant solutions to (9.4.162) then have the form

$$u(S,t) = \int \left(\int \frac{k_i(S)^3}{S} \mathrm{d}S \right) \mathrm{d}S + \tan(\phi)\; t\; S + \frac{\epsilon}{\cos(\phi)} + c_1 S + c_2, \tag{9.4.207}$$

where $c_1,\ c_2 \in \mathbb{R}$, $\phi \in [0, \pi]$, $\phi \neq \pi/2$, $\epsilon = \pm 1$. The expressions for these solutions are rather lengthy so they are omitted here.

Now we have described the complete series of invariant reductions of the RAPM model found in [13]. In each of these cases the partial differential Equation (9.4.162) is reduced to an ordinary differential equation. Using the optimal system of subalgebras (Table 9.6) we were able to present the complete set of the non-equivalent reductions of Equation (9.4.162) up to the transformations of the symmetry group. In both cases, $r \neq 0$ and $r = 0$, we obtain three non-trivial reductions to ODEs. In all six cases it is possible to solve these ODEs and to obtain the explicit or parametric representations of exact invariant solutions to the RAPM model. We deal here with the very seldom case when we can compare structures of non-equivalent invariant solutions since they are given in explicit or parametric forms.

Each of these solutions contains two integration parameters and some free parameters connected with the corresponding subgroup. This reasonable set of parameters allows one to approximate a wide class of boundary conditions.

The RAPM model (9.4.162) possesses a non-trivial analytical and singular-perturbed algebraic structure. There exist rather few methods to study equations of such high complexity. An application of both analytical and numerical methods to singular-perturbed equations is a highly non-trivial task. The RAPM model has been studied in detail using numerical methods in [41] and in [79]. The authors of [41] derive a robust numerical scheme for solving the Equation (9.4.162) and perform extensive numerical testing of the model and compare the results with the real market data. In [79] Ševčovič studies the free boundary problem for the RAPM model and provides a description of the early exercise boundary for American style call options. Using the same numerical method he provides some computational examples of the free boundary approximation for American style of Asian call options with arithmetically average floating strike as well. He proposes a numerical method based on the finite difference approximation combined with an operator splitting technique for a numerical approximation of the solution and a computation of the free boundary condition position.

We hope that Lie group analysis of the RAPM model, which was provided in [13] and is shortly described here, gives us a more general, alternative point of view on the structure of this equation. It opens the possibility to exploit the Lie algebraic structure of the equation and may be helpful to improve other methods to find the solution.

Bibliography

[1] M. Abramowitz and I. A. Stegun. Chapter 10. In *Handbook of Mathematical Functions with Formulas, Graphs, and Mathematical Tables.* National Bureau of Standards, 1964.

[2] I. D. Ado. The representation of lie algebras by matrices. *Uspekhi Matematicheskikh Nauk (in Russian)*, 2(6):159–173, 1947. English translation in Ado, I. D. (1949), *The representation of Lie algebras by matrices*, American Mathematical Society Translations 1949 (2): 21, ISSN 0065-9290, MR 0030946.

[3] K. J. Arrow. *Essays in the Theory of Risk Bearing.* Markham Pub. Co., Chicago, reprint edition, 1965.

[4] P. Bank and D. Baum. Hedging and portfolio optimization in financial markets with a large trader. *Mathematical Finance*, 14(1):1–18, 2004.

[5] J. G. F. Belinfante and B. Kolman. A survey of Lie groups and Lie algebras with applications and computational methods. Technical report, Society for Industrial and Applied Lie mathematics, Philadelphia, 1989.

[6] A. Bihlo and R. O. Popovych. Invariant discretization schemes for the shallow-water equations. *SIAM Journal on Scientific Computing*, 34(6):B810–B839, 2012.

[7] F. Black and M. Scholes. The Pricing of Options and Corporate Liabilities. *Journal of Political Economy*, 81:637–654, 1973.

[8] G. W. Bluman and S. C. Anco. *Symmetry and integration methods for differential equations*, volume 154 of *Applied Mathematical Sciences.* Springer-Verlag, New York, 2002.

[9] G. W. Bluman, A. F. Cheviakov, and S. C. Anco. *Applications of symmetry methods to partial differential equations*, volume 168 of *Applied Mathematical Sciences.* Springer-Verlag, New York, 2010.

[10] G. W. Bluman and S. Kumei. *Symmetries and differential equations*, volume 81 of *Applied Mathematical Sciences*. Springer-Verlag, New York, 1989.

[11] L. A. Bordag. Symmetry reductions and exact solutions for nonlinear diffusion equations. *International Journal of Modern Physics A*, 24(8-9):1713–1716, 2009.

[12] L. A. Bordag. Pricing options in illiquid markets: Optimal systems, symmetry reductions and exact solutions. *Lobachevskii Journal of Mathematics*, 31(2):90–99, May 2010.

[13] L. A. Bordag. Study of the risk-adjusted pricing methodology model with methods of Geometrical Analysis. *Stochastics An International Journal of Probability and Stochastic Processes*, 83(4):333–345, 2011.

[14] L. A. Bordag and A. Y. Chmakova. Explicit solutions for a nonlinear model of financial derivatives. *International Journal of Theoretical and Applied Finance*, 10(1):1–21, February 2007.

[15] L. A. Bordag and R. Frey. Pricing options in illiquid markets: symmetry reductions and exact solutions. In M. Ehrhardt, editor, *Nonlinear Models in Mathematical Finance*. Nova Science Publishers, New York, 2009.

[16] L. A. Bordag and A. Mikaelyan. Models of Self-Financing Hedging Strategies in Illiquid Markets: Symmetry Reductions and Exact Solutions. *Letters in Mathematical Physics*, 96(1-3):191–207, February 2011.

[17] L. A. Bordag and I. P. Yamshchikov. Optimization problem for a portfolio with an illiquid asset: Lie group analysis. 2015.

[18] L. A. Bordag, I. P. Yamshchikov, and D. Zhelezov. Optimal allocation-consumption problem for a portfolio with an illiquid asset. *International Journal of Computer Mathematics*, 2014. http://dx.doi.org/10.1080/00207160.2013.877584.

[19] L. A. Bordag, I. P. Yamshchikov, and D. Zhelezov. Portfolio optimization in the case of an asset with a given liquidation time distribution. *International Journal of Engineering and Mathematical Modelling*, 2(2):31–50, 2015.

[20] L. A. Bordag. On option-valuation in illiquid markets: invariant solutions to a nonlinear model. In A. Sarychev, A. Shiryaev, M. Guerra,

and Grossinho M.R., editors, *Mathematical Control Theory and Finance.* Springer, 2007.

[21] B. J. Cantwell. *Introduction to symmetry analysis.* Cambridge University Press, Cambridge, 2002.

[22] U. Cetin, R. A. Jarrow, and P. Protter. Liquidity risk and arbitrage pricing theory. *Finance and Stochastics*, 8:311–341, 2004.

[23] A. Y. Chmakova. *Symmetriereduktionen und explicite Lösungen für ein nichtlineares Modell eines Preisbildungsprozesses in illiquiden Märkten.* PhD thesis, BTU Cottbus, 2005.

[24] H. T. Davis. *Introduction to nonlinear differential and integral equations.* Dover publications, USA, 2010.

[25] V. A. Dorodnitsyn. Finite-difference models entirely inheriting continuous symmetry of original differential equations. *Int. J. Mod. Physics. C*, 5:723–734, 1994.

[26] V. A. Dorodnitsyn and Winternitz P. Lie point symmetry preserving discretizations for variable coefficient korteweg- de vries equations. *Nonlinear Dynam.*, 22:49–59, 2000.

[27] L. Dresner. *Applications of Lie's Theory of ordinary and partial differential equations.* IOP, 1999.

[28] R. Frey. Perfect option replication for a large trader. *Finance and Stochastics*, 2:115–148, 1998.

[29] R. Frey. Market illiquidity as a source of model risk in dynamic hedging. In R. Gibson, editor, *Model Risk*, pages 125–136. Risk Publications, London, 2000.

[30] R. Frey and P. Patie. Risk management for derivatives in illiquid markets: a simulation study. In K. Sandmann and P. Schönbucher, editors, *Advances in Finance and Stochastics.* Springer, Berlin, 2002.

[31] R. Frey and A. Stremme. Market volatility and feedback effect from dynamic hedging. *Mathematical Finance*, 7(4):351–374, 1997.

[32] G. Gaeta. *Nonlinear Symmetries and Nonlinear Equations.* Kluwer Academic Publishers, Dordrecht, NL Boston, USA London UK, 1994.

[33] R. K. Gazizov and N. H. Ibragimov. Lie symmetry analysis of differential equations in finance. *Nonlinear Dynamics*, 17(4):387–407, 1998.

[34] K. Glover, P. Duck, and D. Newton. On nonlinear models of markets with finite liquidity: some cautionary notes. Technical report, University of Manchester, 2007.

[35] V. Henderson and D. Hobson. Utility Indifference Pricing - An Overview. In R. Carmona, editor, *Indifference Pricing: Theory and Applications*, chapter 2. Princeton University Press, 2009.

[36] S. D. Hodges and A. Neuberger. Optimal replication of contingent claims under transaction costs. *Review of Future Markets*, 8:222–239, 1989.

[37] T. Hoggardt, A.E. Whalley, and P. Wilmott. Hedging option portfolios in the presence of transaction costs. *Advances in Futures and Options Research*, 7:21–35, 1994.

[38] P. E. Hydon. *Symmetry methods for differential equations. A beginner's guide.* Cambridge Texts in Applied Mathematics. Cambridge University Press, Cambridge, 2000.

[39] N. H. Ibragimov. *Lie group analysis of differential equations.* CRS Press, Boca Raton, 1994.

[40] N. H. Ibragimov. *Elementary Lie Group Analysis and Ordinary Differential Equations.* John Wiley&Sons., Chichester, New York, USA Weinheim, Germany Brisbane, USA Singapore, Singapore Toronto Canada etc, 1999.

[41] M. Jandačka and D. Ševčovič. On the risk-adjusted pricing-methodology-based valuation of vanilla options and explanation of the volatility smile. *Journal of Applied Mathematics*, 2005(3):235–258, 2005.

[42] Y. M. Kabanov and M. M. Safarian. On Leland's strategy of option pricing with transactions costs. *Finance and Stochastics*, 250(February 1996):239–250, 1997.

[43] Y. M. Kabanov and M. M. Safarian. *Markets with transaction costs. Mathematical Theory.* Springer, 2010.

[44] I. Karatzas and S.E. Shreve. *Methods of mathematical finance.* Springer, New York, 1998.

[45] D. Kramkov and W. Schachermayer. The asymptotic elasticity of utility functions and optimal investment in incomplete markets. *The Annals of Applied Probability*, 9:904–950, 1999.

[46] M. Kratka. No mystery behind the smile. *Risk*, 9:67–71, 1998.

[47] Y. K. Kwok. *Mathematical models of financial derivatives.* Springer, 1998.

[48] V. I. Lagno, S. V. Spichak, and V. I. Stognii. Symmetry analysis of evolution type equations (in Russian: Simmetrinyi analis uravnenii evolutionnogo tipa). Technical report, Institute of computer sciences, Ishevsk, Moscow, 2004.

[49] H. E. Leland. Option pricing and replication with transaction costs. *Journal of Finance*, 40:1283–1301, 1985.

[50] H. E. Leland. Comments on Hedging errors with Leland's option model in the presence of transactions costs. *Finance Research Letters*, 4(3):200–202, September 2007.

[51] S. Lie. *Vorlesungen über Differentialgleichungen mit bekannten infinitesimalen Transformationen (Bearbeitet und herausgegeben von Dr. G. Scheffers).* Teubner, Leipzig.(Reprinted by Chelsea Publishing Company, New York, 1967), 1891.

[52] S. Lie. *Vorlesungen über Differentialgleichungen mit bekannten infinitesimalen Transformationen.* Teubner, Leipzig, 1912.

[53] S. Lie. *Gesammelte Abhandlungen (Vol. 1-7).* Teubner, Leipzig, 1927.

[54] H. Liu and J. Yong. Option pricing with an illiquid underlying asset market. *Journal of Economic Dynamics and Control*, 29(12):2125–2156, 2005.

[55] R. C. Merton. Optimum consumption and portfolio rules in a continuous-time model. *Journal of economic theory*, 3:373–413, 1971.

[56] M. Musiela and T. Zariphopoulou. An example of indifference prices under exponential preferences. *Finance and Stochastics*, 8(July 2003):229–239, 2004.

[57] V. Naicker, J. G. O'Hara, and P. G. L. Leach. A note on the integrability of the classical portfolio selection model. *Applied Mathematics Letters*, 23(9):1114–1119, 2010.

[58] F. Oliveri. *Program packages ReLie (REDUCE) and SymboLie (Mathematica).*

[59] P. J. Olver. *Application of Lie groups to differential equations.* Springer, New York, 1986.

[60] L. V. Ovsiannikov. *Group Analysis of Differential Equations.* Academic Press, New York, 1982.

[61] J. Patera and P. Winternitz. Subalgebras of real three- and four-dimensional Lie algebras. *Journal of Mathematical Physics*, 18(7):1449–1455, 1977.

[62] E. Platen and M. Schweizer. On Feedback Effects from Hedging Derivatives. *Mathematical Finance*, 8(1):67–84, January 1998.

[63] U. Polte. *On Hedging and Pricing of Derivatives in Incomplete Markets - a PDE Approach.* PhD thesis, Universität Leipzig, 2007.

[64] R. O. Popovich and A. Bihlo. Symmetry preserving parameterization schemes. *Journal of Mathematical Physics*, 53(7):073102, 2012.

[65] J. W. Pratt. Risk aversion in the small and in the large. *Econometrica*, 32(1-2):122–136, 1964.

[66] R. Rebelo and F. Valiquette. Symmetry preserving numerical schemes for partial differential equations and their numerical tests. *Journal of Difference Equations and Applications*, 19(5):738–757, 2012.

[67] V. K. Romanko, N. H. Agakhanov, V. V. Vlasov, and L. I. Kovalenko. Collection of problems to the theory of differential equations and variational calculus (in Russian: Sbornik sadach po differentialnym uravnenijam i variazionnomu ischesleniju). Technical report, Unimediastile, Moscow, 2002.

[68] K Saxe. *Beginning functional analysis.* Springer-Verlag, New York, 2002.

[69] P. Schönbucher and P. Wilmott. The feedback effect of hedging in illiquid markets. *SIAM Journal on Applied Mathematics*, 61:232–272, 2000.

[70] A. N. Shiryaev. *Essentials of Stochastic Finance: Facts, Models, Theory.* World Scientific, New Jersey, London, Singapore, Hong Kong, 2003.

[71] S. E. Shreve. *Stochastic Calculus for Finance II: Continuous–Time Models.* Springer, New York, 2004.

[72] K. R. Sircar and G. Papanicolaou. General Black-Scholes models accounting for increased market volatility from hedging strategies. *Applied Mathematical Finance*, 82:45–82, 1998.

[73] J. Staum. Incomplete Markets. In J.R. Birge and V. Linetsky, editors, *Handbooks in OR & MS*, pages 511–563. Elsevier, 2008.

[74] M. Steiner and B. Brunner. Marktgerechte Bewertung von Power-Optionen, Teil IV. In W. Kürsten and B. Nietert, editors, *Kapitalmarkt, Unternehmensfinanzierung und rationale Entscheidungen Festschrift für Jochen Wilhelm*. Springer, 2006.

[75] H. Stephani. *Differential Gleichungen: Symmetrien und Lösungsmethoden*. Spektrum Akademischer Verlag GmbH, Heidelberg, 1994.

[76] R. G. Tompkins. Power options: hedging nonlinear risks. *Journal of Risk*, 2000.

[77] V. S. Vladimirov, O. A. Ladyzenskajia, and N. N. Ural'tzeva. *Linear and Quasilinear Equations of Parabolic Type (Transl. Math, Monographs, 23)*. AMS, New York, 1968.

[78] J. von Neumann and O. Morgenstern. *Theory of Games and Economic Behavior*. Princeton University Press, Princeton, NJ, 1944.

[79] D. Ševčovič. Transformation methods for evaluating approximations to the optimal exercise boundary for linear and nonlinear Black-Scholes equations. In *Nonlinear Models in Mathematical Finance: Research Trends in Option Pricing*, chapter 6, pages 173–218. Nova Science Publishers, INC., 2009.

[80] P. Wilmott, S. Howison, and J. Dewynne. *The mathematics of financial derivatives. A Student introduction*. Cambridge University Press, 2002.

Index